DAVID J JAMES
HRI EM

BIOTECHNOLOGY
INTELLIGENCE
UNIT

GENOME MAPPING IN PLANTS

Andrew H. Paterson

Department of Soil and Crop Sciences
Texas A&M University
College Station, Texas, U.S.A.

Academic Press

R.G. LANDES COMPANY
AUSTIN

BIOTECHNOLOGY INTELLIGENCE UNIT

GENOME MAPPING IN PLANTS

R.G. LANDES COMPANY
Austin, Texas, U.S.A.

This book is printed on acid-free paper.
Copyright 1996 © by R.G. Landes Company and Academic Press, Inc.

Please address all inquiries to the Publisher:
R.G. Landes Company
909 Pine Street, Georgetown, Texas, U.S.A. 78626
Phone: 512/ 863 7762; FAX: 512/ 863 0081

Academic Press, Inc.
525 B Street, Suite 1900, San Diego, California, U.S.A. 92101-4495

United Kingdom Edition published by Academic Press Limited
24-28 Oval Road, London NW1 7DX, United Kingdom

Library of Congress Catalog Number: 581.87'3282--dc20
International Standard Book Number (ISBN): 0-12-546590-4

Printed in the United States of America

While the authors, editors and publisher believe that drug selection and dosage and the specifications and usage of equipment and devices, as set forth in this book, are in accord with current recommendations and practice at the time of publication, they make no warranty, expressed or implied, with respect to material described in this book. In view of the ongoing research, equipment development, changes in governmental regulations and the rapid accumulation of information relating to the biomedical sciences, the reader is urged to carefully review and evaluate the information provided herein.

Library of Congress Cataloging-in-Publication Data

Paterson, Andrew H., 1960-
    Genome mapping in plants / Andrew H. Paterson.
       p. cm. — (Biotechnology intelligence unit)
    Includes bibliographical references and index.
    ISBN 0-12-546590-4, 1-57059-359-0 (alk. paper)
    1. Plant genome mapping. I. Title. II. Series.
       QK981.45P37 1996
       581.87'3282--dc20

96-17126
CIP

Publisher's Note

R.G. Landes Company publishes six book series: *Medical Intelligence Unit, Molecular Biology Intelligence Unit, Neuroscience Intelligence Unit, Tissue Engineering Intelligence Unit, Biotechnology Intelligence Unit and Environmental Intelligence Unit.* The authors of our books are acknowledged leaders in their fields and the topics are unique. Almost without exception, no other similar books exist on these topics.

Our goal is to publish books in important and rapidly changing areas of bioscience and the environment for sophisticated researchers and clinicians. To achieve this goal, we have accelerated our publishing program to conform to the fast pace in which information grows in bioscience. Most of our books are published within 90 to 120 days of receipt of the manuscript. We would like to thank our readers for their continuing interest and welcome any comments or suggestions they may have for future books.

Deborah Muir Molsberry
Publications Director
R.G. Landes Company

CONTENTS

EDITOR

Andrew H. Paterson, Ph.D.
Department of Soil and Crop Sciences
Texas A&M University
College Station, Texas, U.S.A.
Chapters 1-7, 18, and 20

CONTRIBUTORS

Gregor Benning
Institute of Plant Sciences
Swiss Federal Institute of
    Technology
Zürich, Switzerland
Chapter 12

Stuart M. Brown
U.S. Department of Agriculture
Agricultural Research Service
Plant Genetic Resources
    Conservation Unit
Griffin, Georgia, U.S.A.
Chapter 8

Charlene Chang, M.S.
Department of Soil and Crop
    Sciences
Texas A&M University
College Station, Texas, U.S.A.
Chapter 20

Wing Y. Cheung, Ph.D.
DNA LandMarks
St. Jean, Quebec, Canada
Chapter 16

Jorge Dubcovsky
Instituto de Recursos Biologicos
CIRN-INTA
Buenos Aires, Argentina
Chapter 19

Kenneth A. Feldmann
Department of Plant Sciences
University of Arizona
Tucson, Arizona, U.S.A.
Chapter 13

Martin W. Ganal, Ph.D.
Institute for Plant Genetics and
    Crop Plant Research
Gatersleben, Germany
Chapter 10

Bikram S. Gill
Department of Plant Pathology
Kansas State University
Manhattan, Kansas, U.S.A.
Chapter 11

Erwin Grill
Institute of Plant Sciences
Swiss Federal Institute of
    Technology
Zürich, Switzerland
Chapter 12

Tim Helentjaris
Pioneer Hi-Bred International, Inc.
Research and Product Development,
  Trait and Technology
  Development, Agronomic Traits
Johnston, Iowa, U.S.A.
Chapter 9

Matthew A. Jenks
Arkansas State University
State University, Arkansas, U.S.A.
Chapter 13

Jiming Jiang
Department of Horticulture
University of Wisconsin
Madison, Wisconsin, U.S.A.
Chapter 11

Richard V. Kesseli
Department of Biology
University of Massachusetts
Boston, Massachusetts, U.S.A.
Chapter 15

Gary Kochert, Ph.D.
Department of Botany
University of Georgia
Athens, Georgia, U.S.A.
Chapter 17

Stephen Kresovich, Ph.D.
U.S. Department of Agriculture
Agricultural Research Service
Plant Genetic Resources
  Conservation Unit
Griffin, Georgia, U.S.A.
Chapter 8

Benoit S. Landry, Ph.D.
DNA LandMarks
St. Jean, Quebec, Canada
Chapter 16

Candice B. Lewis
Department of Plant Breeding and
  Biometry
Cornell University
Ithaca, New York, U.S.A.
Chapter 21

Zhongsen Li
Department of Biology
Texas A&M University
College Station, Texas, U.S.A.
Chapter 14

Yann-rong Lin, M.S.
Department of Soil and Crop
  Sciences
Texas A&M University
College Station, Texas, U.S.A.
Chapter 20

Sin-Chieh Liu, Ph.D.
Department of Soil and Crop
  Sciences
Texas A&M University
College Station, Texas, U.S.A.
Chapter 20

Susan R. McCouch
Department of Plant Breeding
  and Biometry
Cornell University
Ithaca, New York, U.S.A.
Chapter 19

Knut Meyer
Institute of Plant Sciences
Swiss Federal Institute of
   Technology
Zürich, Switzerland
Chapter 12

Richard W. Michelmore
Department of Vegetable Crops
University of California
Davis, California, U.S.A.
Chapter 15

David B. Neale
Institute of Forest Genetics
Pacific Southwest Research Station
USDA Forest Service
Albany, California, U.S.A.
Chapter 22

James C. Nelson
Department of Plant Breeding
   and Biometry
Cornell University
Ithaca, New York, U.S.A.
Chapter 19

Andrew N. Nunberg
Department of Biology
Texas A&M University
College Station, Texas, U.S.A.
Chapter 14

Klaus Pillen
Department of Plant Breeding
   and Biometry
Cornell University
Ithaca, New York, U.S.A.
Chapter 21

Omaira Pineda
Department of Plant Breeding
   and Biometry
Cornell University
Ithaca, New York, U.S.A.
Chapter 21

Ronald R. Sederoff
Depts. of Forestry, Genetics
   and Biochemistry
North Carolina State University
Raleigh, North Carolina, U.S.A.
Chapter 22

Stephen Smith
Pioneer Hi-Bred International, Inc.
Research and Product Development,
   Trait and Technology
   Development, Agronomic Traits
Johnston, Iowa, U.S.A.
Chapter 9

Mark E. Sorrells
Department of Plant Breeding
   and Biometry
Cornell University
Ithaca, New York, U.S.A.
Chapter 19

David M. Stelly, Ph.D.
Department of Soil and Crop
   Sciences
Texas A&M University
College Station, Texas, U.S.A.
Chapter 18

Steven D. Tanksley
Department of Plant Breeding
   and Biometry
Cornell University
Ithaca, New York, U.S.A.
Chapter 21

Terry L. Thomas
Department of Biology
Texas A&M University
College Station, Texas, U.S.A.
Chapter 14

Norman F. Weeden, Ph.D.
Department of Horticultural
   Sciences
New York Experiment Station
Cornell University
Geneva, New York, U.S.A.
Chapter 17

Jonathan F. Wendel, Ph.D.
Department of Botany
Iowa State University
Ames, Iowa, U.S.A.
Chapter 18

Nevin D. Young, Ph.D.
Department of Plant Pathology
University of Minnesota
St. Paul, Minnesota, U.S.A.
Chapter 17

Xinping Zhao, Ph.D.
University of Michigan Medical
   Center
Ann Arbor, Michigan, U.S.A.
Chapters 10, 18

BIOGRAPHY

D r. Andrew Paterson is an associate professor of Plant Molecular Genetics at Texas A&M University, teaching courses in genetic and molecular analysis of eukaryotic genomes. Dr. Paterson earned his B.S. in Plant Science (Summa Cum Laude) at the University of Delaware, and his M.S. in Plant Breeding and Ph.D. in Plant Genetics at Cornell University. He worked as a postdoctoral fellow in Plant Molecular Genetics at Cornell University, and a Research Biochemist in Agricultural Biotechnology at E. I. duPont de Nemours (with an adjunct faculty appointment at the University of Delaware), before moving to Texas A&M University in 1991. He is the author or co-author of more than 40 refereed publications, several book chapters, and is a widely-sought speaker. Research in his laboratory encompasses the plant families Malvaceae (cotton), Poaceae (grasses), Cruciferae (*Arabidopsis thaliana*, and *Brassica* spp.), and Fabaceae (peanut). His research team studies the organization, evolution, and function of plant genomes, and the molecular basis of crop productivity.

PREFACE

In the past decade, genome mapping has emerged as a powerful new approach to research in botany, as well as in many other fields. As one whose formal education was largely in the applied field of plant breeding, I am still amazed by the ease with which my colleagues can subvert the genetic machinery of an unsuspecting bacterium into producing large amounts of DNA from maize, sorghum, cotton, or some other plant. While the emergence of genome mapping as a major research area occurred well after the basic technologies of DNA cloning had been reduced to routine practice, a number of new techniques in molecular cloning have been developed as a direct result of the needs of genome analysis.

Many of the concepts associated with genome mapping are more than a century old, however, recent technical advances have afforded description of the structure and function of plant genomes in unprecedented detail. Molecular-level understanding of the inheritance of agriculturally important traits creates new opportunities to streamline plant breeding, the process of altering plant genotypes to better fit the needs of humankind. Further, this understanding provides a channel for communication between the farmer's field and the molecular biology laboratory, in principle enabling the scientist to identify the specific DNA element(s) responsible for particular plant characteristics. Ongoing improvement in our understanding of plant genomes is likely to streamline the process of gene identification. By better documenting the (many) similarities and (few) differences among even distantly-related plants, one can study the course of evolution of particular genes over millions of years, and extrapolate information from one species to another.

In seeking to summarize a large and rapidly-growing body of literature into a succinct but still useful form, it was clear that the task would best be accomplished by enlisting the aid of experts in various aspects of the field. Consequently, this volume has been prepared in two sections. Section 1 (chapters 1-6) summarizes broad principles which I consider to be applicable to genome analysis in many plant (and animal) species, and is intended to bring new students of the field up to speed on basic concepts which represent the foundation of the field, while remaining within reach of the interested lay person with a scientific bent. Section 2 presents in-depth discussions of widely-used tools and techniques (chapters 7-14) and well-studied plant taxa (chapters 15-22), written by appropriate experts for an audience of researchers, and providing noteworthy examples of contemporary results for the student or lay person who has assimilated Section 1.

While public attention to genome analysis is dominated by the "Human Genome Project," plant genome analysis offers equal (if not greater) opportunities to improve human well-being. In days of historically low real prices for agricultural crops, and federal subsidies to grow (or not grow) crops, it is easy to forget that agriculture sustains humanity. A small number of crop plants stand between humankind and famine, converting solar energy into a form usable by humans and other animals. In view of continuing growth in world populations, gains in crop productivity remain essential to international economic and political stability. Further attempts to improve crop productivity by altering the environment incur risk of damage to the ecosystem, from effects of agrichemicals on natural biota, effects of intensive cultivation on soil erosion, and depletion of fresh water resources to sustain short-term gains in crop yield. Intrinsic genetic changes can improve plant productivity at minimal cost to both the grower and the ecosystem. Plant genome analysis will play a major role in accomplishing the genetic changes needed to improve the sustainability of modern agriculture, as well as the quantity and quality of products which agriculture provides to society.

Andrew H. Paterson, Ph.D.

CHAPTER 1

AN HISTORICAL PERSPECTIVE

Andrew H. Paterson

*"...a set of probes for DNA polymorphism...should provide a new horizon in...genetics."** *

1.1. WHAT IS GENOME MAPPING?

Genome mapping, a synthesis of concepts from classical genetics with tools from molecular biology, is an exciting new research area in the life sciences. Since the 1970s when the first genes were cloned and characterized, genetic research in most organisms has shifted to the DNA level. While it has been understood for more than 50 years that there was a cause-and-effect relationship between changes in "genotype" (information encoded by DNA), and changes in ("phenotype") (appearance or behavior of living organisms) the capability to identify the specific DNA element(s) responsible for a particular phenotype has only emerged very recently.

Genome mapping is a generally-applicable approach to studying the repertoire of genetic information which directs the growth and development of extant life forms. Higher organisms are estimated to have 10^4-10^5 "genes"—hereditary instructions for specific steps in biological processes. The vast majority of these genes remain uncharacterized, and their functions remain unknown (although rapid progress is being made toward characterizing all of the genes found in the genomes of selected plant and animal taxa: see chapters 13, 14). A particular strength of genome mapping is that it facilitates isolation of genes based simply on measurement of their effect(s) on phenotype—requiring no information about the specific functions performed by the gene(s).

The word "genome" describes the total repertoire of DNA in a particular organelle. Animals have one genome in the nucleus, and a second, very different genome in the mitochondrion. Plants have yet a third genome, in the chloroplast. The mitochondrial and chloroplast genomes are believed to have once been the nuclear genomes of independent

* Botstein D, White RL, Skolnick M, Davis RW. Construction of a genetic linkage map in man using restriction fragment length polymorphisms. Am J Hum Genet 1980; 32:314-331.

organisms—which found the intracellular matrix hospitable, and conferred advantages to their host cells. In this book, we will focus on the nuclear genome, by far the largest of the three, and the one which imparts the vast majority of characteristics to an organism.

A "genome map" can be thought of much as a road map, reflecting the relative proximity of different landmarks to one another. "Genome mapping" is made possible by the fact that the nuclear genome in higher organisms is organized and transmitted as linear units, called chromosomes. Just as mileposts guide the motorist along a linear highway, molecular tools enable the geneticist to establish specific "DNA markers" at defined places along each chromosome. DNA markers can then be used to delineate when one has reached (or passed by!) a particular gene of interest.

Genome mapping encompasses a wide range of techniques, useful for studying DNA at different levels of magnification. The genomes of many plants, like the genome of human, include more than one billion individual units (nucleotides) of information. Information is conveyed by the "genetic code"[1] based on the order in which the units occur. A gene, the functional unit which encodes a single protein, is often 1000 nucleotides or more in length—even with only four chemically-different nucleotides to choose from, one can envision an enormous number of possible permutations in the composition and order of 1000 nucleotides. However, as few as 10^3-10^4, of the 10^{52} possible exons (defined in the cited analysis as having an average length of 120 nucleotides), may be needed to construct most known proteins, suggesting that there are strong constraints on the diversity of extant genes (see chapter 6 for more discussion, and citations).

1.2. THE PRE-HISTORY OF GENOME MAPPING

Genome mapping was practiced for nearly four decades prior to the demonstration that DNA was the hereditary molecule. The most important targets of genome mapping are Mendel's "factors."[2] However, the beginnings of the field of genome mapping lie not in Mendelism, but in the phenomenon of "genetic linkage" explained by Morgan.[3] The fact that a strict linear organization is imposed on the hereditary information, affords the opportunity to construct ordered "genetic" maps. Whether one considers chromosomal maps of visible markers, local physical maps based on contiguous overlapping megabase DNA clones, or the DNA sequence of a single gene, all modern approaches to characterizing and cataloging genomes and their components rely on the linear organization of genetic information.

Prior to the era of molecular biology, "mileposts" along the genome map were comprised of "visible markers." A visible marker is simply a mutation in a particular gene, which imparts a discrete, easily-identified phenotype to an organism. Today, such "visible markers" can be created at will—and provide the means to quickly and efficiently isolate genes associated with a particular phenotype (see chapter 13). However, in 1910, identification of visible markers was a serendipitous event, at the mercy of Nature.

Visible markers were used to establish the principle of "genetic mapping," based on the hypothesis that the likelihood of co-transmission of any two markers reflected the proximity between the markers (mileposts) along the chromosome (highway). Students of *Drosophila*, bean, and a variety of other organisms had demonstrated the basic principles now associated with genetic mapping, and in fact had constructed partial genetic maps, well before it was established that DNA was the molecule they were studying.

Several limitations of "visible markers" obstructed the progress in genome mapping which could be made by its early practitioners. First, visible markers often had deleterious effects on the study organism—after all, they represented "errors" in the replication of a gene, that served some function in the organism. (For example, consider the competitiveness of wingless fruitflies, in nature). While laboratory strains of organisms were carefully maintained in non-competitive environments, a particular lineage could only survive under the "genetic load" of a modest number of "visible markers." Because hundreds of such markers were needed to provide mileposts for all regions of all chromosomes, extensive genetic mapping experiments simply were not feasible. Moreover, visible markers were rare—genetic linkage analysis required that they be assembled into common lineages by tedious breeding experiments—assembly of such a lineage suitable for conducting a genetic mapping experiment might take longer than the experiment.

1.3. DNA: THE BIRTH OF MOLECULAR BIOLOGY

A 1944 experiment revolutionized our understanding of the transmission of genetic information, and hinted at the possibility of manipulating this information. Avery and colleagues[4] studied two strains of *Streptococcus pneumoniae* which differed in their ability to cause pneumonia in humans. In one of the most significant experiments in biology, DNA from pathogenic (S) strains of *St. pneumoniae* was isolated, and incubated with minimally pathogenic (R) strains of the bacterium in a test tube. When the bacteria in the test tube were evaluated, some were S strains, and transmitted the S phenotype to their progeny. The only substance which could cause this "transformation" (from R-type to S-type), was deoxyribonucleic acid (DNA).

The demonstration that wild-type DNA could be introduced into another organism and complement a mutant phenotype established that the "factors" of Mendel, were comprised of DNA...in other words, that DNA was the hereditary molecule. An understanding of the structure of DNA followed,[5] revealing both the chemical simplicity of DNA (four different nucleotides), and suggesting a means by which DNA replicates (on a "mirror-image template").[6] It had been known since the 1940s that there was a direct correspondence between one gene and one protein.[7,8] However, it was not understood until the 1960s that this correspondence involved "transcription" of DNA into an RNA intermediate, which in turn was "translated" into an enzyme or protein.[9,10]

Moreover, the demonstration that DNA could be removed from its natural host, and become an integral part of the hereditary information of a completely different organism, hinted at means by which specific DNA elements might be produced en masse, and studied in detail. A general understanding of the structure and function of genes, together with clever use of fortuitous properties of specific "molecular shears" called "restriction enzymes,"[11] permitted the isolation and "cloning" (replication of the DNA in a bacterium) of genes such as the human insulin gene.[12]

Molecular cloning has come to be a central component of modern life sciences research, being practiced every day in thousands of laboratories across the globe. The ability to "clone" DNA permits biologists to "grow" large amounts of a specific DNA element, study it in detail, and retain it indefinitely, in a convenient form for future study.

By basing genetic analysis directly on DNA, rather than on naturally occurring "visible markers," the primary obstacle to progress in genome mapping was overcome (chapter 2). No longer did scientists have to endure long and tedious breeding experiments to assemble visible markers into "multiply marked stocks"—since members of a taxon shared a common DNA language. Instead, techniques were devised by which one took advantage of "spelling errors" (mutations) in the language—at the molecular level, such errors were similar to visible mutations, except that most had no effect on the viability or fitness of the organism, thus could accumulate freely in a lineage without impairing viability. Moreover, because one had millions, sometimes billions of "letters" in which "spelling errors" (mutations) might occur, the number of potential "DNA markers" is enormous.

1.4. GENOME MAPPING FITS INTO A CONTINUUM OF TECHNIQUES FOR GENE MANIPULATION

Different techniques are employed to obtain information from different levels in the hierarchy of DNA organization. Domestication and breeding of plants and animals has relied on selection of individuals whose entire genomic complement added up to a superior overall phenotype. The classical discipline of cytogenetics (see chapter 11) enabled manipulation of individual chromosomes, and even parts of chromosomes, even before it was known that DNA was the information-bearing component of the chromosome. However, a chromosome might contain more than 1,000 individual genes, each representing the blueprint for a particular protein.

At the opposite extreme of the hierarchy of DNA organization, chemists devised the means to determine the order of nucleotides along a defined piece of DNA.[13,14] "DNA sequencing" permitted study of the relationship between abiotic (chemical) information and biotic (phenotype) information. The tools of genome mapping bridge this logical, and physical, gap between the resolution at which classical genetic methods could study genes, and the resolution with which genes can be chemically dissected. Today, a continuum of methods exist by which one can observe a trait in nature, determine the number of genes responsible for the trait, find the location(s) of the gene(s) on the chromosomes (chapter 4), identify a contiguous stretch of DNA which includes the location(s) of the gene(s) (chapter 5), and finally identify the gene(s) (chapters 12 and 13). Many of the steps in this process remain difficult, and technological improvements are needed. However, the basic process can be applied to many organisms, and has been successfully applied to human, as well as to several plant and animal species, to clone genes associated with diseases and other traits.

Genome mapping has been especially important in isolation of genes for which the mechanism of action (e.g., biochemical function of the enzyme or protein) is unknown. Cloning of genes based upon genetic map location (see chapter 12) has become an important alternative to cloning genes based on functional assays. Opportunities for "map-based cloning" of genes known only from phenotype have stimulated development of new molecular tools and technologies, capable of in vitro maintenance and replication of intact segments of the chromosomes of higher eukaryotes, often as "artificial chromosomes" which behave much as the native chromosomes of the host cell.

1.4.1. GENES ARE A TINY COMPONENT OF GENOMES

Genes, the contemporary term for the "factors" of Mendel, are the single most important entities studied by genome mapping—curiously, however, in the DNA of many organisms, genes are as rare as the proverbial "needles in a haystack." Much of the DNA, in fact the vast majority of DNA in most organisms, does not encode any instructions which we can recognize at present (see chapter 10). Some DNA elements serve "structural" roles, delineating a point of association for the spindle fibers which guide separation of chromosome pairs during meiosis (centromeres) or delineating the "end" of the chromosome (telomeres). A vast quantity of DNA is thought by many to be "selfish"[15,16] lacking any function essential to growth and development, but persisting because it has evolved the means to replicate itself.

1.5. GENOME MAPPING UNIFIES STUDY OF DIVERSE ORGANISMS

Classical "genetics" is confined to the study of organisms which can be interbred—however, genome mapping reaches beyond this boundary. Most known life forms share very similar means of transmitting hereditary information from one generation to another, suggesting that life on Earth ultimately derives from a single evolutionary lineage. This inference suggests that there may be similarities in the hereditary information of different organisms. In particular, DNA elements which encode instructions essential to the survival and/or reproduction of an organism are likely to remain similar, over long time periods—since "mutations," or changes in such basic instructions may quickly lead to the immediate death of the organism, or to the ultimate extinction of the organism's family as a result of producing few progeny.

A vast number of experiments, across a wide range of taxa, clearly demonstrate that the elements which encode hereditary information are "conserved"—protected from changes which might impact function—over very long periods of time. Some extreme cases are widely-documented in basic biology texts. For example, the DNA elements which encode "ribosomes," the basic protein assembly sites common to virtually all cells, differ only nominally in higher plants such as corn or beans, and higher animals such as human or bovine. A high proportion of genes exhibit similar DNA sequence in organisms which are considered to fall within the same taxonomic "family." Virtually all genes are very similar in organisms which fall within the same taxonomic "genus," or "species."

The commonality among organisms resulting from "conservation" of genes, affords the opportunity to use genes as "DNA markers" common to different organisms. By such "comparative mapping," we have learned that not only the information content of genes, but also the order of genes along the chromosome, tend to persist over long time periods (see chapters 3 and 15-22). Such information creates new learning opportunities, specifically the capability to compare organization and function of genes in different taxa. Further, this information offers tremendous potential efficiencies in genome mapping—because new information about how genes are ordered in one organism is likely to be relevant to many other related organisms. Finally, comparative information helps the scientific community to minimize redundancy in its collective research agenda—a gene need only be cloned once, to understand its function in many different organisms.

1.5. WHAT ARE SOME PRACTICAL APPLICATIONS OF GENOME MAPPING?

There is widespread public awareness of the "human genome projects" underway in several countries. Because of the cause-and-effect relationship between DNA information and the nature of organisms, complete documentation of the genes which comprise an organism affords a first step toward understanding how the organism grows and develops. By associating variation among different organisms with variation in genetic information, the basis of attributes such as athletic prowess, or accidents such as genetic diseases might be understood.

Particularly good examples of how such information might be utilized come from plant breeding, where moral issues associated with altering heredity are minimal. Most crop plants were domesticated within the past 15,000 years, a very short time in the history of plant (and human) evolution.[17] However, the changes effected in this short time are quite dramatic—for example, evolution of the flower of a small Meso-American grass into the "ear" of maize, evolution of the seed-containing berry of wild nightshades into the "fruit" of tomato, and evolution of the flower buds of wild mustard into the "curd" of cauliflower. A high priority of plant genome mapping is the identification of genes associated with evolution of crop productivity, and use of this information to further improve crops to meet the needs of a hungry world. (see chapters 4, 5, 7, and 15-22).

The utility of DNA is by no means limited to altering particular traits. DNA-level investigations have clarified relationships among related genotypes, provided guidance in collection, cataloging, and utilization of genetic resources (see chapter 8), and provided a means for documenting and protecting the products of investment in crop improvement efforts (see chapter 9).

The following chapters will attempt to eludicate basic principles associated with genome analysis (chapters 2-6), provide more detailed information on specific areas of expertise in genome analysis (chapters 7-14), and summarize the state of genome analysis in leading crop plants and model systems (chapters 15-22).

REFERENCES

1. Nirenberg MW and JH Matthei. The dependance of cell-free protein synthesis in *E. coli* upon naturally-occurring or synthetic polyribonucleotides. Proc Nat'l Acad Sci USA 1961; 47:1588-1602.

2. Mendel G. Versuche uber Pflanzen-Hybriden. *Verh. Naturforshung Ver. Brunn* 1865; 4:3-47. Translated by W. A. Bateson as "Experiments in plant hybridization," and reprinted in J A Peters, Classic *Papers in Genetics*, Prentice-Hall, 1959.

3. Morgan TH. Sex-linked inheritance in *Drosophila*. Science 1910; 32:120-122.

4. Avery OT, CM Macleod, M McCarty. Studies on the chemical nature of the substance inducing transformation of pneumococcal types. J Exp Med 1944; 79:137-158.

5. Watson JD and FHC Crick. A structure for deoxyribose nucleic acid. Nature 1953; 171:737-738.

6. Watson JD and FHC Crick. General implications of the structure for deoxyribose nucleic acid. Nature 1953; 171:964-967.

7. Beadle GW and EL Tatum. Genetic control of biochemical reactions in Neurospora. Proc Nat'l Acad Sci USA 1941; 27:499-506.

8. Lederberg J and EL Tatum. Gene recombination in *Escherichia coli*. Nature 1946; 158:588.

9. Jacob F and J. Monod. Genetic regulatory mechanisms in the synthesis of proteins. J Mol Biol 1961; 3:318-356.

10. Brenner S, F Jacob and M Meselson. An unstable intermediate carrying information from genes to ribosomes for protein synthesis. Nature 1961; 190:576-581.

11. Smith HO. Nucleotide sequence specificity of restriction endonucleases. Science 1970; 205:455-462.

12. Goeddel DV, DG Kleid, F Bolivar, HL Heyneker, DG Yansura, R Crea, T Hirose, A Kraszewski, K Itakura, A Riggs. Expression in *Escherichia coli* of chemically synthesized genes for human insulin. Proc Nat'l Acad Sci USA 1979; 76:106-110.

13. Maxam AM and W Gilbert. A new method for sequencing DNA. Proc Nat'l Acad Sci USA 1977; 74:560-564.

14. Sanger F, S Nicklen, AR Coulson. DNA sequencing with chain-terminating inhibitors. Proc Nat'l Acad Sci USA 1977; 74:5463-5467.

15. Doolittle WF and C Sapienza. Selfish genes, the phenotype paradigm, and genome evolution. Nature 1980; 284:601-603.

16. Orgel LE, and FHC Crick. Selfish DNA: the ultimate parasite. Nature 1980; 284:604-607.

17. Harlan JR. Crops and Man. Madison, WI: Crop Science Society of America, 1975.

CHAPTER 2

THE DNA REVOLUTION

Andrew H. Paterson

"The genes are units useful in concise description of the phenomena of heredity. Their place of residence is the chromosome. Their behavior brings about the observed facts of genetics. For the rest, what we know about them is merely an interpretation of crossover frequency. In terms of geometry, chemistry, physics, or mechanics we can give them no description whatever." *

2.1. COMPOSITION OF A GENOME

The beginning of the description which Prof. East sought, came in 1944, with the demonstration by Avery and colleagues that deoxyribonucleic acid (DNA) from one organism was able to confer new characteristics upon other organisms.[1] Subsequent descriptions of the structure of DNA,[2] use of its mirror-image structure to replicate DNA,[3,4] relationship of DNA to protein mediated by RNA,[5,6] "cracking" of the genetic code,[7] and ability to determine the "sequence" of the four "letters" in the genetic code along a piece of DNA,[8,9] provided much of the basic knowledge essential to the evolution of contemporary "biotechnology."

The breakthrough which ushered in the era of biotechnology was the ability to efficiently obtain large quantities of specific DNA elements. The genome of even a relatively simple organism such as the human gut bacterium *Escherichia coli*, is comprised of more than four million "letters" (nucleotides) of DNA. The letters are organized into "words" (genes) and "sentences" (operons)...and finally into a "book" (genome) comprising the single chromosome of *E. coli*. From the above, it was well understood that the letters themselves meant relatively little, and the basic unit of hereditary information was the gene ("word").

Genes (DNA sequences encoding polypeptides) are only a tiny fraction of the DNA of most plants. This can be illustrated by a thumbnail calculation of the approximate amount of DNA which might be required to encode all of the genes in a higher plant. Recent estimates suggest that *Arabidopsis thaliana*, the smallest plant genome known, might proceed

---

* *East EM. The concept of the gene. Proc Int Congr Plant Sci 1926; 1: 889-895.*

Genome Mapping in Plants, edited by Andrew H. Paterson. © 1996 R.G. Landes Company.

through its life cycle using as few as 25,000 genes. Based on an average poly-peptide length of 400 amino acids, it would require about 3×10^7 nucleotides to encode these 20,000 polypeptides. Further, genes include non-coding regions (introns), and both upstream and downstream regu-latory DNA sequences necessary to proper gene expression (promoters, enhancers)—these elements together might typically double the amount of DNA needed to en-code all the genes, to about 6×10^7 nucle-otides. However, the *Arabidopsis* genome, the smallest known among higher plants, has 2.5 times this much DNA (1.5×10^8 nucleotides; see chapter 5). Moreover, this does not begin to account for all of the DNA in the garden lily, or other large-genome plants such as wheat (see chap-ter 5). Genes in the genome are like needles in a haystack!!

While the nature of all the other DNA comprising a plant genome will be ad-dressed elsewhere (see chapter 10), the sa-lient point is that in order to isolate the gene which performs a specific function, one must have an efficient means of sort-ing through an enormous amount of DNA. Because genes are comprised of different permutations of the same four nucleotides (letters in the DNA alphabet), there are few chemical features which distinguish one gene from another, or distinguish genes from non-functional DNA sequences.

The ultimate solution to the problem of isolating genes, came from the combined understanding of genetics derived from the pioneering work cited above, technologi-cal developments which enabled separation of DNA molecules based on size, and some curious properties of certain bacteria.

2.2. MOLECULAR DISSECTION OF A GENOME

2.2.1. ELECTROPHORESIS—A MEANS TO SEPARATE DNA (AND OTHER) MOLECULES BASED ON SIZE

Electrophoresis is the most commonly-used method to separate, identify, and pu-rify nucleic acid (DNA or RNA) molecules.

Electrophoresis is based on the fact that nucleic acid molecules are negatively-charged, and migrate toward the anode in an electrical field. DNA electrophoresis is usually performed in a matrix of agarose or polyacrylamide, immersed in an aque-ous salt buffer which is suitable for estab-lishment of an electrical field.

The basic principles of electrophoresis are illustrated in Figure 2.1. Molecules of linear double-stranded DNA migrate at a rate inversely proportional to their size—large molecules migrate more slowly be-cause of (1) greater frictional drag, and (2) because they "worm" their way through the pores of gel matrices less efficiently than smaller molecules. Factors other than size can influence migration of DNA mol-ecules, such as conformation of the DNA (linear, circular, superhelical), and base composition—however, electrophoretic separation of DNA is based primarily on the size (e.g., number of nucleotides) of DNA molecules.[10]

Electrophoresis can be used to separate DNA molecules which differ in size by as little as one nucleotide, or as much as mil-lions of nucleotides. By varying agarose or acrylamide concentration, buffer composi-tion and/or temperature, or electrical field strength, conditions can be optimized for separation of molecules falling into differ-ent size ranges. A particularly important innovation has been "pulsed-field" gel elec-trophoresis, in which the direction of the electrical field alternates. The technical details of electrophoresis are well-described in excellent lab manuals such as Sambrook, Fritsch, and Maniatis,[10] to which the reader is referred for more details.

In conjunction with electrophoresis, "Southern blotting"[11] affords efficient study of DNA separated by electrophoresis (Fig. 2.1). This technique uses capillary ac-tion to transfer DNA from a delicate "gel" matrix used for electrophoresis, to a more durable DNA-binding membrane (usually made of nylon, although nitrocellulose was widely used in the past). DNA can be tightly bound to the membrane, such that the membrane can be used for 20 or more

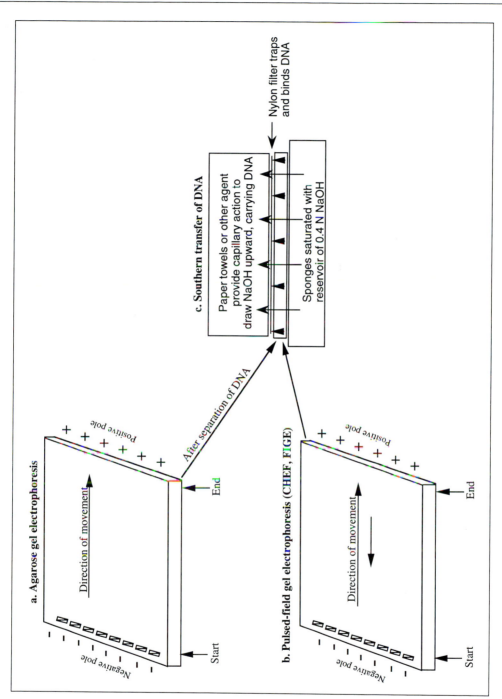

Fig. 2.1. Electrophoresis and Southern blotting of DNA. Agarose gel electrophoresis (a) has long been the most common means for separating DNA fragments, based on differential migration in an electrical field as described in text. The more recent need for manipulation of large DNA fragments motivated the development of techniques such as "pulsedfield" gel electrophoresis (b), which afford separation of large DNA elements by periodic switching of the direction of the electrical field. Once DNA fragments have been separated, Southern blotting (c) simplifies further analysis, by transferring DNA from the delicate agarose gel to a durable nylon membrane. Finally, gel electrophoresis can also be performed in vertical orientation, and using different matrices such as acrylamide, as is presently used in DNA sequencing (see chapter 14).

cycles of analysis, in some applications. The recyclability of the membrane is important to efficient genetic analysis of large populations, at many DNA marker loci distributed throughout the genome.

2.2.2. RESTRICTION ENZYMES: A MEANS TO GENERATE SPECIFIC, REPRODUCIBLE DNA FRAGMENTS

Molecular genetics is based heavily upon the use of "restriction enzymes," highly-specific molecular shears produced by certain bacteria and fungi, which recognize and cleave DNA at specific sites.[12] A particular restriction enzyme, for example the commonly-used *Eco* RI, recognizes a specific "recognition sequence" in the DNA (in the case of *Eco* RI, it is the 6-nucleotide sequence 5' GAATTC), and cleaves both DNA strands in or near the recognition sequence (Fig. 2.2). These recognition sequences are either absent, or chemically modified in the host, but are present in the DNA of viruses which parasitize these organisms—consequently these enzymes "restrict" the growth of parasites in bacteria or fungi, by attacking the DNA of the parasite.

The specificity of "restriction enzymes" for particular recognition sequences is the basis for most DNA cloning strategies. The vast majority of restriction enzymes used in molecular cloning have recognition sequences which are 4, 5, or 6 nucleotides in length, and display 2-fold symmetry. The approximate range of DNA fragment sizes can be estimated based on the number of nucleotides in the recognition sequence—for example a "6-cutter" (an enzyme with six nucleotides in its recognition sequence) would be expected to find its recognition site at average intervals of 4^6, or an average of every 4,096 nucleotides (4 kilobases, or kb). A "4-cutter" would find its recognition sequence more frequently, yielding "restriction fragments" with an average length of 256 nucleotides. For analysis of large DNA fragments, "8-cutters" are often used, which find their recognition sequences only rarely, an average of once in 64 kb. These estimates of restriction fragment size are only approximate, and vary greatly in relation to the ratio of different nucleotides in different genomes (for example, plants tend to be AT-rich, and restriction enzymes recognizing AT-rich sequences cut frequently), sensitivity to DNA methylation, and the number and composition of repetitive DNA elements in various genomes.

Many restriction enzymes generate DNA fragments with "sticky ends"—each DNA strand of the double helix is cleaved at similar locations on opposite sides of the axis of symmetry (Fig. 2.2), creating DNA fragments that carry protruding single-stranded termini. Because of the symmetry of the recognition site, the protruding ends are complementary to those of other DNA fragments generated by cleavage with the same restriction enzyme—this provides a basis for adhesion between DNA fragments from different sources, and is a key to many molecular cloning experiments.

2.2.3. CLONING VECTORS: A MEANS TO REPLICATE SPECIFIC DNA SEQUENCES

Complementing the ability to create well-defined DNA fragments with restriction enzymes, and to separate DNA fragments based on size by electrophoresis, the third component needed to "clone" DNA was a means to isolate and propagate specific DNA fragments. A wide variety of "vectors" have been developed, for carrying exogenous DNA into a prokaryotic or eukaryotic cell, then assuring that the exogenous DNA is replicated along with endogenous cellular DNA. Cloning vectors vary in the size of the DNA fragment which might be cloned (from a few nucleotides to millions of nucleotides), the nature of the vehicles used to propagate the DNA (bacteria, phage, yeast), and the behavior of the vectored DNA (extrachromosomal plasmid, infective phage, or artificial chromosome), as well as many fine points associated with how the DNA is cloned and maintained.

The original vectors used in molecular cloning were bacterial "plasmids,"[13] double-

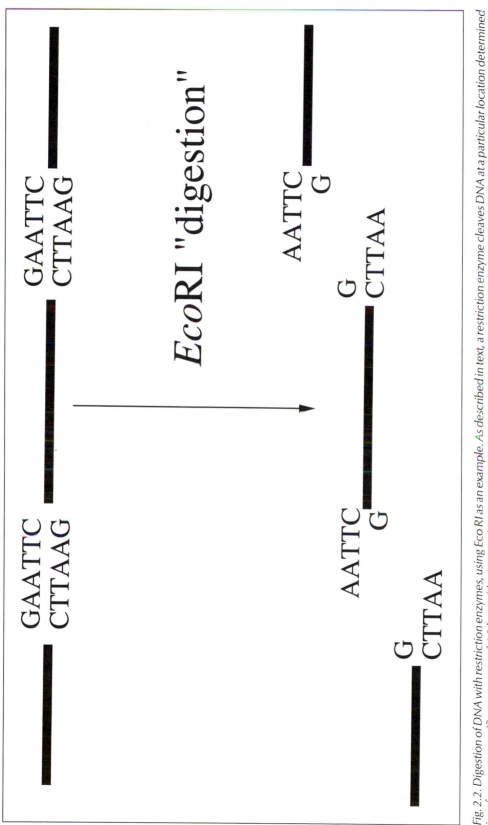

Fig. 2.2. Digestion of DNA with restriction enzymes, using Eco RI as an example. As described in text, a restriction enzyme cleaves DNA at a particular location determined in reference to specific sequences of 4-8 bases (the "recognition sequence"). Restriction enzyme cleavage generates reproducible DNA fragments, which usually have "sticky" (complementary) ends, and can, reanneal to form a single fragment—this property is crucial to DNA cloning (see Fig. 2.4).

stranded closed circular DNA molecules ranging in size from 1 kb to more than 200 kb.[10] They are found in many bacterial (and other) species, and are inherited and replicate independently of the bacterial chromosome. However, the proteins and enzymes necessary to replicate plasmids or express plasmid-encoded genes are the same as those which replicate and express genes on the bacterial chromosome, and are encoded by the bacterial chromosome.

Over the years, genetic engineering has produced plasmids which have become progressively more sophisticated, as vehicles for cloning DNA.[10] Extraneous DNA not necessary to replicate the plasmid (and the "insert" of exogenous DNA) has been removed, recognition sequences for multiple restriction enzymes have been introduced, antibiotic resistance genes have been added to assure that plasmids are maintained in the bacterial cell (by growing the cells in the antibiotic), colorimetric systems have been developed to reveal the presence of exogenous DNA in the plasmid. Ancillary sequences have been added to permit generation of single-stranded DNA templates from the plasmid (for DNA sequencing), transcription of the foreign DNA, or expression of the foreign protein. These specialized functions are described both in many lab manuals, and in excellent publications provided by commercial biotechnology suppliers.

Because plasmids had a relatively small DNA-carrying capacity,[10] other vectors were designed to permit cloning of larger quantities of DNA (Fig. 2.3). The first of these was bacteriophage λ.[14] The genome of λ is a double-stranded DNA molecule about 50 kb in length, and is contained within a protective protein coat. The DNA of λ is injected into appropriate bacterial host strains, and subverts the bacterial cellular machinery to replicate itself, as well as to produce protein coats, ultimately killing the bacterium. Only about 30 kb (60%) of the λ DNA is actually essential to replication—yet the viability of bacteriophages declines dramatically if the length of their genome is less than 78%

or more than 105% of the physical size of the wild-type genome—accommodating as much as 20 kb of exogenous DNA. Recently-engineered strains of λ can accommodate up to 90 kb of DNA.[15] As is true of plasmids, λ vectors of increasing sophistication have been developed in recent years, and the reader seeking detail is referred to lab manuals or commercial product literature.

The next step in propagation of larger DNA elements was the development of "cosmid" vectors, modified plasmids that contain the DNA sequences (*cos* sites) required for packaging them into bacteriophage λ particles.[14] Because the vital parts of the plasmid are much smaller than the vital parts of the bacteriophage λ genomic DNA, the cosmid can carry more exogenous DNA—from 33-47 kb per cosmid. The basic features of cosmid vectors are similar to those of plasmids.[10]

The state of the art in cloning vectors is "artificial chromosomes," which are fundamentally different from plasmids, cosmids, or phage. Artificial chromosomes contain the basic features of a eukaryotic chromosome, and replicate along with the chromosomal DNA. The first (and still largest) artificial chromosomes were propagated in yeast cells.[16] Exogenous DNA, usually restriction fragments of 100,000 or more nucleotides (e.g., "megabase" DNA, 100 kb or more), are cloned into a vector comprised primarily of a yeast centromere, and two flanking telomeres. This element behaves as a chromosome, and becomes a part of the nuclear genome of the cell which contains it. As described above, drug resistance is used to help maintain this extra chromosome in the host cell, and colorimetric markers are used to verify that the extra chromosome carries exogenous DNA. Yeast artificial chromosome (YAC) vectors can carry very large amounts of DNA—individual clones containing more than 1,000,000 nucleotides (1000 kb) have been reported.

A recent modification has simplified megabase DNA cloning, although not improved the size of the cloned DNA.

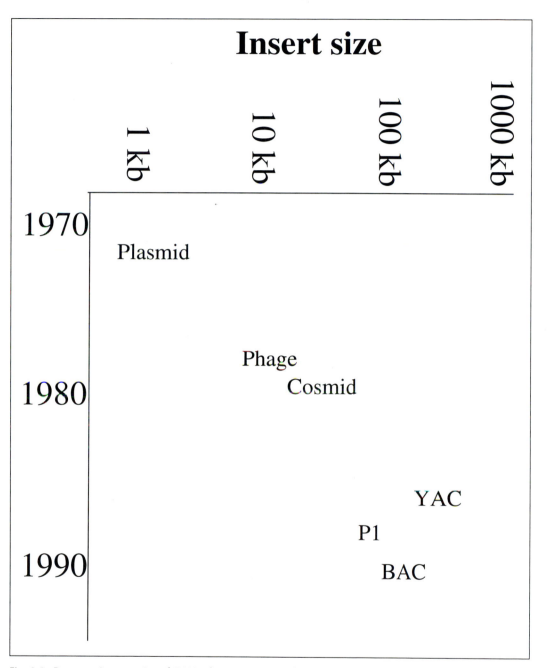

Fig. 2.3. Progress in capacity of DNA cloning vectors. The needs of molecular genetics have impelled development of DNA cloning vectors with progressively larger capacity, starting from plasmids in the early 1970s and ranging upward to YACs in the late 1980s. The next logical step may be the construction of artificial chromosomes directly in plants, further increasing the potential size of DNA which can be cloned. However, it is important to remember that physical maps of DNA based on large inserts have relatively low resolution—while it is faster to assemble physical maps based on large DNA clones, some advantages accrue to maps based on smaller inserts, in that genes can be more precisely delineated (see chapter 5).

Bacterial artificial chromosomes (BACs), using extrachromosomal DNA ("F factors") carried in some bacterial strains, are becoming increasingly popular as a supplement to YACs.[17] Because bacteria grow very quickly, and bacterial DNA cloning, now in its third decade, has become very sophisticated, use of bacterial systems to replicate exogenous DNA has numerous advantages over yeast. Moreover, BACs occur in only one copy per bacterial cell—a feature which confers a high degree of stability to the exogenous DNA, minimizing its opportunity for recombination and/or rearrangement.[17] While the size of BACs is somewhat smaller than YACs (the largest BACs reported are in the range of 350 kb), the facility of bacterial systems makes it likely that BACs will continue to be an important complement to YACs.

2.2.4. MOLECULAR CLONING: CREATION OF RECOMBINANT DNA, USING RESTRICTION ENZYMES AND CLONING VECTORS

Molecular cloning is the "subversion" of an organism's genetic machinery to produce many copies of a piece of extraneous DNA. The extraneous DNA may come from any organism (chemically, all DNA is essentially identical), may or may not be known to have a particular function, and may be of a size ranging from a few "letters" to literally millions of "letters" in their original order. The extraneous DNA is first made an integral part of a DNA "vector." Just as a mosquito carries malaria to an unsuspecting human, the DNA vector carries extraneous DNA into its host cell.

Molecular cloning has three basic requirements.
(1) A DNA molecule worthy of being cloned.
(2) A DNA molecule capable of serving as a vector.
(3) The means to integrate the exogenous DNA into the vector.

Construction of a DNA library, individual recombinant bacteria representing a target population of DNA molecules, is

illustrated in Figure 2.4. DNA from the organism under study, in this case a maize plant, is isolated (by a series of physical and chemical manipulations, tailored to the peculiarities of each plant species: see chapters 15-22). The DNA is exposed to a restriction enzyme, in this case one derived from *E. coli* and called *Eco* RI, which is the most commonly-used (and cheapest) of many such enzymes available. The enzyme locates all "recognition sequences" (in the case of *Eco* RI, sequences of GAATTC), and cuts the DNA at the same place within this sequence, wherever it is found. The plasmid has been "engineered" to have only 1-2 *Eco* RI recognition sequences, which will hereafter be referred to as the "cloning site." (Many other enzymes could be used equally well—modern cloning vectors have genetically-engineered "multiple cloning sites," which contain restriction site for many different restriction enzymes, in a short stretch of DNA).

After *Eco* RI "digestion," both the maize genomic DNA and the vector DNA have short single-stranded overhangs at each end of the "restriction fragments." Because of the specificity of *Eco* RI for restriction sites of identical sequence, together with the "mirror-image" nature of DNA (A pairs only with T; C pairs only with G), the single-stranded ends of different DNA fragments are complementary, and can associate to form double-stranded DNA. By providing an excess of maize DNA, relative to the amount of vector DNA, one increases the likelihood that "recombinant" DNA molecules are formed, inserting a piece of maize DNA into the bacterial plasmid. Finally, by incubating the plasmids in the presence of large numbers of bacteria, one can stimulate bacteria to "take up" a plasmid, and become "transformed."

To be representative of a defined population of DNA molecules, DNA libraries need to contain some redundancy. The likelihood that a DNA library will contain a specific sequence can be estimated from the number of "primary recombinants" (Recombinant bacteria or phage derived from

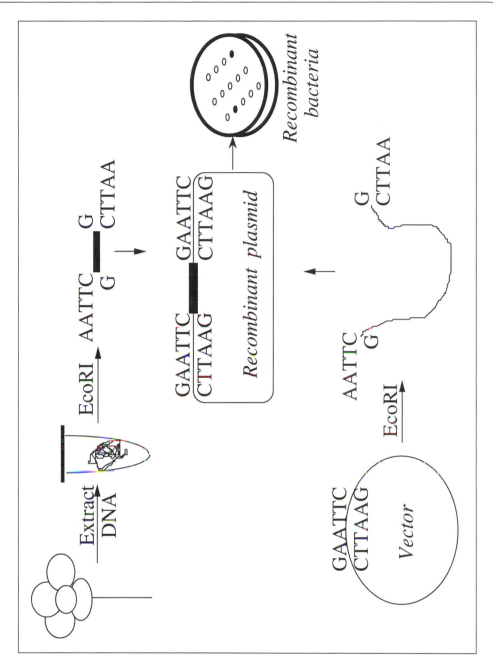

Fig. 2.4. Development of a DNA library. Construction of DNA libraries in a wide range of cloning vectors, and for a wide range of purposes, is based on similar principles. DNA is isolated from a target organism (in this case a flowering plant), and cleaved with a restriction enzyme. Simultaneously, DNA of the cloning vector is cleaved with the same restriction enzyme. The plant DNA and vector DNA are mixed, with a high ratio of plant DNA to vector DNA favoring annealing of plant DNA to vector DNA (in addition, the vector DNA has usually been chemically treated to prevent it from self-annealing). The population of DNA molecules formed are transformed into receptive ("competent") bacteria, which are then grown in the presence of an antibiotic for which a resistance gene is encoded by the vector DNA (assuring that the vector DNA is present in the surviving cells). Cells in which the vector has been disrupted by insertion of plant DNA, are revealed colorimetrically, as they fail to produce a particular enzyme encoded by the intact vector. Cells lacking inserted plant DNA produce the enzyme, which acts on compounds in the medium to produce a colored product (such as the two dark colonies on the plate).

the initial construction of the library, prior to "replication" of the library.) in the library. Succinctly, to be 99% confident that a particular DNA molecule is in a library, the library must contain about five times (5x) the aggregate length of DNA molecules in the target population. A smaller "3x" library affords about 95% confidence. For example, if one is seeking to represent the entire genome of *E. coli*, about 4,000,000 nucleotides (4000 kb), in a plasmid library with average insert size of 1 kb, about 20,000 randomly-chosen plasmids must be obtained to be 99% confident that any specific region of the genome is represented in the library.[10]

Many steps in this schematic have been tremendously streamlined over the past 20 years.[10] Pitfalls such as re-circularization of plasmid are prevented by phosphatase treatment, or by clever use of multiple restriction enzymes. By locating the cloning site in a gene which produces a colored (usually blue) product, one can identify recombinant (white) clones visually on an appropriate growth medium. Moreover, similar procedures are used to engineer bacteriophage and yeast. Schemes have been devised to clone not just random genomic DNA, but genes which produce a mRNA product (see chapter 14), DNA fragments which contain unique features such as short arrays of tandem repeats (see chapter 10), and many other variants. Finally, procedures for growing out all individual clones within a DNA library, and utilizing DNA, RNA, or antibody probes to identify clones containing a target element of known features, are well-established.[10]

Moreover, the principles of molecular cloning are similar in a wide range of circumstances—for example, construction of "artificial chromosome" libraries differs primarily in that the exogenous DNA must be very carefully handled to maintain it in long intact segments, usually embedding it in agarose beads or plugs to avoid shear forces. Such artificial chromosome vectors, using yeast[16] or bacteria[17] as host cells, have permitted large-scale physical mapping of complex genomes (see chapter 5).

2.2.5. THE POLYMERASE CHAIN REACTION: IN VITRO DNA SYNTHESIS MAKES DNA ANALYSIS FASTER AND MORE EFFICIENT

The fundamental power of molecular cloning is that it affords a rapid and efficient means to obtain a specific, defined DNA element in large enough quantity to study it in extraordinary detail. For example, one might first sequence the element, to study association between the chemical information content (genetic code) of the element and a suspected function (see chapter 14). Next, one might engineer modifications in the molecule, and transform the modified version back into a null mutant lacking a functional molecule, to study "mutant" forms of the gene...learning about function by comparison to the function of the wild-type form (see chapters 12 and 13).

In the past five years, a new technique has simplified and accelerated many routine procedures associated with genome mapping, and led to creation of several new systems for identifying DNA markers. The "polymerase chain reaction (PCR),"[18,19] has revolutionized virtually all aspects of DNA manipulation, and yielded the 1993 Nobel Prize to its discoverer, Kary Mullis.

It has long been known that the requirements of DNA synthesis reactions could be emulated in a test tube. Chemically, the requirements for synthesis of DNA are simple: an ample supply of the four nucleoside triphosphates, a single-stranded DNA template with a double-stranded region upstream...the free hydroxyl group of the last paired nucleotide providing a starting point for DNA synthesis, a DNA polymerase (one of a class of enzymes capable of "reading" one strand of DNA and synthesizing the complement), and the cofactors and abiotic conditions (temperature, pH) necessary for the chosen DNA polymerase to function.

However, prior to PCR, most techniques were limited to a single cycle of DNA synthesis, using a single-stranded DNA "template" to make a double-stranded product. While PCR works much the same way, it is cyclic, first making DNA, then using this product as template to make more product...hence polymerase chain reaction. Consequently, the quantity of a specific DNA element in solution grows exponentially, rather than linearly. After 20-30 cycles of "amplification," a specific DNA product can be amplified by 10^5 or more.

The mechanics of PCR are illustrated in Figure 2.5. The requirement for a free hydroxyl group, provided by double-stranded DNA upstream of the "start site" for DNA synthesis, is used to confer specificity to the PCR. In the PCR, this requirement is met by annealing of a synthetic, single-stranded DNA "primer," usually of 20-25 bases in length, to the single-stranded template DNA. A specific DNA sequence of 20 bases is likely to occur in only one location (if at all) in the genome of a higher organism, since 4^{20} is much larger than the number of nucleotides in the largest known genome. PCR requires basic ingredients common to all DNA synthesis reactions including buffers and salts which permit DNA polymerase to function, an ample supply of the four deoxynucleoside triphosphates, a thermophilic DNA polymerase, and a DNA template for the polymerase to read and copy. In theory, the quantity of target DNA might be increased by 100% in each cycle, however, actual yields are more often in the range of 50-60% per cycle and tend to decline as the reaction proceeds.

Moreover, in most protocols (but see RAPD, AP-PCR, and DAF, below), a second primer delineates the start of DNA synthesis for the mirror-image DNA strand, conferring further specificity. Only the DNA lying between these two primers is "amplified." In the early history of PCR, obtaining DNA sequence information was tedious and expensive, and identifying primers was an important limiting factor. This limitation partly motivated development of several DNA amplification methods which relied upon primers of arbitrary sequence (see chapter 3). However, the advent of automated DNA sequencing (see

chapter 14) has remedied this potential limitation.

An elegant refinement made PCR the single most-widely used technique in molecular biology. In its initial implementation, PCR required continuous supervision. The three phases of each PCR "cycle" require very different temperatures (template DNA denaturation at 95°C, primer-template hybridization typically at 37-55°C, DNA synthesis at 72°C), and the early practitioners of PCR were often junior students, sitting in front of three waterbaths with a stopwatch. Further, since the DNA polymerases available at the time did not retain activity at temperatures of 95°C, fresh enzyme had to be added frequently. The discovery that DNA polymerase from thermophilic organisms (mostly bacteria which live in hot springs) retained activity even after prolonged exposure to temperatures of 95°C[20] meant that the reaction could be mechanized—the enzyme need only be added once, and the reaction can proceed by itself, requiring only pre-programmed changes in temperature.[21] Today, laboratory instruments for programmed, cyclical variation in sample temperature are widely available.

PCR has found an enormous number of applications in genome analysis and other areas of molecular biology. Many once-tedious laboratory procedures have been greatly streamlined by PCR, including DNA sequencing, isolation of novel alleles at well-defined genetic loci, production of large quantities of specific cloned DNA probes, and many others. Much detail regarding such shortcuts is readily available from primary literature, laboratory manuals, and vendor publications, and will not be addressed in detail herein.

More to the immediate topic, PCR has enabled us to ask questions and pursue learning opportunities which previously were not accessible, largely due to the newfound ability to work with very small quantities of DNA, and sometimes DNA in poor condition. DNA forensics is becoming an increasingly important area, and is largely made possible by the ability of PCR

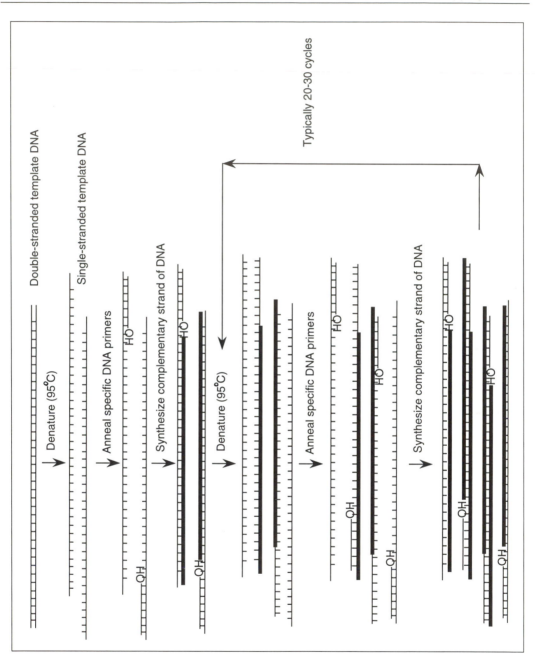

Fig. 2.5. The Polymerase Chain Reaction. Double-stranded DNA (for example, total genomic DNA from a particular plant) is heat-denatured, and permitted to anneal to specific DNA primers of a sequence complementary to the target site of the genome. Starting from the 3' (hydroxyl) end of this short double-stranded region, new complementary DNA is synthesized by a DNA polymerase. The process is then repeated, and since the newly-synthesized DNA can now serve as template for further synthesis, DNA is propagated exponentially. Amplification of 10⁵ or more can be accomplished, permitting visualization of specific DNA sequences which were represented in the initial reaction by as little as a single molecule (such as a single human sperm cell—see text).

to "amplify" specific informative DNA molecules from exceedingly small amounts of tissue (e.g., a single human hair[22] or a single sperm nucleus[23]). PCR has afforded access to sequence information from other exceedingly-rare DNA samples, such as DNA from fossilized or amber-embedded tissues.[24-27]

2.3. DNA MARKERS FOR GENOME ANALYSIS

It was quickly realized that the utility of restriction enzymes, for providing local landmarks in the DNA, extended far beyond molecular cloning. In 1980, a landmark paper[28] outlined the basic principles which would represent the foundations of an explosion in molecular genetic analysis of plants and animals. The concept was that mutations in restriction sites, or mutations which altered the distance between adjacent restriction sites, could be visualized as "DNA markers."

This technique, referred to as restriction fragment length polymorphism (RFLP), has been the basis of an explosion of genetic mapping activity. RFLPs were used to construct the first molecular map of the human genome, although the large quantities of DNA and relatively low allelic richness of RFLPs in humans led to subsequent implementation of new techniques in humans and experimental animals. In plants, which regenerate organs plucked for DNA, RFLP has remained a relatively more important technique, and is the basis for detailed genetic maps of most major crops (see chapters 15-22).

A DNA basis for genome mapping represented a breakthrough in two important ways: (1) It provided a virtually unlimited potential supply of markers (mileposts) in the hereditary information, and (2) The mileposts were components of the same molecule as the functional genes, thus provided a means to gain direct access to specific genes.

Finally, the commonality among organisms as a result of "conservation" of genes, affords the opportunity to use genes as "DNA markers" (mileposts) common to different organisms. By such "comparative mapping," we have learned that not only the information content of genes, but also the order of genes along the chromosome, tend to persist over long periods of evolution. Development and utilization of comparative maps will be addressed in chapter 3, and examples are provided in chapters 15-22. Comparative mapping offers tremendous potential efficiencies in genome analysis—because new information about both location and function of genes in one organism are likely to be relevant to other related organisms.

The increasing capacity of DNA cloning vectors has enabled development of "physical maps" comprised of contiguous overlapping megabase DNA clones spanning large regions of plant chromosomes,[29] and represents a major step toward complete sequencing of entire plant chromosomes.

2.4. SUMMARY

From modest beginnings, less than two decades ago, the next few years of DNA-based research will yield the complete DNA sequences of several higher organisms, probably including at least two plants (*Arabidopsis* and rice). We will soon have in our hands (or more accurately, in our computers) the entire genetic complement of these (and other) organisms—opening a vast number of doors to study and understand the molecular basis of plant growth and development.

REFERENCES

1. Avery OT, CM Macleod, M McCarty. Studies on the chemical nature of the substance inducing transformation of pneumococcal types. J Exp Med 1944; 79:137-158.
2. Watson JD and FHC Crick. A structure for deoxyribose nucleic acid. Nature 1953; 171:737-738.
3. Watson JD and FHC Crick. General implications of the structure for deoxyribose nucleic acid. Nature 1953; 171:964-967.
4. Meselson M and F Stahl. The replication of DNA in *E. coli*. Proc Nat'l Acad Sci USA 1958; 44:671-682.

5. Jacob F., and J Monod. Genetic regulatory mechanisms in the synthesis of proteins. J Mol Biol 1961; 3:318-356.

6. Brenner S, F Jacob, and M Meselson. An unstable intermediate carrying information from genes to ribosomes for protein synthesis. Nature 1961; 190:576-581.

7. Nirenberg MW and JH Matthei. The dependance of cell-free protein synthesis in *E. coli* upon naturally-occurring or synthetic polyribonucleotides. Proc Nat'l Acad Sci USA 1961; 47:1588-1602.

8. Maxam AM, and W Gilbert. A new method for sequencing DNA. Proc Nat'l Acad Sci USA 1977; 74:560-564.

9. Sanger F, S Nicklen, AR Coulson. DNA sequencing with chain-terminating inhibitors. Proc Nat'l Acad Sci USA 1977; 74:5463-5467.

10. Sambrook J, EF Fritsch, T Maniatis. Molecular Cloning: A Laboratory Manual. (Second edition). Cold Spring Harbor Laboratory Press, 1989, Cold Spring Harbor, NY.

11. Southern EM. Detection of specific sequences among DNA fragments separated by gel electrophoresis. J Mol Biol 1975; 98:503-517.

12. Smith HO. Nucleotide sequence specificity of restriction endonucleases. Science 1970; 205:455-462.

13. Cohen SN, ACY Chang, HW Boyer, RB Helling. Construction of biologically functional bacterial plasmids in vitro. Proc Nat'l Acad Sci USA 1973; 70:3240-3244.

14. Hohn B. In vitro packaging of λ and cosmid DNA. In Methods of Enzymology: Recombinant DNA (volume 68), Ed. R. Wu, Academic Press, 1979, New York.

15. Sternberg N. Bacteriophage P1 cloning system for the isolation, amplification, and recovery of DNA fragments as large as 100 kilobase pairs. Proc Nat'l Acad Sci USA 1990; 87:103-107.

16. Burke DT, G Carle, MV Olson. Cloning of large segments of exogenous DNA into yeast by means of artificial chromosome vectors. Science 1987; 236:806-812.

17. Shizuya H, B Birren, U-J Kim, V Mancino, T Slepak, Y Tachiiri and M Simon. Cloning and stable maintenance of 300-kilobase-pair fragments of human DNA in *Escherichia coli* using an F-factor-based vector. Proc Nat'l Acad Sci USA 1992; 89:8794-8797.

18. Mullis K, F Faloona, S Scharf, R Saiki, G Horn, H Erlich. Specific enzymatic amplification of DNA in vitro: The polymerase chain reaction. Cold Spring Harbor Symp Quant Biology 1986; 51:263.

19. Mullis K, and F Faloona. Specific synthesis of DNA in vitro via a polymerase-catalyzed chain reaction. Methods Enzymol 1987; 155:335.

20. Chien A, DB Edgar, JM Trela. Deoxyribonucleic acid polymerase from the extreme thermophile *Thermus aquaticus*. J Bacteriol 1976; 127:550.

21. Saiki RK, DH Gelfand, S Stoffel, SJ Scharf, R Higuchi, GT Horn, KB Mullis, HA Erlich. Primer-directed enzymatic amplification of DNA with a thermostable DNA polymerase. Science 1988; 239:487.

22. Gill P, AJ Jeffreys, DJ Werrett. Forensic applications of DNA fingerprints. Nature 1985; 318:577-579.

23. Li H, UB Gyllensten, X Cui, RK Saiki, HA Erlich, N Arnheim. Amplification and analysis of DNA sequences in single human sperm and diploid cells. Nature 1988; 335:414.

24. Golenberg EM, DE Giannasi, MT Clegg, CJ Smiley, M Durbin, D Henderson, G Zurawski. Chloroplast DNA sequence from a Miocene Magnolia species. Nature 1990; 344:656-658.

25. Soltis PS, DE Soltis, CJ Smiley. An RBCL sequence from a miocene Taxodium bald cypress. Proc Natl Acad Sci USA 1992; 89:449-451.

26. DeSalle R, J Gatesy, W Wheeler, D Grimaldi. DNA sequences from a fossil termite in oligo-Miocene amber, and their phylogenetic implications. Science 1992; 257:1933-1936.

27. Cano RJ, HN Poinar, NJ Pieniazek, A Acra, GO Poinar, Jr. Amplification and sequencing of DNA from a 120-135 million year old weevil. Nature 1993; 363:536-538.

28. Botstein D, RL White, M Skolnick, RW Davis. Construction of a genetic linkage map in man using restriction fragment length polymorphisms. Am J Hum Genet 1980; 32:314-331.

29. Schmidt R, J West, K Love, Z Lenehan, C Lister, H Thompson, D Bouchez, C Dean. Physical map and organization of *Arabidopsis* chromosome 4. Science 1995; 270:480-483.

CHAPTER 3

MAKING GENETIC MAPS

Andrew H. Paterson

"...the main practical limitation of the technique seems to be the availability of suitable markers, and the time that can be given to the considerable work involved."*

3.1. INTRODUCTION

A "genetic map" can be thought of much as a road map. Just as mile posts guide the motorist along a linear highway, molecular tools enable the geneticist to establish specific "genetic markers" at defined places along each linear chromosome. Genetic markers can then be used as mileposts, to determine when one has reached (or passed by!) a particular gene of interest.

"Genetic markers," differences in the DNA sequence of chromosomes derived from different progenitors, can be visualized in several different ways. Morphological mutations, sometimes called "visible markers" (see chapter 1), can be visualized by a casual glance at an individual. Isozymes, or protein variants[1] require separation by electrophoresis, and are visualized by a colorimetric "activity assay" for the relevant enzyme in crude extracts from living tissues. DNA markers are visualized either by use of radioactivity (autoradiography), fluorescence, or by direct chemical staining of the DNA itself.

To be useful as a genetic marker, a trait must meet two criteria: (1) the trait must differentiate between the parents, and (2) the trait must be precisely reproduced in the progeny. For example, blood type is a good genetic marker. If a parent with blood type A has a spouse with blood type O, their children are of either blood types A or O. In contrast, stature (height) is often different between two parents, but it is usually a poor genetic marker, because the stature of progeny is often intermediate between parents, as a joint result of genetic factors ("nature") and environmental factors ("nurture").

While genetic markers were being used before it was known that DNA was the hereditary material, DNA-based technology revolutionized

---

* *Thoday JM. Location of polygenes. Nature 1961; 191:368-370.*

Genome Mapping in Plants, edited by Andrew H. Paterson. © 1996 R.G. Landes Company.

genetic mapping in many organisms. Visible markers have been used in genetic studies since 1910,[2] and isozymes since 1959.[1] However, as discussed in chapter 1, these technologies proved inadequate for assembling high-density complete genetic maps efficiently.

In 1980, it was suggested that large numbers of genetic markers might be found by studying differences in the DNA molecule itself, revealed as restriction fragment length polymorphisms (RFLPs).[3] Plants have about 10^8-10^{10} nucleotides of DNA[4]—if even a small fraction of these are different between two individuals, an enormous number of potential DNA markers result. This makes DNA markers suitable for developing high-density genetic maps. Moreover, the fact that all higher organisms share DNA as the hereditary molecule, makes DNA-based technology generally applicable to molecular mapping of most organisms.

3.2. MOLECULAR BASIS OF DNA MARKERS

DNA markers arise as a result of several different classes of mutations. The simplest event is substitution of as little as a single nucleotide, differentiating two genotypes. When a base substitution eliminates a restriction site, it changes the length of DNA fragments detected by the relevant assay (see Fig. 3.2, below), and thus represents a discrete marker which is directly representative of an individual's genotype. In assays based on the polymerase chain reaction (PCR), base substitution within the region to which a PCR "primer" would (otherwise) bind has exactly the same effect.

Alternatively, rearrangements in the DNA intervening between two restriction sites, or two priming sites, can generate DNA markers. Such rearrangements might include insertion or excision of mobile DNA elements (such as transposable elements: see chapter 13), or errors in replication of arrays of tandemly-repeated DNAs (see chapter 10). Rearrangements tend to create DNA markers that can be detected by several different restriction enzymes, or different PCR primers flanking the same region, while base substitutions are specific to particular restriction enzymes or PCR primers. The common finding that DNA markers detected with one restriction enzyme can often be detected with other restriction enzymes suggests that the majority of DNA markers in most plant species result from localized rearrangements in DNA.

3.3. VISUALIZATION OF DNA MARKERS

Several techniques are available for visualization of DNA markers. In plants, the most widely-used has been restriction fragment length polymorphisms (RFLPs: Fig. 3.1). This technique uses cDNAs or other cloned DNA elements as probes (labeled using radioactivity, or with conjugated enzymes which yield a colored substrate) to detect DNA-level differences. The RFLP technique remains the single most-widely used DNA marker assay in plants, and mapped DNA probes are available in many plant species (see chapters 15-22). Moreover, the technique is relatively robust, and readily transferable between different labs.

Although it remains widely-used, two basic limitations of the RFLP technique have motivated development of several alternative technologies. The first limitation is the quantity of DNA required—RFLP analysis typically requires 50-200 micrograms of DNA per individual, to generate a DNA fingerprint of the entire genome. Large-scale DNA extraction is tedious and laborious—use of smaller amounts of DNA would increase sample throughput. In contrast to RFLPs, PCR-based techniques require only about 10% of this amount of DNA, as template for de novo synthesis of large quantities of the target sequence. As a result of these limitations, a wide range of other techniques for detecting DNA markers have been described—these are discussed below, and summarized in Figure 3.2.

PCR-based techniques for detecting DNA markers require development of a

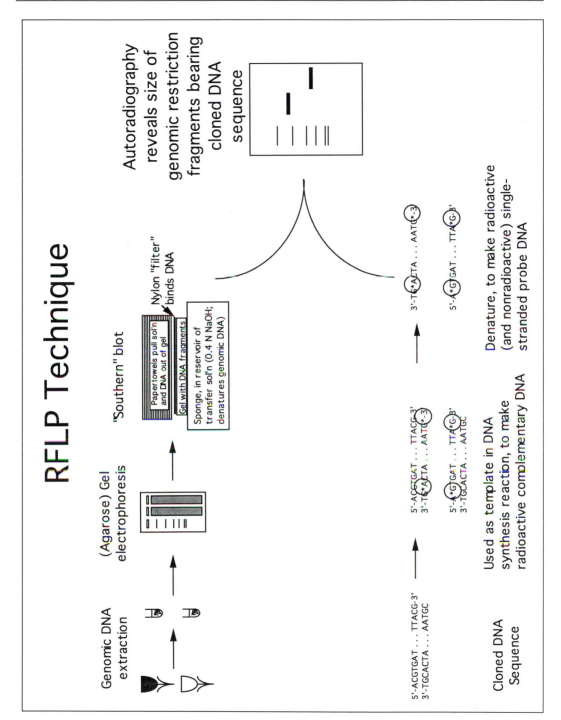

RFLP Technique

Genomic DNA extraction

(Agarose) Gel electrophoresis

"Southern" blot

Nylon "filter" binds DNA

Paper towels pull sol'n and DNA out of gel

Gel with DNA fragments

Sponge, in reservoir of transfer sol'n (0.4 N NaOH; denatures genomic DNA)

Autoradiography reveals size of genomic restriction fragments bearing cloned DNA sequence

Cloned DNA Sequence

5'-ACGTGAT . . . TTACG-3'
3'-TGCACTA . . . AATGC

5'-ACGTGAT . . . TTACG-3'
3'-TC*ACTA . . . AATC(*-3

5'-(*GTGAT . . . TTA*G-3'
3'-TGCACTA . . . AATGC

Used as template in DNA synthesis reaction, to make radioactive complementary DNA

3'-TC*ACTA . . . AATC(*-3

5'-A(*GTGAT . . . TTA(*G-3'

Denature, to make radioactive (and nonradioactive) single-stranded probe DNA

Fig. 3.1. Restriction fragment length polymorphism (RFLP) technique. Total genomic DNA is extracted from two (or more) different genotypes, and subjected to "digestion" with a restriction endonuclease (usually, several restriction endonucleases are screened, but only one has been shown here). The resulting "restriction fragments" are separated by electrophoresis, transferred to a nylon membrane by Southern blotting (see chapter 2) and hybridized to a specific DNA probe which has been "labeled" with radioactivity or fluorescence, to permit visualization of the restriction fragments. In a subset of cases, the combination of DNA probe and restriction enzyme reveal a "restriction fragment length polymorphism," as shown. Such markers have been widely used in genetic mapping, as described in text.

specific DNA primer to serve as the "start" site for amplification. Prior to the advent of automated DNA sequencing, it was tedious and expensive to obtain the large quantities of DNA sequence information needed to devise PCR primers. Consequently, researchers were receptive to techniques using PCR primers of arbitrary sequence, in the absence of a priori knowledge of homology to a specific site in the genome. The most widely-used of these, RAPD (randomly amplified poly-

morphic DNA,[5] with a virtually identical technique referred to as AP-PCR[6]), relies on a single 10-base primer of largely-arbitrary sequence, except that primers were selected to have 60% or more G+C content, to obtain stronger binding to template (G-C pairs involve three hydrogen bonds, while A-T pairs involve only two). PCR amplification would only be expected to occur when the priming site (complementing the 10-mer) occurred twice, in opposite orientation, within about 2,000

Fig. 3.2. Molecular basis of DNA markers. DNA marker alleles arise as a result of two basic mechanisms, illustrated in part 1 and parts 2-3. (1) Different DNA marker alleles resulting from a single base substitution which has occurred since divergence of the two genotypes (a, b) from their most recent common ancestor. Although the illustration shows base substitution within a restriction site, used to reveal RFLPs, it could equally well occur within the binding site for a PCR primer such as a RAPD decamer. (2) Insertion of a new genomic element between consecutive restriction sites, thus increasing the length of the restriction fragment (or PCR amplification product). This might happen as a result of a transposon visitation. (3) is a special case of part 2, in which an array of tandem repeats of a DNA element falls between the restriction/priming sites. Such tandem arrays can change in copy number by a variety of mechanisms (see chapter 10), thus changing the length of the restriction fragment (or PCR amplification product). This is the basis of the widely-used "sequence-tagged microsatellite (STM)" or "simple-sequence length polymorphism (SSLP)" technique.

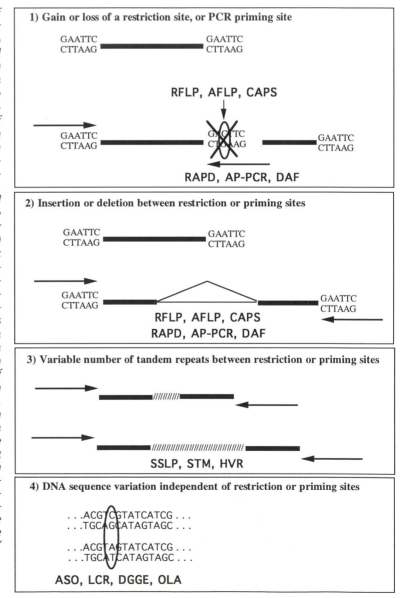

bases of one another along a chromosome. In theory, these conditions are met about five times in a higher eukaryotic genome (such as that of human). A similar technique, "DNA amplification fingerprinting (DAF)" is based on arbitrary primers of as little as six bases.[7]

RAPDs have not been widely-embraced to construct primary genetic maps, probably due to the unpredictable behavior of short primers (10-mers) in PCR reactions.[8] However, arbitrary-primer techniques have found much application in experiments which can afford a large amount of repetition to test reliability (see chapter 4). In addition to the volatility of short primers, numerous other reaction conditions were reported to influence RAPD amplification patterns,[5] including DNA quality, salt (especially Mg) concentrations, DNA polymerase sources and concentrations, and template and nucleotide concentrations. Perhaps as a consequence of these or other variables, amplification patterns tend to be difficult to repeat, especially between different labs.

The advent of automated DNA sequencing (see chapter 14) dramatically increased the availability of sequence information for designing DNA primers. DNA probes previously mapped as RFLPs have been sequenced in several different organisms, then used to amplify corresponding genomic elements. By cutting these amplification products with a restriction enzyme, DNA markers can be revealed. These CAPS, or "cleaved amplified polymorphic sequences"[9,10] retain the basic advantages of PCR, that only nanogram (rather than microgram) quantities of DNA are required, and that immediate detection by fluorescence accelerates results. The longer (15-25 nucleotide) PCR primers used for CAPS tend to generate more reproducible results than RAPDs or other arbitrary-primer techniques. Finally, CAPS can be selected at known genetic map positions by virtue of prior mapping as RFLPs.

However, CAPS fail to remedy the second limitation of RFLPs—that closely-related individuals usually contain the same alleles. This has been a more serious constraint in mammals than in plants—because plant geneticists enjoy greater flexibility to make crosses between highly divergent genotypes. In 1989, mammalian geneticists described a new technique based upon rapidly-evolving regions of the genome.[11] It had been previously known that arrays of short DNA elements repeated in tandem (as per Fig. 3.1) tend to be imprecisely replicated during DNA synthesis, and frequently generate new alleles with different numbers of repeating units.[12] By identifying PCR primers which immediately flank such repeat arrays, one could amplify the intervening region, and detect DNA length polymorphisms. Such "simple sequence length polymorphisms (SSLPs)," or "sequence-tagged microsatellites (STMs)," occur very frequently, and have proven extremely useful in detecting DNA markers which discern closely-related plants and animals. Further, because STMs use long PCR primers which are specific to a single genetic locus (like CAPS), they enjoy the reliability of CAPS. Although STMs are costly to identify, they have become the DNA marker of choice in mammalian systems, and are second only to RFLPs in plant systems (see chapter 10). SSLPs or STMs are closely-related to an earlier form of DNA marker, "hypervariable repeats," (HVRs)[13] which differed primarily in that the repeating unit contained a few more nucleotides, a "minisatellite" rather than a microsatellite. However, the short repeating element of SSLP/STMs is easily detected in a wide range of genomes using short synthetic DNA primers, while the longer HVR elements must first be isolated from a genome, as they are too long to find by chance screening of a few simple sequences.

Finally, several techniques are available to take advantage of DNA sequence differences which are as small as a single nucleotide, but do not fall conveniently in a restriction site or PCR priming site. A pioneering application of synthetic DNA oligonucleotides, even preceding PCR, is the "allele-specific oligonucleotide (ASO)" technique.[14] ASOs are synthetic oligonucleotides of about 20 bases in length, which are a perfect complement to one allele, and a mismatch of one or more bases with an alternative allele. Because a mismatch of 1 base in 20 often represents a substantial difference in DNA melting temperature (perhaps as much as 5°C), DNA hybridization followed by washing at an appropriate temperature can be used to differentiate between alleles. Hybridization between "match" and "mismatch" of as little as 1 nucleotide in 500 can be detected using denaturing gradient gel electrophoresis (DGGE)[15]—a technically demanding procedure, but very effective. Some protocols use DNA ligation-based assays[16] in conjunction with thermostable ligases[17] to detect single-nucleotide differences.

3.4. RELATIONSHIPS AMONG "DNA MARKERS" ARE DETERMINED BY THE PRINCIPLES OF GENETIC LINKAGE

Genetic mapping is made possible by the fact that the nuclear genome of higher organisms is organized and transmitted as linear units, called chromosomes. "Genetic linkage," or co-transmission from parent to progeny of genetic markers which are close together on the same chromosome, provides a means for determining the order of DNA markers along the chromosome.

For example, Figure 3.3 depicts four genetic markers (A-D), assayed on 20 progeny of a "backcross" of a heterozygous individual to one of its homozygous parents. It is clear from a casual inspection that genotype at marker B is unrelated to genotype at any of the other markers—e.g., marker B is "unlinked" to A, C, or D. However, genotype at markers A, C, and D show some similarities. Only four individuals (1, 3, 10, 13) show different genotypes at markers A and C. Two individuals (18, 19) show different genotypes at markers C and D. Five individuals (1, 3, 10, 11, 13) show different genotypes at markers A and D. This suggests that

marker C lies between markers A and D, somewhat closer to D than to A.

By extending such analysis to hundreds, or even thousands of markers, one can build up a "linkage map" describing relationships among the markers. Moreover, by extending upon the principle shown in this simple "backcross" population, other population structures can be used for linkage analysis, including F2 populations derived by selfing or intermating of heterozygous individuals, "recombinant inbred" populations or "intermated" populations which have been through several cycles of recombination, or even outcrossed populations (Fig. 3.4, also see ref. 18). Fortunately, several excellent computer programs have been developed to aid in the assembly of genetic maps. These programs have been well-described in published literature, or in documentation available from the creators. Prominent names include MapMaker,[19] Linkage-1,[20] GMendel,[21] Cri-Map,[22] JoinMap,[23] and MapManager.[24] Readers who are embarking on a new genetic mapping excursion are encouraged to become familiar with the current versions of these programs as their interests warrant.

3.5. CHOICE OF GENETIC MAPPING POPULATIONS

Genetic maps of plants are constructed based on several different kinds of populations, with each population structure having unique strengths and weaknesses. Figure 3.4 illustrates the different kinds of mapping populations, and Table 3.1 summarizes the differences among them. Most genetic mapping populations in plants have been derived from crosses between largely-homozygous parents, however several plant species (e.g., sugarcane, potato, alfalfa) which are intolerant of inbreeding must rely on crosses between heterozygous individuals—genetic analysis of such plant populations relies on segregation of "single-dose restriction fragments," and has been described in detail.[25]

By crossing the F1 hybrid to one of the parents (the "recurrent"), a "backcross" population can quickly be generated, which segregates for alleles derived from the "donor" (non-recurrent) parent. Each backcross individual is recombinant along only one of each pair of homologous chromosomes, and has undergone only one cycle of meiosis—so associations between marker alleles from a common parent ("linkage disequilibrium") remain very strong. Backcross populations are most often made out of necessity—many interspecific crosses in plants produce backcross progeny which are fertile, but self-pollinated progeny which are sterile and have little promise for genetic studies.

By selfing the F1 hybrid, or intermating it to its sibling, an F2 population can be made. Each F2 individual is recombinant along both of each pair of homologous chromosomes, but like the backcross has undergone only one cycle of meiosis.

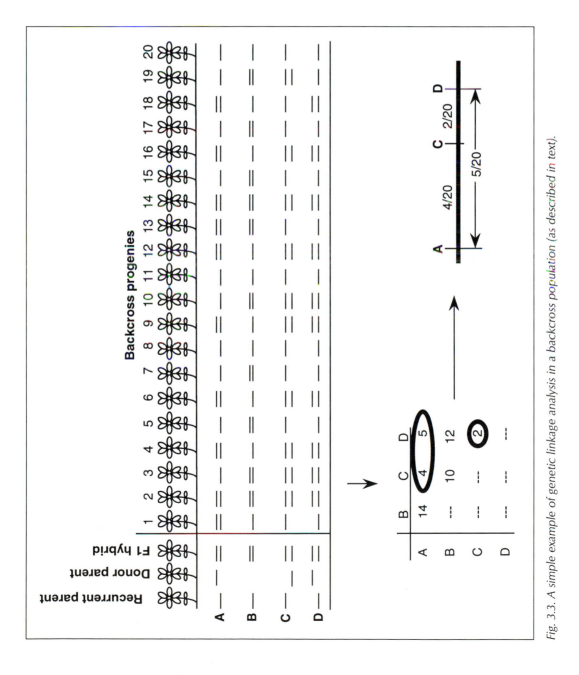

Fig. 3.3. A simple example of genetic linkage analysis in a backcross population (as described in text).

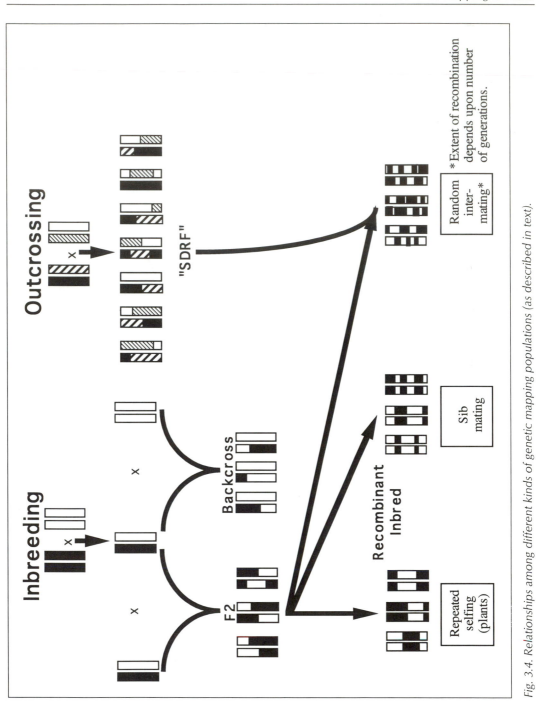

Fig. 3.4. Relationships among different kinds of genetic mapping populations (as described in text).

Table 3.1. Features of different genetic mapping populations

	(Per chromosome)		(Per locus)		
	# informative gametes per indiv.	# recombination events per gamete	# possible genotypes per locus	# generations to make	Replication?
DH-Backcross:	1	x	2	2	clonal only
F2	2	x	3	2	clonal only
RI (SSD)	1	2x	2	6-8	yes
Intermated	2	x+(xn/2)	3	n*	clonal only
"CEPH" (human)	2	2x	10	2	no

The product of the # informative gametes per individual and the # recombination events per gamete is the # recombination events per individual (per chromosome), and expresses the relative amount of information available for discerning recombination between closely-linked points.

*Typically n = 4, however other numbers may be used under certain circumstances. It is worthy of note that intermated populations require large numbers of crosses at each generation to make, and consequently require considerably more time and effort than other forms of mapping populations.

As a result of the fact that each homologue contains unique recombination events, an F2 individual harbors 2x as much information than a backcross individual, for resolving the order of genetic markers along a chromosome. Because of this added information, and also because F2 populations harbor all possible combinations of parental alleles (i.e., AA, Aa, aa), the F2 population has been the most widely used of genetic linkage mapping populations in plants.

"Doubled haploids" are commonly-used in some plant species, which are amenable to anther culture. Doubled haploids (hereafter DH) are made by regenerating plants from single pollen grains, and inducing chromosome doubling. Because the plant has two identical homologs, the amount of recombinational information is exactly equivalent to a backcross. However, DH individuals are completely homozygous, and can be self-pollinated to produce large numbers of progeny which are all genetically identical. This permits replicated testing of phenotypes, and also facilitates distribution of identical DH populations to many different researchers, so that detailed information about common genotypes can be accumulated.

Homozygous populations can also be made by traditional means—the most widely-used genetic mapping population in the mouse (a key mammalian model system) has been the "recombinant inbred (RI)" population,[26] in which the progeny of an F2 cross are inbred to homozygosity. In the mouse, inbreeding is accomplished by recurrent mating between siblings, until homozygosity is reached. The many cycles of meiosis needed to reach homozygosity result in additional opportunities for recombination, so that RI mouse populations harbor about 4x as much recombinational information per homologue as F2 mouse populations. However, RI individuals have two identical homologues, while F2 individuals have two homologues which each contribute unique information, consequently the overall gain in recombinational information for RI mice is 2x.

Recombinant inbred populations have also been used in plants—with a few key differences. Because many plants can be "self-pollinated," homozygosity can be reached more quickly than in mice, and most RI plant populations have been developed by recurrent selfing, also known as "single-seed descent."[27] Selfing is a much more intensive form of inbreeding than is sib mating. The more rapid achievement of homozygosity has the consequence that heterozygosity is lost more quickly, and there are fewer opportunities to obtain new recombinants. RI plant populations harbor only about 2x as much recombinational information per homologue as F2 plant populations—however, as noted above, RI individuals have two identical homologues while F2 individuals have two homologues which each contribute unique information. Consequently, RI plants offer approximately the same amount of information as F2 plants, for resolving close linkages.[18]

Plant geneticists seeking the gains in resolution which are afforded by recombinant inbreds in mice, do have several alternatives. In principle, recombinant inbred plant populations can be made by sib mating, just like mice, thereby increasing the amount of recombinational information per individual. However, a preferred option is to use random intermating among different F2-derived lineages for several generations, which permits the accumulation of 50% more recombinational information each generation. An intermated population of *Arabidopsis* has recently been described, and intermated populations of several other taxa are being developed.[18]

Different objectives can be expedited by the choice of different genetic mapping populations. For initial construction of a primary genetic linkage map, in an organism which has not been previously studied, the strong linkage disequilibrium of the F2 or backcross population permit one to detect linkage between widely-scattered markers. For resolving the order of closely-linked markers in detailed genetic maps, the intermated population is clearly the preferred population. For genetic analysis of complex phenotypes which are subject to the vagaries of environment, choice of recombinant inbred populations permits one to conduct replicated trials and minimize the influence of non-genetic factors on the data. Some of the advantages and disadvantages of different mapping populations are summarized in Table 3.1.

3.6. GENETIC DISTANCE

"Distance" along a genetic map is measured in terms of the frequency of recombination between genetic markers. For example, reconsider markers C and D in Figure 3.3. Among the 20 progeny assayed, only two, or 10% differed from the parental allele combinations. Such a value is said to reflect a "recombination fraction" of 10%. Usually, genetic maps adjust this distance for the possibility that two simultaneous recombination events occurred within the same interval, thereby making the individuals appear to represent a parental type, but in fact contain two recombination events. For example, if one considers only the genotypes at markers A and D, there appear to have been five recombination events among the 20 progeny examined. By including marker C, it becomes apparent that individuals 18 and 19 have each had two recombination events, one between A and C, and the other between C and D. The likelihood of such "double recombinants" is proportional to the square of the recombination distance between A and D—and is accommodated by adjustment of "recombination fraction" to units called centiMorgans, in honor of T.H. Morgan. The most commonly-used algorithm for adjustment of recombination fraction into centiMorgans was described by Kosambi,[28] and shown by Perkins[29] to approximately reflect the genetic behavior of several plant and animal taxa. Direct calculation of centiMorgan distances from recombination fraction is incorporated into several computer programs for genetic mapping.

The precision with which genetic distance is measured, is directly related to the number of individuals which are studied. In the example, only 20 individuals were studied. If no recombinants were found between two markers, one would infer a recombination distance of 0 cM. However, study of an additional 80 individuals might reveal a recombinant, and suggest a recombination distance of 1 cM. The number of individuals studied in a genetic mapping experiment is determined jointly by the level of precision needed, and the diminishing return afforded by linear increases in the amount of work and cost. Typically, primary genetic maps are constructed based on 50-100 individuals, permitting one to detect recombination between markers 1-3 cM apart. For experiments requiring a high degree of precision, additional individuals must be studied. The mathematics of genetic map resolution are well-established, and have been addressed in detail in several excellent publications, to which the reader is referred for more detail, if needed.[18,30-34]

3.6.1. GENETIC DISTANCE VERSUS PHYSICAL DISTANCE

"Genetic distance" is only loosely related to the physical quantity of DNA between genetic markers. Recombination frequency is influenced by a wide range of genetic, epigenetic, and environmental factors—progeny from the exact same cross, made in different environments, can show marked differences in recombination frequency between common genetic markers. Moreover, corresponding genetic markers in different taxa often show similar recombinational distances, despite dramatic differences in physical DNA content of the chromosomes. This suggests that different components of the DNA do not participate equally in recombination—in particular the repetitive DNA elements which account for much of the difference in genome size between different taxa (see chapter 10) may be relatively inert in recombination. Methods for precisely determining the physical quantity of DNA intervening between two DNA markers will be discussed in chapter 5.

Although genetic distance, estimated by recombination frequency, is highly variable, the order of genetic markers along the chromosomes as determined by genetic mapping, is highly conserved. "Structural rearrangements" in the order of genes along the chromosomes of different individuals, are major evolutionary events which are usually lethal (and hence go undetected).

Differences in the order of genes along the chromosomes among members of the same species are rare (save exceptional cases such as balanced structural polymorphisms). In fact, structural rearrangements are indicative of speciation—closely-related species such as tomato and potato differ by relatively few rearrangements (see chapter 21), while more distantly-related species such as wheat and rice differ by more rearrangements (see chapter 19).

3.7. WHEN IS THE MAP COMPLETE?

How does one know that a genetic map is "complete"—for example, that the map represents all regions of all chromosomes in the genome? Development of a genetic map involves detection of a sufficient number of genetic markers to be statistically confident (see below) that all regions of all chromosomes are represented on the map. The specific number of DNA markers needed to make a "complete" genetic map varies with the number, and genetic length, of chromosomes in the organism—for example, a complete genetic map of the 26 gametic chromosomes of cotton requires far more markers than does a complete genetic map of the five gametic chromosomes of *Arabidopsis*. To evaluate the relative completeness of genetic maps in taxa with different chromosome number and length, one can think in terms of the average spacing of genetic markers along the chromosomes, again measured in terms of recombination. As a rule of thumb, if genetic markers are spaced at an average density of about 5% recombination, then only about 1% of intervals between markers are predicted to measure more than 25% recombination—a distance beyond which genetic linkage becomes difficult to discern.

Two features are diagnostic of a genetic map which is complete. First, the number of "linkage groups," assemblages of DNA markers which show genetic linkage to one another but not to any markers in other such groups, is equal to the number of gametic chromosomes in the organism.

(The latter is usually known from prior cytological studies). Second, newly-mapped genetic markers invariably show genetic linkage to existing markers, suggesting that all regions of the chromosomes are covered by the genetic map.

In a few cases, physical criteria have been used to test the completeness of genetic maps. Telomeres, long tandem arrays of short repetitive DNA elements, delineate the physical ends of eukaryotic chromosomes. By determining the recombinational distance between the telomeres and the terminal marker on a genetic linkage group, one can determine whether the end of the linkage group is near the end of the chromosome. This has been done in both tomato and maize,[35-36] in both cases finding that genetic maps considered to be complete by the criteria of marker density and number of linkage groups, did in fact cover the vast majority of the chromosomes that could be evaluated.

3.8. HOW IS THE MOLECULAR MAP TIED TO THE CYTOLOGICAL MAP?

In many plants, different chromosomes are distinguishable under the microscope, either by size, by the relative lengths of different chromosomes arms, by specific features such as "constrictions" or "knobs," or by chemically-detectable "banding patterns." Prior to the availability of genetic maps based upon DNA markers, study of the appearance of genetic stocks with abnormal dosage of individual chromosomes had yielded a large body of information about the repertoire of genes on these chromosomes (see chapter 11).

To better utilize this information, it has been important to relate "linkage groups" of molecular maps to cytologically-distinguishable chromosomes. The most common approach to this objective has been use of aneuploid genetic stocks, either lacking a particular chromosome, or with abnormal dosage of a particular chromosome. For example, in tomato, a series of genetic stocks which were each trisomic (e.g., had three copies rather than two) for

one of the 12 chromosomes, were studied with DNA markers mapping to different "linkage groups." DNA markers hybridizing to a trisomic chromosome revealed 50% higher signal intensity than those hybridizing to normal (disomic chromosomes), relative to an internal standard—and afforded assignment of linkage groups to chromosomes (see chapter 21). Plants which are polyploid can often tolerate absence of a chromosome, and in wheat it was possible to assign most DNA markers to chromosomes based on such "nullisomic" stocks (see chapter 19). In cotton, genetic stocks were available in which a chromosome from one species had been substituted for a single one of the 26 chromosomes from the cultivated species (*Gossypium hirsutum*), and DNA markers falling on the substituted chromosome detected an RFLP (see chapters 1-7, 18 and 20).

A relatively new approach is providing the means to correlate molecular maps to chromosomes which are not readily distinguished by cytological features. "In situ hybridization," application of DNA probes directly to chromosomes, is a powerful means for investigating the distribution of a DNA element across the chromosomes of an organism (see chapter 11). The technique is routinely applied to determining the locations of tandemly-repetitive DNA elements, which occur in thousands of copies at one location—thus generate a very intense signal. While difficult, in situ hybridization has also been used for locating elements which occur in single copies. Not only is this technique useful for correlating particular chromosomes to genetic linkage groups, but it is also very powerful for studying the distribution of repetitive DNA elements across the genome.

3.9. HOW ARE THE GENETIC MAPS OF DIFFERENT TAXA RELATED TO ONE ANOTHER ("COMPARATIVE MAPPING")?

Over the past two decades, molecular genetics has revealed fundamental similarity in gene repertoire between taxa (species, genera, or higher divisions in the evolutionary tree) that have been reproductively-isolated for millions of years. Similarity of genes in different taxa thus provides a means to evaluate similarities and differences in the organization of genes along the chromosomes in taxa such as corn and sorghum, which cannot interbreed (thus cannot be subjected to genetic linkage analysis).

"Comparative mapping," the study of similarities and differences in gene order along the chromosomes of taxa which cannot be hybridized, has recently been used to demonstrate that a surprisingly small number of chromosomal rearrangements (inversions and/or translocations) distinguish many major crops and model systems. Moreover, comparative maps provide a conduit for communication—permitting information gathered during study of one species to be quickly and efficiently applied to related species.

Comparative mapping involves analysis of a common ("orthologous") set of genetic markers in two different taxa, then evaluation of the respective arrangement of these markers along the chromosomes. The genetic markers to be used must meet two specific criteria (Fig. 3.5). First, comparative mapping is based upon analysis of "orthologous" genetic loci in different plant species. To be "orthologous," genetic loci must be directly derived from a common ancestral locus. Orthology is similar to allelism—except that it refers to corresponding loci in taxa which cannot hybridize sexually, thus a direct test of allelism by classical means is impossible. For example, if a brown-eyed man and a blue-eyed woman have two brown-eyed children, the gene(s) conditioning brown eye color in the different children are almost certainly "orthologous." However, the gene(s) may or may not be orthologous to the gene(s) which impart brown eye color to children of a different family, or to the family dog. The most common reagent for comparative mapping is DNA sequences that occur at only one place in the genomes of each of the respective taxa to be studied. It is readily determined that a DNA sequence

meets this criterion, simply by "labeling" the DNA sequence with radioactivity or fluorescence, and applying it to immobilized total genomic DNA from each taxon (usually on a nylon membrane—the "Southern blotting technique[37]"). If the labeled probe only reveals a single genomic DNA fragment, it is likely to occur at only a single location in the genome.

The second constraint on genetic markers useful for comparative mapping is that they must be "conserved," e.g., they must retain recognizably similar form in divergent taxa, so that "orthologous" locations in the genome can be evaluated. cDNAs (DNA complementary to sequences which encode an mRNA product; see chapter 2) are particularly well-suited to establishment of orthology. Since expressed genes are responsible for specific functions in plants, mutations in genes are quickly revealed by a change in phenotype. Existing alleles of particular genes have persisted because they were able to stand the test of time, conferring a selective advantage to the organism that carried them. Because mutation is largely a random process, most mutations are less well-adapted than the original allele, and are thus quickly eliminated from the gene pool by natural (or artificial) selection. Consequently, expressed sequences tend to evolve slowly, and are particularly useful for comparative mapping.

Only a subset of DNA marker technologies are suitable for comparative analysis. Most comparative maps made to date have relied on RFLP analysis, using cDNAs as the DNA probes. The PCR-based technique best suited to comparative mapping is the "CAPS" technique (described above). CAPS markers can be derived from cDNAs, thus meeting both the criterion of orthology and of conservation.

PCR-based techniques using locus-specific DNA primers to detect allele-rich "simple-sequence repeats," or "sequence-tagged microsatellites"[11] are sometimes useful for comparative mapping, but less-frequently than RFLPs or CAPS. Such systems are locus-specific, thus in principle orthology can be established.

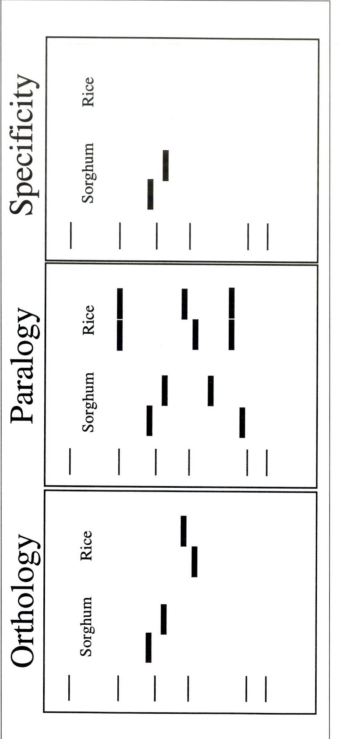

Fig. 3.5. Suitability of DNA markers for comparative mapping: the criteria of "orthology," and "conservation." Genetic loci which are "orthologous" are known to be directly descended from a common ancestral locus—such loci are revealed by DNA markers which detect only a single locus in each of two taxa (such as sorghum and rice), and are the preferred type of loci for comparative mapping. Genetic loci which are paralogous are descended from a common ancestral locus, but with intervening duplication events—it is more difficult to establish correspondence between such loci in different taxa, since there are several possible combinations of orthologous loci. Finally, genetic loci which are only found in one taxon are not useful for comparative mapping.

However, unlike CAPS, the primer sequences are usually not located in conserved sequences, thus are less likely retain homology to DNA of disparate taxa.

PCR-based techniques which use "arbitrary primers," such as RAPD,[5] AP-PCR,[6] DAF,[7] and AFLP[38] are poorly-suited to comparative mapping. Difficulty in using these techniques for comparative mapping stems from the fact that each simultaneously reveals many genetic loci which are independent of each other, except for sharing DNA sequence homology over 1-2 priming sites of 10 nucleotides or less. Since such limited homology occurs in many different places in large genomes, it does not necessarily reflect correspondence of genetic loci—thus orthology is difficult to establish. It is important to recall that the original basis for creating "arbitrary-primer" techniques was avoidance of the need for DNA sequence information—by using PCR primers which would bind to many different sites in different eukaryotic genomes.

In principle, however, individual RAPD amplification products can be rendered more suitable for comparative analysis by developing locus-specific PCR primers ("SCARs," or sequence-characterized amplified regions[39]). With regard to comparative mapping, SCARs are similar to microsatellites, in that the cloned element is usually not a cDNA, thus may or may not be conserved across taxa.

3.9.1. STATUS OF COMPARATIVE MAPPING IN PLANTS

Many different plant families, each including major crops, enjoy comparative maps which are illustrated in detail in subsequent chapters (see chapters 15-22). These include monocots and dicots, diploids, disomic polyploids, and polysomic polyploids, autogamous and allogamous crops, tropical and temperate crops, and uncultivated experimental models. Within each of these plant families, a recurring picture of conservation of gene order, punctuated by occasional chromosome rearrangement, bodes well for cross-utilization of genetic information in closely-related taxa. The utility and accuracy of molecular maps are reflected by corroboration of predictions from classical genetics, regarding gene and genome duplication (maize, soybean, cotton, *Brassica*), subgenomic origin and composition (cotton), and similarity in gene order of closely-related taxa (mungbean and cowpea, tomato and potato).

3.10. SUMMARY

Genetic maps, based largely upon DNA markers, afford the modern crop scientist with a powerful array of tools for analyzing the inheritance of agriculturally important traits (see chapter 4), for monitoring the transmission of specific genes or genomic regions from parent to progeny, and for cloning specific genes relevant to plant development and/or agricultural productivity (see chapter 12). Genetic maps have been developed for most major crops, and can readily be constructed for additional taxa as needed.

Moreover, establishment of correspondence between the chromosomes of different taxa at the DNA level by comparative mapping opens new avenues for scientific inquiry, and affords marked efficiencies in DNA marker-assisted plant breeding. Vavilov's "law of homologous series in variation"[40] was perhaps the earliest recognition of fundamental similarity between different cultivated species. Most plant breeders recognize similarities between their target species and related taxa. Working groups associated with a wide range of crops that are included in common plant families, e.g., the Crucifer Genetics Cooperative, recognize that the importance of specific genes and/or traits transcends the boundaries which nature has imposed on sexual gene exchange. Genetic mapping techniques which facilitate comparative analysis provide a direct conduit for exchange of information between taxa which cannot interbreed.

REFERENCES

1. Markert CL and F Moller. Multiple forms of enzymes: tissue, ontogenetic, and species-specific patterns. Proc Natl Acad Sci USA 1959; 45:753-763.
2. Morgan TH. Sex-linked inheritance in *Drosophila*. Science 1910; 32:120-122.
3. Botstein D, RL White, M Skolnick, RW Davis. Construction of a genetic linkage map in man using restriction fragment length polymorphisms. Am J Hum Genet 1980; 32:314-331.
4. Arumunganathan K and ED Earle. Nuclear DNA content of some important plant species. Plant Mol Biol Rptr 1991; 9:208-218.
5. Williams JGK, Kubelik, AR, Livak, KJ et al. 1990. Oligonucleotide primers of arbitrary sequence amplify DNA polymorphisms which are useful as genetic markers. Nucl Acids Res 18:6531-6535
6. Welsh J and M McClelland. Fingerprinting genomes using PCR with arbitrary primers. Nucl Acids Res 1990; 18:7213-7218.
7. Baum TJ, PM Gresshoff, SA Lewis, RA Dean. DNA amplification fingerprinting (DAF) of isolates of four common Meloidogyne species, and their host races. Phytopathology 1992; 82:1095.
8. He Q, M Marjamaki, H Soini, J Mertsola, MK Viljanen. Primers are decisive for sensitivity of PCR. BioTechniques 1994; 17:82-87.
9. Konieczny A and FM Ausubel. A procedure for mapping Arabidopsis mutations using co-dominant ecotype-specific PCR-based markers. Plant J. 1993; 4:403-410.
10. Jarvis P, C Lister, V Szabo, C Dean. Integration of CAPS markers into the RFLP map generated using recombinant inbred lines of Arab*idopsis thaliana*. Plt Mol Biol 1994; 24:685-687.
11. Weber JL and PE May. Abundant class of human DNA polymorphisms which can be typed using the polymerase chain reaction. Am J Hum Genet. 1989; 44:388-396.
12. Tautz D. Hypervariability of simple sequences as a general source for polymorphic DNA markers. Nucl Acids Res 1989; 17:6463-6471.
13. Jeffreys AJ, V Wilson, SL Thein. Hypervariable minisatellite regions in human DNA. Nature (London) 1985; 314:4251-4255.
14. Conner BJ, AA Reyes, C Morin, K Itakura, RL Teplitz, RB Wallace. Detection of sickle-cell β^S—globin allele by hybridization with synthetic oligonucleotides. Proc Natl Acad Sci USA 1983; 80:278-282.
15. Fischer SG and LS Lerman. DNA fragments differing by single-base substitutions are separated in denaturing gradient gels: Correspondence with melting theory. Proc Natl Acad Sci USA 1983; 80:1579-1583.
16. Nickerson DA, Kaiser R, Lappin S et al. Automated DNA diagnostics using an ELISA-based oligonucleotide ligation assay. Proc Natl Acad Sci USA 1990; 87:8923-27.
17. Barany F. Genetic disease detection and DNA amplification using cloned thermostable ligase. Proc Natl Acad Sci USA 1991; 88:189-193.
18. S-C Liu, SP Kowalski, T-H Lan, KA Feldmann, AH Paterson. Genome-wide High-Resolution Mapping by Recurrent Intermating, Using *Arabidopsis thaliana* as a Model. Genetics 1996; 142:247-258.
19. Lander ES, P Green, J Abrahamson, A Barlow, MJ Daly, SE Lincoln, L Newburg. MAPMAKER: An interactive computer package for constructing primary genetic linkage maps of experimental and natural populations. Genomics 1987; 1:174-181.
20. Suiter KA, JF Wendel, JS Case. Linkage-1: a Pascal computer program for the detection and analysis of genetic linkage. J Hered 1983; 74:203-204.
21. Liu B-H, SJ Knapp. G Mendel: A program for Mendelian segregation and linkage analysis of individual or multiple progeny populations using log-likelihood ratios. J Hered 1992; 81:407.
22. Weaver R, C Helms, SK Mishra, H Donis-Keller. Software for analysis and manipulation of genetic linkage data. Am J Hum Genet 1992; 50:1267-1274.
23. Stam P. Construction of integrated genetic linkage maps by means of a new computer package: JoinMap. Plant J 1993; 3:739-744.
24. Manly KF. New functions in Map Manager, a microcomputer program for genomic mapping. Proceedings of Plant Genome III, San Diego, CA, 15-19 Jan 1995:61.

25. Wu KK, W Burnquist, ME Sorrells, TL Tew, PH Moore, SD Tanksley. The detection and estimation of linkage in polyploids using single-dose restriction fragments. Theor Appl Genet 1992; 83:294-300.

26. Taylor B. Recombinant inbred strains: use in gene mapping. In: Origins of *Inbred Mice*, edited by H. Morse. New York: Academic Press, 1978: 423-438.

27. Brim CA. A modified pedigree method of selection in soybeans. Crop Sci 1966; 6:220.

28. Kosambi DD. The estimation of map distance from recombination values. Ann Eugen 1944, 12:172-175.

29. Perkins DD. Crossing-over and interference in a multiply-marked chromosome arm of Neurospora. Genetics 1962; 47:1253-1274.

30. Allard RW. Formulas and tables to facilitate the calculation of recombination values in heredity. Hilgardia 1956; 24:235-278.

31. Silver J and CE Buckler. Statistical considerations for linkage analysis using recombinant inbred strains and backcrosses. Proc Natl Acad Sci USA 1986; 83:1423-1427.

32. Neumann PE. Two-locus linkage analysis using recombinant inbred strains and Bayes' theorem. Genetics 1990; 126:277-284.

33. Ritter E, C Gebhardt, F Salamini. Estimation of recombination frequencies and construction of RFLP linkage maps in plants from crosses between heterozygous parents. Genetics 1990; 125:645-654.

34. Churchill G, JJ Giovannoni, SD Tanksley. Pooled-sampling makes high-resolution mapping practical with DNA markers. Proc Natl Acad Sci USA 1993; 90:16-20.

35. Ganal MW, Broun P, Tanksley SD. Genetic mapping of tandemly repeated telomeric DNA sequences in tomato *Lycopersicon esculentum*. Genomics 1992; 14:444-448.

36. Burr B, FA Burr, EC Matz, J Romero-Severson. Pinning down loose ends: Mapping telomeres, and factors affecting their length. Plant Cell 1992; 4:953-960.

37. Southern EM. Detection of specific sequences among DNA fragments separated by gel electrophoresis. J Mol Biol 1975; 98:503-517.

38. Zabeau M and P Vos. 1993. European Patent Application, # EP0534858.

39. Paran I and RW Michelmore. Development of reliable PCR-based markers linked to downy mildew resistance genes in lettuce. Theor Appl Genet 1993; 85:985-993.

40. Vavilov NI. The law of homologous series in variation. J Genet 1922; 12(1).

CHAPTER 4

MAPPING GENES RESPONSIBLE FOR DIFFERENCES IN PHENOTYPE

Andrew H. Paterson

*"...differences...may be effected by the independent action of...factors in different linkage groups. These factors, when combined, have a cumulative effect. The...factors in different chromosomes may not be equal in their effect."** *

4.1. WHY MAP A GENE?

Genetic mapping permits study of any morphological, physiological, or developmental processes in which genetic variants exist, with a minimum of a priori information.[1] Genetic mapping provides a direct means for investigating the number of genes influencing a trait, the locations of these genes along the chromosomes, and the effects of variation in dosage of these genes. Genetic mapping provides the information needed to implement "DNA marker-assisted selection," an approach of growing importance in plant and animal improvement (chapter 7). Using genetic maps, one can evaluate the correspondence of genes regulating phenotype in different populations, or even different taxa. Finally, genetic mapping is the first step toward "map-based cloning" of genes responsible for specific phenotypes—which opens up many new avenues for investigation of gene function, as well as new opportunities for engineering novel traits.

Genetic mapping has been particularly important in higher plants, in which molecular genetics is not nearly so elegant as that possible for simple organisms such as *Escherichia coli* (gut bacterium), or *Drosophila melanogaster* (fruitfly). The entire genome of *E. coli*, for example, can be

---

* Sax K. The association of size differences with seed-coat pattern and pigmentation in Phaseolus vulgaris. *Genetics 1923; 8:552-560.*

Genome Mapping in Plants, edited by Andrew H. Paterson. © 1996 R.G. Landes Company.

represented by 4200 1 kb fragments of DNA—however, the smallest plant genome (*Arabidopsis*) would require 150,000 such fragments. Further, higher plants have relatively long generation times, months to years (versus a few minutes for *E. coli*). Finally, many plant phenotypes reflect the effects of numerous independent genes, acting at different times during growth and development. Such "multigenic" or "polygenic" traits are especially cumbersome to manipulate by classical breeding methods. While genes can be cloned readily in bacteria simply by transformation of a mutant population with wild-type DNA and growth under selective conditions, such an experiment is prohibitively difficult in plants.

Having a complete genetic map is important in locating genes, because one can be confident that all regions of all chromosomes are searched—no stones have been left unturned. As described in chapter 3, assembly of detailed molecular maps for higher plants is now routine, and has largely been completed for most leading crops and model systems (see chapters 15-22).

4.2. MONOGENIC VERSUS POLYGENIC TRAITS, AND DISCRETE VERSUS CONTINUOUS VARIATION

The number of genes accounting for variation in a phenotype can vary widely, in different pedigrees, taxa, or even environments. Phenotypic variation which is due to segregation at a single genetic locus is very simple to map—in fact, if such variation is discrete, the trait is a "visible marker." By contrast, if the effects of the locus are obscured by environmental factors (such as differences in temperature, moisture, or nutrition), genetic mapping becomes more difficult, requiring a mathematical approach which accommodates the fact that phenotype is only a quantitative (rather than qualitative) measure of genotype. Finally, if a phenotype is influenced not just by one such "quantitative trait locus" (QTL), but by many such loci segregating in the same population, genetic mapping becomes even more complicated. The remainder of this chapter will discuss approaches to efficient genetic mapping, under each of these scenarios.

4.2.1. MAPPING SIMPLY-INHERITED TRAITS

The chromosomal location of a mutation is determined by identifying nearby genetic markers which are co-transmitted from parent to progeny with the mutant phenotype. Extending upon the example of genetic mapping used in chapter 3, it is clear how a discrete phenotype might be associated with a specific genetic locus (Fig. 4.1). Many discrete mutations are important to agriculture, evolution, or development, and have been placed on genetic maps (see chapters 15-22).

In recent years, several approaches have been described to simplify, and accelerate, the mapping of discrete mutations with molecular markers. One early approach took advantage of "near-isogenic lines," genetic stocks created by plant breeders, often for the purpose of introducing a valuable gene from a wild species to a crop plant. Many genes for disease resistance, color, reduced plant height, early flowering, or other traits have been introduced into cultivated germplasm by way of near-isogenic lines. Near-isogenic lines are developed by backcrossing, e.g., crossing a "donor genotype" carrying a specific trait of interest, to a "recipient" with generally desirable attributes. By recurrently selecting for the trait of interest, and repeatedly crossing to the recipient, donor chromatin is progressively eliminated except for a small amount which is closely-linked to the trait under selection (Fig. 4.2). By comparing the backcross-derived stock to the original recurrent parent, one can determine the likely location of the target gene simply by identifying DNA markers which reveal the donor allele in the backcross-derived stock.[2]

While near-isogenic stocks are efficient mapping tools, they require many breeding cycles to make—a recent molecular

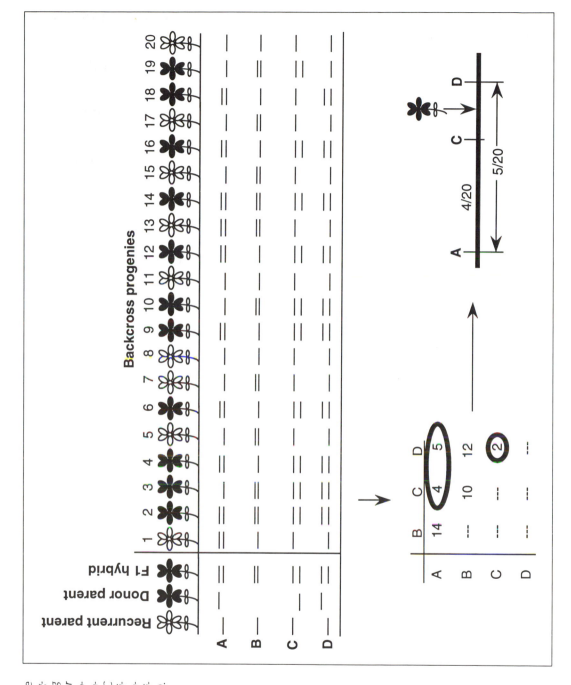

Fig. 4.1. Mapping discrete genes. Re-examination of Figure 3.4, with a segregating phenotype added (flower color), shows that the phenotype is separated by 1 recombinant each from markers C and D. The most parsimonious location of the gene controlling this phenotype is equidistant between loci C and D.

Fig. 4.2. Near-iso-
genic line develop-
ment. The F1 hybrid
of a cross between
two different geno-
types is crossed re-
currently to one par-
ent (the "recurrent"
parent), while selec-
tion is practiced for
an attribute from the
other ("donor") par-
ent. After several gen-
erations of back-
crossing, the major-
ity of the "donor"
chromatin (in black)
has been eliminated,
except for a small
segment immedi-
ately surrounding the
gene under selec-
tion. Based on iden-
tification of DNA
polymorphisms with-
in this small chro-
mosome segment,
which distinguish the
new genotype from
the recurrent parent,
the gene can be
placed on a genetic
map.

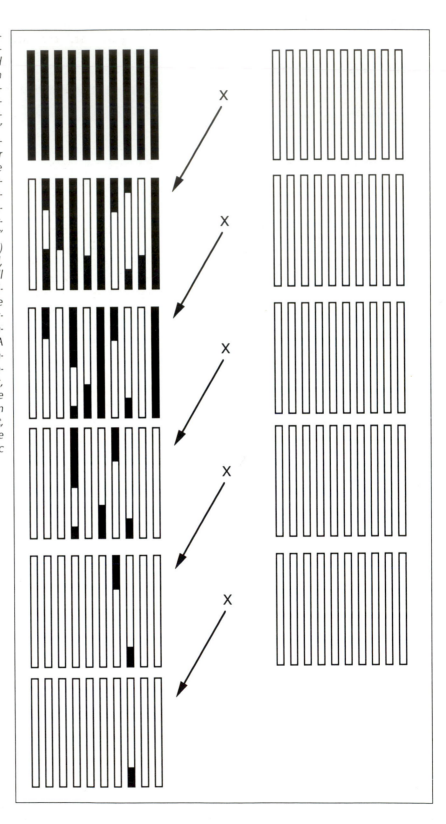

approach called "bulked segregant analysis" (BSA) provides a faster alternative (Fig. 4.3). The salient feature of near-isogenic lines is that they are comprised of different chromatin in one specific chromosomal region, carrying the target gene. Two groups[3,4] have described approaches to creating this same scenario in vitro, simply by mixing DNA from many different siblings which share a trait. The principle of DNA pooling strategies is the grouping of informative individuals together so that a particular genomic region can be studied against a randomized genetic background of unlinked loci. The first example of this approach involved genes for resistance to lettuce downy mildew. A lettuce population was created which segregated for a gene conferring resistance to lettuce downy mildew. Several individual sibling plants were identified which were resistant to the disease. A different set of siblings were identified which were susceptible to the disease. Because response to the pathogen segregated as a single genetic locus which completely accounted for resistance versus susceptibility, it was anticipated that the "resistant" group all shared a common genotype in the region of the gene, and the "susceptible" group all shared an alternative genotype in the same region. By pooling many individual resistant (or susceptible) plants into a "resistant pool" and a "susceptible pool," random segregation at unlinked genetic loci homogenizes the DNA elsewhere in the genome. The gene can then be mapped by identifying DNA markers which show the allele of the resistant parent in the resistant pool, and the allele of the susceptible parent in the susceptible pool. This method was used to identify three RAPD markers in lettuce linked to a gene for resistance to downy mildew in an F_2 population.[3] The result of sensitivity tests indicated that the rare allele in DNA pools was barely detected if it represented 10% of the mixture and was never detected if it constituted a proportion of 4% or less of the total.

In principle, "bulked segregant analysis" can be performed even if the phenotype is not discrete. By performing a "progeny test," evaluating several progeny from each plant tentatively selected for a pool, one can re-evaluate the reliability of single-plant phenotype measurements. However, this requires another generation, and if time is more limiting than genetic mapping resources, one might simply choose to map the trait based on the traditional approach of characterizing individual plants with DNA markers spanning all of the chromosomes.

DNA pooling can be very useful even when the location of the gene is already known. Map-based cloning of a gene is aided by having a large number of different DNA markers near the gene, to pinpoint it to as small a region as possible. One can use DNA markers flanking the gene to develop alternative DNA pools with different genotype near the gene, thereby targeting a search for additional DNA markers to a very specific chromosomal region. Such pools need not even be made from populations segregating for the target gene, but can use highly-polymorphic well-mapped populations. Using an existing mapping population and DNA marker-based pools, two RAPD markers were identified which were tightly linked to loci affecting ripening and jointless stem in tomato.[4]

Use of DNA pools to map genes minimizes the number of independent DNA samples which must be analyzed. To further streamline this process, DNA marker assays which screen many DNA fragments simultaneously, such as RAPD, AP-PCR, or AFLP, are frequently used (see chapter 2 for descriptions). This type of experiment has proven uniquely well-suited to "arbitrary-primer" (AP)-PCR, as the problems with reproducibility can be overcome by replication.[5] By screening hundreds of PCR primers for putative polymorphisms, one can quickly eliminate 90-95% which do not provide useful information, then repeat the putative positives several times to test reliability. While it would be prohibitively time-consuming to have to replicate every datapoint involved in construction of a

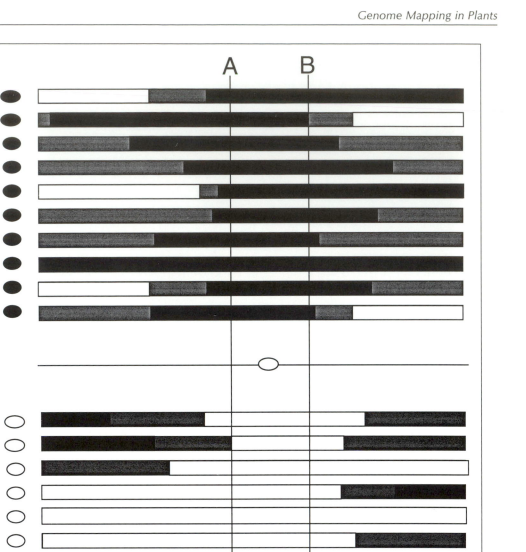

Fig. 4.3. Mapping genes by DNA pooling. Two variations on this approach have been described. The simpler approach, "bulked segregant analysis (BSA)," requires no a priori information about the location of the target gene, and DNA pools are developed from plants which share a common phenotype for a simply-inherited trait (such as seed color, at left).[3] BSA represents a simple method for primary genetic mapping of simply-inherited traits. An alternative approach has the objective of isolating additional DNA markers near a gene for which the genetic map location is known, and pools are constructed based on genotype at DNA markers flanking the gene (A and B, near top).[4] In each of these approaches, the size of the chromosomal "window" targeted is determined by the number of individuals in the pools. More individuals in the pools means that a smaller window is targeted and that fewer markers will be found, however those markers found will tend to be closer to the target. Too few individuals in the pool would risk the possibility of false positive "hits" at unlinked genetic loci, due to chance homogeneity at other loci.

primary genetic linkage map, it is reasonable to replicate a small fraction of putative positives from a "screening" experiment, and identify those which are repeatable.

4.2.2. COMPLEX TRAITS, INFLUENCED BY MANY GENES, REQUIRE LARGE POPULATIONS TO MAP

Mutations which do not impart discrete phenotypes, but merely contribute to a continuous range of phenotype, can also be placed on genetic maps. Using "complete" genetic maps, together with populations typically including 200 or more individuals, such "quantitative trait loci" (QTLs) can be assigned to "likelihood intervals," typically spanning 10-30 cM.[6,7] A large number of phenotypes important to agriculture, evolution, and medicine are influenced by quantitative trait loci, and "QTL mapping" is becoming a powerful new research tool across the life sciences. When necessary, the location(s) of QTLs can be delineated more precisely by techniques described below.

Most commonly, individual QTLs are described by their chromosomal location, the magnitude of their phenotypic effect, the effect of gene dosage at the locus, and their interactions with other QTLs or unlinked genetic loci. Each of these four parameters will be discussed in detail.

4.2.2.1. Chromosomal location of QTLs

Algorithms have been developed for determining the approximate chromosomal locations of QTLs in a wide range of pedigrees and experimental designs, and have been recently reviewed[8]—however, all share the basic principle of testing correlation between marker genotypes and quantitative phenotypes (Fig. 4.4). Using "complete" genetic maps, contemporary analytical methods are able to use information from multiple markers which flank a QTL, in contrast to earlier methods which were limited to information from single markers at unknown distance and direction from the QTL. This fundamental difference permits

algorithms to make accurate estimates of location, and phenotypic effect, of individual QTLs.

Two types of mapping population, F2 and recombinant inbred, have been the most widely used for QTL mapping in plants. F2 populations, derived by selfing or intermating F1 hybrids usually from crosses between homozygous parents, have two primary advantages. Specifically, they are quick to make (two generations), and contain all three possible combinations of two alleles at a locus (e.g., AA, Aa, aa)—thus can be used to estimate mode of gene action and test for complex interactions between loci. The mode of gene action has been determined for QTLs with reasonable success, however, interactions between loci have proven elusive (see below). Recombinant inbred populations, derived by repeated selfing and selection of single individuals derived from different F2 individuals, enjoy the advantage of reproducibility. Because each individual in a recombinant inbred population is homozygous, its genotype can be replicated many times with near-perfect fidelity. This permits replica sets of individuals to be distributed to many different researchers, allowing a large body of data to accumulate on a single population. Moreover, QTL mapping can take advantage of phenotypes measured on replicated plots, improving the ratio of genetic signal to environmental noise. However, recombinant inbred populations take many generations to make, and do not permit one to estimate effects of gene dosage (since only two genotypes at a locus are represented).

In principle, other types of mapping populations, such as backcross, doubled-haploid, intermated, and single-dose restriction fragment (heterozygous parents) can also be used for QTL mapping; the strengths and weaknesses of each are discussed in chapter 3.

4.2.2.2. Phenotypic effect of individual QTLs

Geneticists have long debated the degree of complexity of quantitative traits.[9]

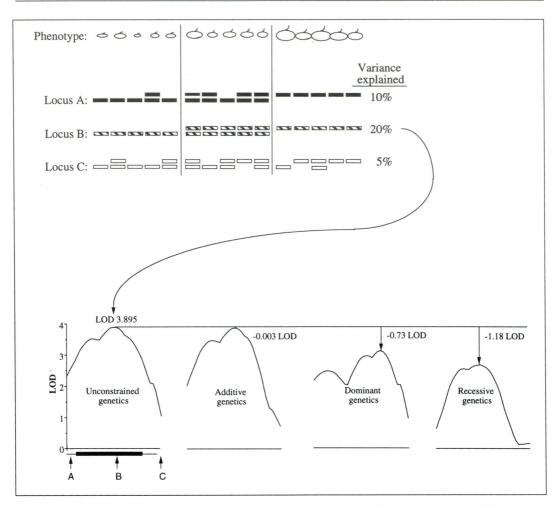

Fig. 4.4. QTL mapping. A quantitatively-measured phenotype is evaluated for association with different DNA markers. The strongest evidence for a gene or genes affecting the phenotype is at the marker most closely correlated with the phenotype. Based on inference from multiple markers, several statistical techniques are available to develop likelihood plots, describing the probable location of the gene(s). Further, by placing different constraints on the null hypothesis being tested, different modes of gene action (additivity, dominance, recessiveness) can be evaluated, and those which are deemed unlikely (recessiveness) can be rejected.

Many theories, ranging from "virtually infinite numbers of genes with tiny effects," to "few genes with large effects" have been proposed, championed, questioned, revised, rejected, and reincarnated. Geneticists realized that some assumptions invoked to simplify quantitative models, such as equality of gene effects, and strict additivity of gene action, were unlikely to precisely describe individual quantitative trait loci. Such assumptions were tolerated so long as models were reasonably predictive of the behavior of study populations. It was no

particular surprise that QTL mapping showed such assumptions to be violated regularly. However, it has remained controversial whether the results of QTL mapping experiments reflect the true complexity of quantitative inheritance, or simply detect a subset of (relatively large) gene effects.

A large body of data from QTL mapping have supported the concept that relatively few genes explain large portions of the phenotypic variance in a trait, with increasing numbers of genes explaining progressively smaller fractions of phenotypic variance.[8,10] Curiously, the number of QTLs observed in "first-generation mapping" (for example, of a BC1 or F2 population[6]) often are similar to the "effective number of factors" predicted by classical models. However, if genes explaining large portions of phenotypic variance are rendered homozygous, the significance threshold is reduced, and yet additional genes explaining even smaller portions of phenotypic variance are revealed.[11]

While the "end" of this stepwise process of revealing genes of smaller and smaller effect is not clear, a large and growing body of data continue to suggest that relatively few genes account for the bulk of variation in many populations, with ever-larger numbers of genes contributing ever-smaller portions of variance.[12] This result will be further supported in discussion of comparative analysis, with the finding that genes which account for the bulk of variation in a trait appear to correspond in different reproductively-isolated species.

Perhaps the single most important consideration in analysis and interpretation of QTL data is the threshold employed for inferring statistical significance. Because QTL mapping involves many analyses of independent (unlinked) genetic markers throughout a genome, there are many opportunities for false-positive results to arise. Significance thresholds which often seem unduly stringent (relative to other applications) must be employed to avoid these false-positive results. Nominal significance criteria of 99.8% or more for any single QTL are usually necessary to assure an "experiment-wide" confidence level of 95% for all QTLs reported across a genome. Appropriate criteria are often described accompanying publication of an analytical method.[7] Alternatively, criteria appropriate to particular data sets can be calculated empirically.[13-15] As the published body of QTL mapping information grows, and opportunities for "comparative analysis" of previously-published QTLs emerge,[16,17] it becomes ever more important that the literature of QTL mapping be based upon

statistical criteria which assure a minimal likelihood of false-positive results.

4.2.2.3. Effects of gene dosage

The "gene action" at QTLs is determined by the same principles as are employed for monogenic traits. Specifically, the goodness of fit to individual genetic models (additivity, dominance, or recessivity) is tested.[10] Models to which the experimental data show very poor fit are deemed unlikely, while models to which the data show relatively good fit are retained as candidates. Virtually all possible modes of gene action have been associated with individual QTLs, including additivity, dominance, recessivity, and all possible intermediates (see Fig. 4.4).

4.2.2.4. Epistasis, or interactions between QTLs

Almost universally, the collective activities of mapped QTLs explain only a portion of the phenotypic difference between parents, even with nearly complete genome coverage by DNA markers. Classical evidence has strongly suggested the importance of epistasis, or nonlinear interactions between unlinked genetic loci, in quantitative inheritance.[18-22]

Until recently, QTL mapping experiments have shown very little evidence in support of the importance of epistasis, with nonlinear interactions among DNA marker loci reaching statistical significance at approximately the frequency which would be expected to occur by chance.[6,10,23] Hints of epistasis among QTLs have derived from the demonstration of "genetic background effects,"[24] and from the discovery of occasional loci reported to show interaction with multiple unlinked sites in a genome.[10]

Modified experimental designs may reconcile QTL mapping results with the importance attributed to epistasis in classical studies. A recent study[25] evaluated genetic stocks differing by two QTLs suspected to interact epistatically, but otherwise uniform in genetic background—and found strong evidence for epistasis between the loci. Another study[26] utilized recombinant inbred lines to reduce the complexity of interactions, and replicate phenotypic measurements—and found evidence of epistasis between QTLs, in genetic control of several agronomic traits. Each of these results suggest that the absence of epistasis in prior QTL mapping studies may have been due to minimal replication, and/or minimal statistical resolution to detect interactions, in the presence of many QTLs with large main effects. While the effects of some QTLs appear independent of interacting loci,[11] in at least some cases it is becoming clear that "the whole" is, indeed, greater than the sum of the parts. Improved documentation of epistasis may account for a portion of the "genetic difference between parents" which was previously unexplained by QTL mapping.

4.2.2.5. Streamlined approaches to QTL mapping

The fact that large populations are used in QTL mapping provides strong incentive to seek experimental approaches which can reduce the labor and cost involved in these experiments. Since DNA pooling strategies have been successfully used to identify simply-inherited traits in F2 populations, it was of interest to determine their potential utility for analysis of complex (polygenic) traits. When DNA pooling strategies are used to tag QTLs, individuals with the highest and lowest phenotypic value would be selected to form DNA pools regardless of dominance, segregation distortion, epistatic effects or environmental noise. The fundamental difference, relative to mapping of monogenic traits, is that a number of different gene combinations, together with the vagaries of environment, could in principle be responsible for an extreme phenotype.

Post hoc analysis of previously-mapped populations suggests that DNA pooling may permit a few QTLs with relatively large effects on phenotype (≥ 1.0 SD) to be identified, if pools are developed from a sufficiently large population—however, QTLs which account for much of the genetic variation in a trait will probably es-

cape detection.[27] Most of the QTLs identified by RFLP mapping in three prior experiments in tomato and rice could not have been identified using DNA pools. To increase the likelihood of mapping QTLs using DNA pooling strategies, several precautions are suggested,[27] including: (1) use of homozygous populations, i.e., RI or DH lines; (2) use of large populations; (3) carrying out replication or progeny testing whenever possible; and (4) using crosses in which extreme variation is observed.

A compromise between the independent genotyping of hundreds of individuals, and the pooling of extreme individuals into two samples, offers significant improvements of efficiency in mapping QTLs. Lander and Botstein[7] proposed "selective genotyping," a method in which only the subset of individuals of a population with extreme phenotypes (high, and low respectively) are mapped. Selective genotyping increases mapping efficiency, often by five-fold or more, for traits such that growing and phenotyping additional progeny requires less effort than completely genotyping individuals.[7] However, selective genotyping only affords improved efficiency in analysis of a single (independent) trait at a time—in situations where multiple independent traits are of importance, such as many plant breeding experiments, the advantages of selective genotyping are diminished.

4.2.2.6. Fine-mapping of QTLs

By using a progeny testing approach, QTLs can be mapped to intervals of a size determined by the number and spacing of previously-mapped markers in the region surrounding the QTL. "Substitution mapping"[11] is similar to development of near-isogenic stocks, in that selective breeding is used to develop populations which are segregating for only a small number of QTLs (preferably just one). The phenotypic effects of different overlapping recombinant chromosome segments are assayed by studying large numbers of progeny derived from plants containing the different segments. By comparison of the effects of different segments, the locations of individual QTLs can be inferred. The limit to resolution of QTLs by this method is determined jointly by the ability to identify recombinants in as small genomic region, and by the number and spacing of markers available to show that these recombinants are different. Using this method, several QTLs have been mapped to regions of as small as 3 cM.[11] In principle, this approach could be used iteratively, in conjunction with isolation of new markers, to delineate QTLs to intervals small enough to span by chromosome walking.

4.2.3. COMPARATIVE ANALYSIS OF QTLs

"Comparative mapping," alignment of the chromosomes of related taxa based on genetic mapping of common DNA markers, affords many benefits to genome analysis. Many DNA probes can be cross-utilized among different species in the same taxonomic family, increasing the number of genetic markers available for many genera simultaneously. In one well-studied plant family, the Poaceae, extensive conservation of gene repertoire and order[28-31] has led to the suggestion that the cultivated cereals (diverse genera within the family Poaceae) might be treated as essentially a "single genetic system."[32-34]

Comparative mapping is also useful in molecular dissection of quantitative traits. Close correspondence among QTLs affecting complex traits such as seed size, as well as traits of varying complexity in different taxa such as disarticulation ("shattering") of the inflorescence, and day-neutral flowering, has been shown for sorghum, sugarcane, maize, wheat, barley, and rice."[35] Correspondence among an unexpectedly high proportion of genes affecting height and flowering of maize, sorghum, and other grasses has also been reported.[36,37] Correspondence of QTLs in different species of the plant genera *Lycopersicon*[10] and *Vigna* [38] had previously been reported, however, such correspondence spans rather short periods of genetic divergence, and the relative promiscuity of plant "species" makes it difficult to preclude the possibility of recent gene flow.

Correspondence of QTLs on duplicated chromosome segments within a particular species has also been suggested, indicating that chromosome duplication may contribute to polygenic inheritance. Specifically, pairs of loci affecting shattering of the maize inflorescence,[35,39] maize height and flowering, and sorghum height,[37] fall on corresponding duplicated chromosome segments. Chromosome duplication in these taxa occurred millions of years before domestication. Domestication, in the past 10,000 years, most likely imparted strong selective advantages to reduced shattering, reduced height, and early flowering. Consequently, if duplicated genes retained common functions, it is intuitive that QTLs might be found on homoeologous (duplicated) chromosomal sites.[35] Ongoing research in these and other polyploids is likely to reveal additional cases of putatively duplicated QTLs.

4.2.3.1. Implications of QTL correspondence

The suggestion that mutations in corresponding genes may account for phenotypic variation in taxa reproductively-isolated for millions of years, has many implications. Perhaps first and foremost, QTL analysis in one taxon may be predictive of results in other taxa. Such predictive value would afford broader utility of QTL mapping results than was previously envisioned, enabling research on facile systems to be extrapolated to more difficult ones.

Correspondence of QTLs across diverse taxa also provides strong empirical support for model systems research on complex phenotypes. For example, the facility of genetic analysis possible in rodents and domesticated mammals has permitted mapping of genes associated with diabetes, hypertension, obesity, alcohol/drug addiction, and other medically-important phenotypes. The inherent difficulties associated with mapping complex traits in humans are partly ameliorated by the possibility of cloning QTLs from mouse or other mammals which account for phenotypic variation in humans. In a similar manner, crop plants which grow particular organs of extraordinary size, such as the enlarged root of turnip, inflorescence ("curd") of cauliflower, or fruit of tomato, might be used to isolate genes important to particular aspects of plant growth and development.

Finally, correspondence among QTLs in different taxa strongly supports the hypothesis that a relatively small number of mutations (genes) account for a large portion of phenotypic variation in many populations. If the possibilities for such mutations were infinite and all equally probable, such correspondence seems unlikely. By contrast, if a few genes play disproportionately large roles in genetic control of a trait, then mutations in (one of) these genes are more likely to have effects sufficiently large to drive the mutant allele to fixation.

It remains unclear whether the correspondence among QTLs which is found in interspecific crosses, is paralleled by correspondence of QTLs in more closely-related genotypes such as elite crop cultivars.[40] Crop gene pools may be homogeneous at loci with large phenotypic effects, fixed during the initial stage of crop domestication, and the remaining variation within elite gene crop pools is a result of subsequent mutations at other loci with smaller effects.

4.3. GENETIC MAPS AS TOOLS FOR CLONING GENES

Three general approaches are employed to clone plant genes. The "candidate gene approach" requires much a priori information about the function of the target gene, permitting one to assay for proteins which perform that function, and work backwards to the gene. Alternatively, one can scan rapidly-growing databases of cloned, sequenced DNA to identify genes in other organisms which perform the function (see chapter 14).

The "mutagenesis" approach involves destroying function of a gene—resulting in a marked difference in phenotype. Molecu-

lar tools have made this a powerful approach to cloning genes—by using specific DNA elements as mutagenic agents, genes responsible for specific phenotypes can be quickly and efficiently isolated (see chapter 13).

"Map-based cloning" provides a means to isolate naturally-occurring mutations, even when obscured by the presence of multiple mutations, or by the vagaries of environment. Genetic linkage analysis is employed to locate a gene to a region of perhaps 0.1% or less of a genome, then physical mapping is used to clone a contiguous stretch of DNA including the gene (see chapter 5). Transcribed sequences from this region are isolated by a number of means (see chapter 5) and ultimately the target gene is identified by mutant complementation (see chapter 12).

Although map-based cloning is the only strategy wholly dependent upon it, genetic mapping is a key component of many gene isolation strategies. In fact, the conceptual lines between different gene cloning approaches are becoming increasingly blurry—most successful attempts to isolate genes based upon phenotype derive information from several different approaches. Mapping of candidate cDNAs can show that many do not map to the same location as the phenotype, ruling out many implausible candidates. Cosegregation of a mutagenic DNA element and a putatively-mutant phenotype is a key test that the underlying gene has been isolated.

4.4. SUMMARY

A "complete" genetic map, including "genetic markers" for all regions of all the chromosomes in a genome, serves as a framework for determining the location of genes responsible for variation in plant growth and development. Strategies for mapping both discrete and continuous phenotypes are well-established, as are efficient experimental approaches which minimize the time and cost of genetic mapping. Genetic map information represents a powerful tool in studying the basis of variation in plant growth and development, and in testing hypotheses about possible candidate genes. Moreover, genetic mapping serves as a starting point for cloning genes, the function(s) of which are known only from mutant phenotypes (see chapter 12).

REFERENCES

1. Paterson AH and RA Wing. Genome mapping in plants. Current Opinions in Biotechnology 1993; 4:142-147.
2. Young ND, D Zamir, MW Ganal et al. Use of isogenic lines and simultaneous probing to identify DNA markers tightly linked to the *Tm-2a* gene in tomato. Genetics 1988; 120:579-586.
3. Michelmore RW, I Paran, RV Kesseli. Identification of markers linked to disease-resistance genes by bulked segregant analysis: A rapid method to detect markers in specific genomic regions by using segregating populations. Proc Natl Acad Sci USA 1991; 88:9828-9832.
4. Giovannoni JJ, RA Wing, MW Ganal and SD Tanksley. Isolation of molecular markers from specific chromosomal intervals using DNA pools from existing mapping populations. Nucl Acids Res 1991; 19:6553-6558.
5. Martin GB, JGK Williams, SD Tanksley. Rapid identification of markers linked to a Pseudomonas resistance gene in tomato using random primers and near-isogenic lines. Proc Natl Acad Sci USA 1991; 88:2336-2340.
6. Paterson AH, ES Lander, JD Hewitt, S Peterson, SE Lincoln and SD Tanksley. Resolution of quantitative traits into Mendelian factors by using a complete map of restriction fragment length polymorphisms. Nature 1988; 335:721-726.
7. Lander ES and D Botstein. Mapping Mendelian factors underlying quantitative traits using RFLP linkage maps. Genetics 1989; 121:185-199; and Corrigendum. Genetics 136:705.
8. Paterson AH. Molecular dissection of quantitative traits: progress and prospects. Genome Research 1995; in press.
9. Dove WF. The gene, the polygene, and the genome. Genetics 1993; 134:999-1002.
10. Paterson AH, S Damon, JD Hewitt, D Zamir, HD Rabinowitch, SE Lincoln, ES Lander and SD Tanksley. Mendelian factors underlying quantitative traits in tomato: comparison across species, generations, and environments. Genetics 1991; 127:181-197.
11. Paterson AH, JW Deverna, B Lanini and SD Tanksley. Fine mapping of quantitative trait loci using selected overlapping recombinant chromosomes, in an interspecies cross of tomato. Genetics 1990; 124:735-742.
12. Lande R and R Thompson. Efficiency of marker-assisted selection in the improvement of quantitative traits. Genetics 1990; 124:743-756.
13. Churchill GA and RW Doerge. Empirical threshold values for quantitative trait mapping. Genetics 1994; 138:963-971.
14. Rebai A, B Goffinet, B Mangin. Comparing power of different methods for QTL detection. Biometrics 1995; 51:87-99.
15. Rebai A, B Goffinet, B Mangin. Approximate thresholds of interval mapping tests for QTL detection. Genetics 1994; 138:235-240.
16. Lin YR, KF Schertz and AH Paterson. Comparative analysis of QTLs affecting plant height and maturity across the Poaceae, in reference to an interspecific sorghum population. Genetics 1995; 141:391-411.
17. Paterson AH, YR Lin, Z Li, KF Schertz, JF Doebley, SRM Pinson, SC Liu, JW Stansel, JE Irvine. Convergent domestication of cereal crops by independent mutations at corresponding genetic loci. Science 1995; 269:1714-1718.
18. Falconer DS. Introduction to Quantitative Genetics. 2nd ed. London and New York: Longman Press, 1981.
19. Mather KP and JL Jinks. Biometrical Genetics. 3rd ed. London: Chapman and Hall, 1982.
20. Pooni HS, DJ Coombs, and PS Jinks. Detection of epistasis and linkage of interacting genes in the presence of reciprocal differences. Heredity 1987; 58:257-266.
21. Spickett SG and JM Thoday. Regular response to selection 3. Interaction between located polygenes. Genet Res 1966; 7:96-121.
22. Allard RW. Genetic changes associated with the evolution of adaptedness in cultivated

plants and their wild progenitors. J Hered 1988; 79:225-238.

23. Edwards MD, C.W. Stuber and J.F. Wendel. Molecular-marker-facilitated investigations of quantitative-trait loci in maize. I. Numbers, genomic distribution and types of gene action. Genetics 1987; 116:113-125.

24. Tanksley SD and JD Hewitt. Use of molecular markers in breeding for soluble solids in tomato: a re-examination. Theor Appl Genet 1988; 75:811-823.

25. Doebley J, A Stec, C Gustus. teosinte branched 1 and the origin of maize: evidence for epistasis and the evolution of dominance. Genetics 1995; 141:333-346.

26. Lark KG, K Chase, F Adler, LM Mansur, JH Orf. Interactions between quantitative trait loci in soybean in which trait variation at one locus is conditional upon a specific allele at another. Proc Natl Acad Sci USA 1995; 92:4656-4660.

27. Wang G and AH Paterson. Prospects for using DNA pooling strategies to tag QTLs with DNA markers. Theor Appl Genet, 1994; 88:355-361.

28. Hulbert SH, TE Richter, JD Axtell, JL Bennetzen. Genetic mapping and characterization of sorghum and related crops by means of maize DNA probes. Proc Natl Acad Sci USA 1990; 87:4251-4255.

29. Ahn S, and SD Tanksley. Comparative linkage maps of the rice and maize genomes. Proc Natl Acad Sci USA 1993; 90:7980-7984.

30. Ahn S, JA Anderson, ME Sorrells, and SD Tanksley. Homeologous relationships of rice, wheat, and maize chromosomes. Mol Gen Genet 1993; 241:483-490.

31. Kurata N, G Moore, Y Nagumara, T Foote, M Yano, Y Minobe, M Gale. Conservation of genome structure between rice and wheat. Bio/technology 1994; 12:276-278

32. Bennetzen JL and M Freeling. Grasses as a single genetic system: genome composition, collinearity and compatibility. Trends Genet 1993; 9:259-261.

33. Helentjaris T. Implications for conserved genomic structure among plant species. Proc Natl Acad Sci USA 1993; 90:8308-8309.

34. Shields R. Pastoral synteny. Nature 1993; 365:297-298.

35. Paterson AH, YR Lin, Z Li, KF Schertz, JF Doebley, SRM Pinson, SC Liu, JW Stansel, JE Irvine. Convergent domestication of cereal crops by independent mutations at corresponding genetic loci. Science, 1995; 269:1714-1718.

36. Pereira MG, M Lee, and PJ Rayapati. Comparative RFLP and QTL mapping in sorghum and maize. Poster 169 in the Second Internal Conference on the Plant Genome, New York: Scherago International, Inc., 1994.

37. Lin YR, KF Schertz and AH Paterson. Comparative analysis of QTLs affecting plant height and maturity across the Poaceae, in reference to an interspecific sorghum population. Genetics 1995; 141:391-411.

38. Fatokun CA, DI Menacio-Hautea, D Danesh, ND Young. Evidence for orthologous seed weight genes in cowpea and mungbean, based upon RFLP mapping. Genetics 1992; 132:841-846.

39. Doebley J, A Stec, J Wendel, and M Edwards. Genetic and morphological analysis of a maize-teosinte F2 population: implications for the origin of maize. Proc Natl Acad Sci USA 1990; 87:9888.

40. Beavis WD, D Grant, M Albertsen, and R Fincher. Quantitative trait loci for plant height in four maize populations and their associations with qualitative loci. Theor Appl Genet 1991; 83:141-145.

CHAPTER 5

PHYSICAL MAPPING AND MAP-BASED CLONING: BRIDGING THE GAP BETWEEN DNA MARKERS AND GENES

Andrew H. Paterson

"Although the discovery of linked markers (for a gene) greatly limits the length of DNA sequence to be analyzed relative to the entire genome, the distances involved may still be extremely large..."*

5.1. PHYSICAL DISTANCE BETWEEN DNA MARKERS IN A "COMPLETE" GENETIC MAP

What is the correspondence between "genetic distance" as measured by recombination, and "physical distance" as measured by the quantity of DNA intervening between two genetic markers? This relationship is crucial to making the transition between in vivo study of the behavior of DNA during meiosis, and in vitro study of the specific informational properties of particular genes.

The total quantity of DNA in the nuclear genome of a representative sample of higher plants is presented in Table 5.1. While only one recent estimate for each of these species is presented, and differences in techniques for estimating DNA content impart some variation, as do differences in DNA content within a taxon, these estimates provide a sound basis for considering the approximate relationship between genetic and physical distance.

---

* *Wicking C and Williamson B. From linked marker to gene. Trends Genet 1991; 7:288-292.*

Genome Mapping in Plants, edited by Andrew H. Paterson. © 1996 R.G. Landes Company.

By using a complete genetic map to estimate the total "genetic length" of a genome, one can calculate the average quantity of DNA corresponding to a "genetic distance" of 1% recombination (i.e., 1 centiMorgan). The physical quantity of DNA corresponding to 1 centiMorgan varies widely among higher plants, from about 280 kb in *Arabidopsis* to more than 7000 kb in barley (and probably several times higher than this in tulip, however, there is not yet a tulip map available to estimate genetic length).

Curiously, despite gross differences in the average amount of DNA per centiMorgan in different taxa, the genetic (recombinational) distance between orthologous loci in different taxa tends to be remarkably similar (see chapters 15-22). This tends to suggest that the largely-repetitive DNA elements which account for the differences in physical size of plant genomes (see chapter 10) may be relatively inactive in recombination—a hypothesis which has been tested particularly elegantly in cotton (see chapter 1-7, 18, and 20).

The correspondence between genetic and physical distance varies widely at different locations within a genome. In tomato, an organism with an average of about 750 kb per centiMorgan, particular regions have been estimated to show as little as 50 kb per cM, to as much as > 4000 kb per cM (see chapter 21). It has long been known that centromeric regions tend to be subject to "recombination suppression,"[3-8] and many genetic maps show "clustering" of DNA markers near the centromere. The tandem array of 45s ribosomal DNA in tomato comprises most of one chromosome arm, but represents essentially a single genetic locus (see chapter 21). Factors other than repetitive DNA, such as introgressed chromatin[9,10] or "recombinational hotspots" can also markedly influence the relationship between genetic and physical distance.

Genetic mapping of a locus influencing phenotype has a practical limit of about 0.1-1.0 cM resolution.[10] Even in the smallest plant genome known (*Arabidopsis*), this corresponds to 28,000-280,000 nucleotides (28-280 kb), and much more in most plant genomes. Only recently has it been possible to manipulate such large segments of DNA in vitro, as a result of new megabase DNA cloning technology developed to meet the needs of genome mapping applications. The challenges of analyzing such large segments of DNA remain considerable. Global approaches to assembling "physical maps" of selected small-genome plants will expedite megabase DNA analysis in the near future. However, at present,

Table 5.1. Approximate genome sizes of various plants,[1] mammals, and bacteria[2]

	Amt. DNA per nucleus (pg)	Amt. DNA per cM	# 150 kb clones for 5x coverage
Bacteria	~ 0.002-0.01	N.A.	127
Human	~ 6.0	1000 kb	92,100
Plants:			
Arabidopsis thaliana	~ 0.3	280 kb	4,600
Rice (*O. sativa*)	~ 0.9	300 kb	13,800
Sorghum (*S. bicolor*)	~ 1.3	400 kb	19,950
Tomato (*L. esculentum*)	~ 1.9	750 kb	29,160
Cotton (*G. hirsutum*)	~ 4.5	400 kb	69,100
Barley (*H. vulgare*)	~10	7000 kb	154,000
Tulip	~64	N. A.	982,400

Note that 1 picogram (pg) = 0.965 x 10^9 bp.

the challenges of megabase DNA manipulation are a major consideration in evaluating prospective approaches to cloning genes known only from phenotype, and are one reason for using alternative techniques such as insertion mutagenesis (see chapter 13) which are less dependent upon the correspondence between genetic and physical distance.

5.2. CONTIGS

Attempts to interface genetic (recombinational) and physical (cloned DNA) maps date back to the early 1980s.[11] The principle is simple—start with a single DNA element at a location of interest, use the terminal part of the first element to identify a second element, and "walk" down the chromosome (Fig. 5.1). By identifying two DNA markers which flank a target locus, one can obtain the genic DNA corresponding to the target locus by walking from one marker to the other.

This simple principle, of "walking" along a chromosome to identify a series of contiguous DNA clones ("contigs") is fundamental to a tremendous amount of research. On a local scale, chromosome walks in particular regions provide the means to isolate genes which have been assigned to the region by genetic mapping (see chapter 12). On a global scale, assembly of contigs for entire chromosomes can efficiently provide a resource which will, in the future, reduce the need for small-scale investigations of particular locations.

A primary consideration in contig assembly is the size of each "step"—i.e., the amount of DNA which can be held by the cloning vector used to construct the library. This consideration is a "two-edged sword"—larger steps afford faster progress in assembling the contig, but yield lower resolution because target genes must be identified from a larger DNA segment. Early attempts at "chromosome walking" used bacteriophage lambda (λ) as a cloning vector, which could carry only 10,000-20,000 nucleotides (10-20 kb) of exogenous DNA.[11]

"Cosmid" vectors, which could carry up to 35 kb of DNA (see chapter 2), quickly came to be preferred over lambda for the purpose of assembling physical maps,[12] and are still widely-used to improve the resolution of physical maps in both plants and animals. However, it quickly became apparent that large genomes such as those of human and many crop plants, would be characterized much more quickly (albeit at lower resolution) by using larger clones.

The development of "artificial chromosome" vectors revolutionized physical mapping and contig assembly, and represents a key component in the "Human Genome Project," as well as smaller genome projects for various plants and animals.[13] Essentially, it was found that a segment of cloned DNA, spliced to a cloned centromere and flanked by two telomeres, could behave as an autonomous chromosome in yeast cells, being replicated along with the normal genomic complement of yeast. Because the DNA in compacted, chromosomal form is protected from shear forces in the cellular environment, it is possible to propagate very long pieces of DNA by this approach—in principle, pieces of DNA at least as long as yeast chromosomes (which range from 100 kb on up). Development of yeast artificial chromosomes (YACs) represented a turning point in the practice of physical mapping, making it truly practical to consider large-scale projects such as assembly of contigs representing entire chromosomes of higher eukaryotes. Early YAC libraries were comprised of DNA clones with average insert size of ca. 150 kb, and modern libraries have pushed the average size up to 400-700 kb.

Technology for cloning megabase DNA continues to improve, most recently with the development of "bacterial artificial chromosomes"[14] (BACs), exploiting the facile nature of bacterial cloning systems to accelerate preparation and characterization of clones. While BAC inserts are not as large (so far) as YAC inserts (early BAC libraries average 150 kb, and range up to

350 kb[14]), the relative ease of cloning DNA into bacteria has made BACs a popular system.

Megabase DNA libraries, like other DNA libraries described in chapter 2, must include considerable redundancy of cloned DNA, in order to be confident that virtu-ally all regions of the genome are repre-sented at lease once in the library. The number of DNA clones needed for a cer-tain probability of finding a target clone, is calculated by the formula:

n=ln (1-probability required)/ln(1-DNA insert size/haploid genome size).

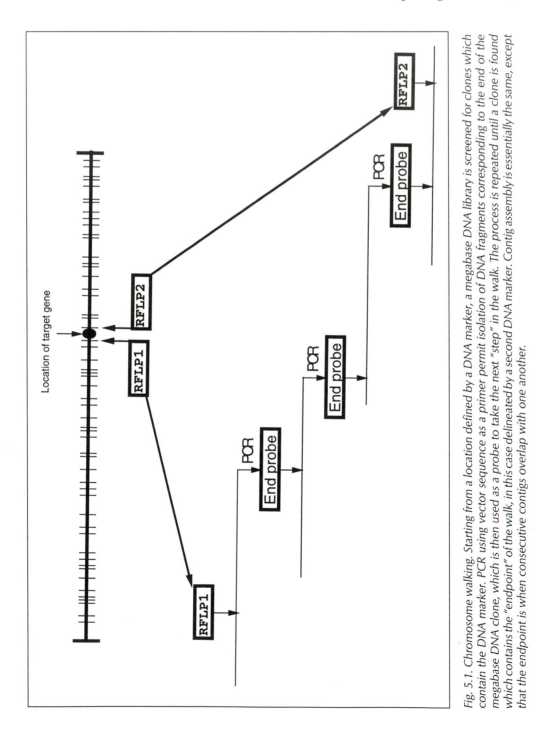

Fig. 5.1. Chromosome walking. Starting from a location defined by a DNA marker, a megabase DNA library is screened for clones which contain the DNA marker. PCR using vector sequence as a primer permit isolation of DNA fragments corresponding to the end of the megabase DNA clone, which is then used as a probe to take the next "step" in the walk. The process is repeated until a clone is found which contains the "endpoint" of the walk, in this case delineated by a second DNA marker. Contig assembly is essentially the same, except that the endpoint is when consecutive contigs overlap with one another.

As a rule of thumb, a library which contains DNA inserts which collectively add up to three (3x) the amount of DNA in a single gamete of the organism, will afford about 95% confidence that any DNA element in the genome is represented at least once in the library. A library with "5 genome-equivalents" (rather than three), will afford about 99% confidence of including the target element. The number of BACs of average size 150 kb for 5x coverage of several plant genomes are shown in Table 5.1.

In several organisms, contig assembly is proceeding very quickly. In human, a YAC library of 33,000 clones, with average insert size 900 kb, has been fingerprinted by several techniques to permit assembly of a sets of overlapping cloned genomic DNA fragments that span each of the human chromosomes.[15] In *Arabidopsis*, 374 YAC clones have been organized into a physical map of four contigs which span the vast majority of chromosome 4 (1 of the 5 chromosomes in the genome).[16] It seems likely that within a few years, it may be possible to identify a relatively small number of YAC (or other) clones which represent the vast majority of these genomes. These clones, comprising a physical map of the genome, can readily be integrated with existing genetic maps, simply by applying mapped DNA markers as labeled probes to the array of YACs. In this manner, new DNA markers might be quickly mapped to a level of resolution which otherwise would require study of several hundred recombinant plants.[17]

5.3. LOCAL ENRICHMENT FOR DNA MARKERS HELPS TO FILL THE GAPS

Economic considerations are likely to limit construction of complete contig maps to a select group of plant taxa, at least for the near future. *Arabidopsis* and rice enjoy sizable efforts in large-scale physical mapping, by virtue of the small genome size of each, and the worldwide economic importance of the latter. By virtue of comparative maps integrating *Arabidopsis* with

cultivated vegetables,[18] and rice with most of the world's cereal grains (see chapter 19), the contig maps of these facile systems may also be applied to other important crops.

To expedite cloning of high-priority genes in large-genome crop plants, where construction of complete contig maps remains prohibitively costly (circa 1996), high-density maps of DNA markers have proven a valuable aid.[19] To obtain the large numbers of DNA markers needed in a specific chromosomal region, molecular maps based on semi-random DNA probes (see chapters 15-22) are often supplemented by PCR-based screens for additional DNA markers in targeted chromosomal regions. As described in chapter 4, near-isogenic lines, and/or DNA pools can provide DNA polymorphisms which are diagnostic of a particular chromosomal segment closely-linked to a target gene. Because near-isogenic lines take several years to develop, synthetic DNA pools have become the method of preference for "enriching" a specific genomic region for DNA markers.[20]

A large number of DNA markers in a target region are more useful, if one also has a sufficient number of genetic recombinants to determine the orientation of the markers along the chromosome. Recently, a second DNA pooling technique was described which affords major efficiencies in high-resolution mapping of specific chromosomal regions.[21] By combining "marker enrichment" with high-resolution mapping, the process of "chromosome walking" can be avoided in many taxa—rather, one can identify markers so close to a target gene that it is possible to directly "land" on a megabase DNA clone containing the gene.[19]

5.4. STRATEGIES FOR GENE IDENTIFICATION

Contig assembly, or chromosome walking/landing, facilitate map-based cloning—the isolation of genes based on genetic map information. Map-based cloning is a means for isolating genes based upon segregation for naturally-occurring (or induced) mutant alleles, and permits isolation of genes for which a minimum of a priori information exists. Map-based cloning has proven an effective means of isolation of genes in higher plants (see chapter 12), but can be complicated by physically large genomes, prominent repetitive DNA fractions, and polyploidy. Consequently, many developmentally important plant genes are also being cloned by alternative means such as insertional mutagenesis, which will be discussed in detail (see chapter 13).

Map-based cloning has several basic requirements:

(1) Delineation of a target gene to a small chromosomal interval, preferably flanked by two DNA markers, and spanned by a single megabase DNA clone, or by a contig of several megabase DNA clones.

(2) A means for identifying transcripts in the megabase DNA clone.

(3) An efficient transformation system for introducing exogenous DNA into the plant species of interest, permitting identification of the target gene by mutant complementation.

As discussed above, delineation of the target gene to a small chromosomal interval is largely a matter of numbers. Screening large numbers of candidate markers, in conjunction with large numbers of recombinant genotypes, affords a high-resolution map to guide a search for megabase DNA clones containing the target gene. If a contig map already exists, the search for megabase DNA clones is made more efficient—and if not, a local contig must be assembled by chromosome walking.

However, even if a target gene is assigned to a single YAC, it is probably accompanied on the YAC by 5-10 other genes which are not related to the target phenotype, in most plant genomes. Consequently, these transcripts must be isolated, sequenced, evaluated for the presence of mutations, and perhaps tested for complementation of the mutant phenotype by the wild-type allele. Isolation of transcripts from megabase DNA clones frequently poses many technical difficulties. To date, successful isolation of transcribed sequences

from a target region of DNA has usually involved combinations of several approaches, with no single approach currently being considered the technique of choice for all situations.[22] In genomes as large as tomato, the traditional approach of direct screening of a cDNA library with the megabase DNA clone has been successful.[23] However, in cases where the target transcript is expressed only at a very low level, or only in a particular tissue or developmental stage, it is easy to miss it in a cDNA library. Recently, PCR has facilitated development of alternative strategies, using individual megabase DNA clones to selectively bind cDNAs, which are then eluted and amplified by PCR.[24-28] This approach has proven effective in isolating numerous cDNAs from human and other mammals.[22]

PCR has also simplified approaches to gene isolation which utilize the fact that genes tend to be hypomethylated, largely free of the chemical modification that is commonly found associated with repetitive DNA elements. A technique was recently described which employs PCR to amplify hypomethylated DNA upstream from an interspersed repetitive element found very frequently in human DNA (Alu; see chapter 10).[29] Post hoc analysis revealed that this means could have been used to isolate genes corresponding to some already-known transcripts, and holds promise for isolating unknown transcripts in new target regions.

As a result of large-scale cDNA sequencing projects (see chapter 14), genetic maps are coming to include large numbers of transcribed sequences—and may guide a search for candidate genes. Efforts in two plant species, rice[30] and *Arabidopsis*,[31,32] to develop sequence databases representing the majority of genes in these representative monocot and dicot plants, together with high-throughput mapping efforts,[33] may lead to near-comprehensive genetic maps of expressed sequences in the foreseeable future. Such a tool would reduce the need for region-specific searches for expressed sequences, such as described above.

5.5. PROOF OF IDENTITY OF A CANDIDATE GENE

Mutant complementation remains the most common means of proving that a candidate gene confers the target phenotype. The target gene, isolated from a true-breeding line showing the dominant phenotype, should confer the dominant phenotype to transformants when introduced into the recessive genotype. The concept of this experiment is very simple, dating back to the pioneering experiments of Avery and colleagues (discussed in chapter 2), in which it was demonstrated that DNA is the hereditary material. However, implementation of the experiment can meet with varying degrees of difficulty, because different plant taxa are transformed with different degrees of efficiency. Models such as *Arabidopsis* and tobacco are easily transformed, while many monocots remain difficult to transform. Moreover, the expression of transformed genes can be influenced markedly by "position effects," ill-defined properties of flanking DNA near the transformation site. Consequently, multiple transformation events are often required to evaluate the function (or lack thereof) of a candidate gene. Numerous volumes have been written on the topic of plant transformation, and the reader requiring more detail is referred to these. Several are cited in chapter 13.

5.6. SUMMARY

In partial summary, the integration of recombination-based genetic maps with physical distance along the DNA double helix is becoming increasingly detailed. Well-developed tools and technology routinely permit cloning and characterization of intact DNA segments spanning millions of nucleotides, and containing tens to hundreds of genes. A variety of techniques can be employed to identify the specific gene, among the many found in a megabase DNA segment, that is responsible for a particular phenotype. Rapid progress in megabase DNA analysis will soon yield contiguous sets of DNA clones that collectively span the chromosomes of selected model plants such as *Arabidopsis* and rice. The utility of such contigs will be quickly extended to other related taxa, by the principles of comparative mapping.

REFERENCES

1. Arumuganathan K and ED Earle. Nuclear DNA contents of some important plant species. Plt Mol Biol Rptr 1991; 9:208-218.
2. Lewin, B. Genes II (chapter 18). New York: John Wiley and Sons, 1985.
3. Beadle GW. A possible influence of the spindle fiber on crossing-over in Drosophila. Proc Natl Acad Sci USA 1932; 18:160-165.
4. Mather K. Crossing-over and heterochromatin in the X chromosome of *Drosophila melanogaster*. Genetics 1938; 24:413-435.
5. Roberts PA. Difference in the behavior of eu- and hetero-chromatin: crossing over. Nature 1965; 205:725-726.
6. Khush GS and CM Rick. Studies on the linkage map of chromosome 4 in tomato and on the transmission of induced deficiencies. Genetica 1967; 38:74-94.
7. Lambie EJ & GS Roeder Repression of meiotic crossing over by a centromere (Cen3) in *Saccharomyces cerevisiae*. Genetics 1986; 114:769-789.
8. Tanksley SD, Ganal MW, Prince JP et al. High density molecular linkage maps of the tomato and potato genomes. Genetics 1992; 132:1141-1160.
9. Ganal MW, Young ND, Tanksley SD. Pulsed field gel electrophoresis and physical mapping of large DNA fragments in the *Tm-2a* region of chromosome 9 in tomato. Mol Gen Genet 1989; 215:395-400.
10. Paterson AH, JW Deverna, B Lanini and SD Tanksley. Fine mapping of quantitative trait loci using selected overlapping recombinant chromosomes, in an interspecies cross of tomato. Genetics 1990; 124:735-742.
11. Steinmetz M, K Minard, S Horvath, et al. A molecular map of the immune response region from the major histocompatibility complex of the mouse. Nature 1982; 300:35-42.
12. Coulson A, J Sulston, S Brenner, J Karn. Toward a physical map of the nematode Caenorhabditis elegans. Proc Natl Acad Sci USA 1986; 83:7821-7825.

13. Burke DT, G Carle, MV Olson. Cloning of large segments of exogenous DNA into yeast by means of artificial chromosome vectors. Science 1987; 236:806-812.

14. Shizuya H, B Birren, U-J Kim, V Mancino, T Slepak, Y Tachiiri and M Simon. Cloning and stable maintenance of 300-kilobase-pair fragments of human DNA in *Escherichia coli* using an F-factor-based vector. Proc Natl Acad Sci USA 1992; 89:8794-8797.

15. Cohen D, I Chumakov, J Weissenbach. A first-generation physical map of the human genome. Nature 1993; 366:698-701.

16. Schmidt R, J West, K Love, et al. Physical mapping and organization of Arabidopsis thaliana chromosome 4. Science 1995; 270:480-483.

17. Liu S-C, SP Kowalski, T-H Lan, KA Feldmann, AH Paterson Genome-wide high-resolution mapping by recurrent intermating, using *Arabidopsis thaliana* as a model. Genetics; 1996: in press.

18. Kowalski SD, T-H Lan, KA Feldmann, AH Paterson. Comparative mapping of *Arabidopsis thaliana* and *Brassica oleracea* chromosomes reveals islands of conserved gene order. Genetics 1994; 138:499-510.

19. Tanksley SD, Ganal MW, Martin GB. Chromosome landing: A paradigm for map-based gene cloning in plant species with large genomes. Trends Genet 1995; 11:63-68.

20. Giovannoni JJ, Wing RA, Ganal MW et al. Isolation of molecular markers from specific chromosomal intervals using DNA pools from existing mapping populations. Nucleic Acids Res 1991; 19:6553-6558.

21. Churchill GA, JJ Giovannoni and SD Tanksley. Pooled-sampling makes high-resolution mapping practical with DNA markers. Proc Natl Acad Sci USA 1993; 90:16-20.

22. Gardiner K., R. Mural. Getting the message: identifying transcribed sequences. Trends Genet 1995; 11:77-79.

23. Martin GB, SH Brommonschenkel, J Chunwongse, A Frary, MW Ganal, R Spivey, T Wu, ED Earle, SD Tanksley. Map-based cloning of a protein kinase gene con-ferring disease resistance in tomato. Science 1994; 262:1432-1436.

24. Parimoo S, SR Patanjali, H Shukla, DD Chaplin, SM Weissman. cDNA selection: Efficient PCR approach for the selection of cDNAs encoded in large chromosomal DNA fragments. Proc Nat Acad Sci USA 1991; 88:9623-9627.

25. Lovett M, J Kere, LM Hinton. Direct selection: A method for the isolation of cDNAs encoded by large genomic regions. Proc Nat Acad Sci USA 1991; 88:9628-9632.

26. Morgan JG, GM Dolganov, SE Robbins, LM Hinton, M Lovett. Nucl Acids Res 1992; 20:5173-5179.

27. Fan W, X Wei, H Shukla, S Parimoo, H Xu, P Sankhavaram, Z Li, SM Weissman. Application of cDNA selection techniques to regions of the human MHC. Genomics 1993; 17:575-581.

28. Lovett M. Fishing for complements: finding genes by direct cDNA selection. Trends Genet 1995.

29. Valdes JM, DA Tagle, FS Collins. Island rescue PCR: A rapid and efficient method for isolating transcribed sequences from yeast artificial chromosomes and cosmids. Proc Natl Acad Sci USA 1994; 91:5377-5381.

30. Sasaki T, J Song, Y Koga-ban, E Matsui, et al (21 authors). Toward cataloguing all rice genes: large-scale sequencing of randomly chosen rice cDNAs from a callus cDNA library. Plant J 1994; 6:615-624.

31. Hofte H, et al (28 authors). An inventory of 1152 expressed sequence tags obtained by partial sequencing of cDNAs from *Arabidopsis thaliana*. Plant J 1993; 4:1051-61.

32. Newman T, FJ deBruijn, P Green, K Keegstra, H Kende, L McIntosh, J Ohlrogge, N Raikhel, S Somerville, M Thomashow, E Retzel, C Somerville. Genes Galore: A summary of methods for accessing results from large-scale partial sequencing of anonymous *Arabidopsis* cDNA clones. Plant Physiol 1994; 106:1241-1255.

33. Kurata N, et al (28 authors). A 300 kilobase interval genetic map of rice including 883 expressed sequences. Nature Genetics 1994; 8:365-372.

CHAPTER 6

AND WITH WHAT CONSEQUENCES? NEW OPPORTUNITIES IN LIFE SCIENCES RESEARCH, USING A GENOME-BASED APPROACH

Andrew H. Paterson

*"...further research of the intensity given to genetics will eventually provide man with the ability to describe with completeness the essential features that constitute life."**

6.1. UNIFICATION OF BIOLOGICAL SCIENCES THROUGH THE CONTINUITY OF DNA

Much of the progress in "molecular biology" and "biotechnology" of the past two decades was made possible by the fact that higher organisms share a common mechanism of heredity. Research at the DNA level transcends taxonomic boundaries. DNA from virtually any organism can be isolated, studied and manipulated using similar techniques. Tools and techniques developed to meet a specific application in bacterial genetics are often transferable to study of fruitflies, plants, or humans.

Commonality in the hereditary information of diverse organisms is not merely a general similarity in the chemical composition of DNA, but conservation of specific protein-encoding sequences (exons) over millenia of evolution. As few as 10^3-10^4, of a possible 10^{52} exons (based on an average length of 120 nucleotides), may be needed to construct

---

* Watson JD. The Molecular Biology of the Gene. The Benjamin/Cummings Publishing Company, 1973.

Genome Mapping in Plants, edited by Andrew H. Paterson. © 1996 R.G. Landes Company.

most naturally-occurring proteins.[1] In other words, the naturally-occurring population of exons may represent as few as 0.000000000000000000000000000000-000000000000000001% of the possible arrangements of amino acids.

While the precision of such estimates of exon number is modest,[2] the principle is clear that existing genes comprise only a tiny fraction of the number of possible genes. Two basic constraints limit the diversity of extant genes:

(1) Heredity. The continuity of life, with contemporary biota derived by descent from ancestral forms, places a historical constraint on the gene content of extant organisms. Various means of gene duplication afford some latitude for divergence, but the need to faithfully transmit hereditary information restricts the rate at which divergence can occur.

(2) Selection. Evolution, defined as change, in the genetic information, is constrained by its impact on the fitness of the organism. Changes which reduce survival and/or reproduction tend to be eliminated from the gene pool. Since there are vastly more arrangements of nucleotides which encode "nonsense" than "sense" (an amino-acid product), most changes in the genetic information are likely to reduce fitness—and become extinct.

The likelihood that different organisms may share many genes (with only minor modifications), points to genome analysis as a means by which study of diverse taxa can be unified. A unified approach to the study of heredity affords new learning opportunities—permitting us to extend our study of genetics beyond transmission of genes from one generation to the next within a single taxon, to study of the organization and function of genes in different taxa, which cannot interbreed, and which have been diverging for millenia. Further, unification of genetic maps affords efficiency: by using facile model systems to identify the genetic determinants of a particular phenotype, both a conceptual framework and molecular tools can be established that simplify analysis of less-favorable systems. However, extension of new tools and concepts to less-favorable systems cannot be given short shrift—for it is through study of a cross-section of biota, that levels and patterns of variation (or its reciprocal, conservation) are revealed.

6.1.1. CHROMOSOME ORGANIZATION AS A FRAMEWORK FOR UNIFYING PLANT GENETICS

Comparative study of the organization of genes along the chromosomes of different taxa unifies the study of heredity. It is well-established that diverse taxa within the grasses, legumes, crucifers, and others, have retained common gene order along substantial portions of their chromosomes, over tens of millions of years of evolution (see chapters 3, and 15-22).

For example, in the Poaceae (grasses), extensive conservation of gene repertoire and order (see chapters 1-7, 18, 19 and 20) has stimulated the suggestion that this taxonomic family might be treated essentially as a single genetic system.[3-5] The physically small quantity of DNA comprising the genomes of rice or sorghum may provide an efficient template for identification of genes relevant to important characteristics in maize or wheat, taxa in which an abundance of repetitive DNA complicates genome analysis. In several cases,[6-10] common genes have already been suggested to control traits shared by diverse Poaceae taxa.

In the short term, the greatest benefits from comparative analysis of chromosome organization may accrue to plants in the family Brassicaceae, which includes the species *Arabidopsis thaliana*.[11] The tiny genome of *Arabidopsis* has become a focal point for plant molecular genetics. By virtue of a detailed genetic map, vast array of molecular tools, and rapid progress toward assembly of contigs for entire chromosomes, *Arabidopsis* now occupies a place in genetics research comparable to that of *Escherichia coli* or *Drosophila melanogaster*. *Arabidopsis* is a close relative of the genus

Brassica, and a partial map has been developed which describes the comparative organization of the *Arabidopsis* chromosomes, and those of one widely-cultivated *Brassica* species (See ref. 9; also see chapter 16). Virtually all low-copy DNA elements in *Brassica* are also found in *Arabidopsis*, and share an average 87% DNA sequence correspondence. Genes from *Brassica* can complement mutations in *Arabidopsis* (see chapter 12). By understanding the comparative organization of the *Arabidopsis* and *Brassica* chromosomes, a contig map of the entire *Arabidopsis* genome will essentially serve also as a contig map of the genes found in *Brassica*.

Comparative mapping is important not only because it helps us to identify the similarities among the genomes of diverse taxa, but also because it helps us to study the differences. Conserved, expressed sequences provide a "skeleton" of anchor loci for aligning the chromosomes of different taxa—but they represent islands of conservation in a sea of more rapidly-evolving DNA. Not all genes (expressed sequences) retain sequence conservation across taxa. Study of the intervening DNA between conserved elements, offers opportunities to shed light on rates and processes of genic divergence, seeing what has happened to the genes which are not conserved. Further, identification of individual genetic elements which have become repetitive in derived genomes, affords an opportunity to study the evolution and spread of repetitive DNA elements across a genome.

Analysis of individual DNA sequences by computer is extending the reach of comparative genome analysis. Identification of corresponding genes by traditional means such as cross-hybridization of DNA probes, PCR with degenerate primers, or mutant complementation are laborious, and have a diminishing success rate with increasing genetic distance between taxa. Rapidly-expanding databases of expressed-sequence tags ("ESTs") for numerous organisms, offer a more rapid and more sensitive alternative approach to quickly identify homologous genes in disparate taxa.

Pioneering studies in yeast and mammals[12] point the likely direction for plants, and large-scale EST sequencing projects in several key taxa (see chapter 14) are creating the infrastructure necessary for such analyses.

6.1.2. EVOLUTIONARY SIGNIFICANCE OF CHROMOSOMAL REARRANGEMENTS

Finally, before moving on to a discussion of phenotype, a few words are warranted regarding the significance of particular chromosomal rearrangements. Classical genetics has frequently suggested that associations of particular groups of genes may offer a selective advantage, and that such groups may be kept together by balanced polymorphisms for structural rearrangements, or other restrictions on recombination.[13-15]

A logical question, which is suggested by the identification of chromosomal segments sharing common gene order between taxa is: "Why was it preserved? Does the order itself have importance?" Rates of chromosomal rearrangement can be readily modelled.[16] Comparison of the extent of chromosomal rearrangement between taxa diverged for similar time periods suggests that there are marked differences in the rate of chromosomal rearrangement between taxa.[11] The basis for such differences in genome stability is a fascinating, and under-explored research area, which is becoming increasingly tractable.

"Static" comparison of the number of rearrangements distinguishing the chromosomes of extant taxa does not reveal the temporal pattern of rearrangement. Does chromosomal rearrangement happen as a "burst," accompanying (or causing) speciation? Or is it a continuous process, with rearrangements gradually permitted to accumulate over time, as a result of fluctuating population sizes and genetic drift? Recent data have suggested that specific chromosomal rearrangements may be a direct cause of speciation, rather than an indirect consequence of speciation.[17] This exciting possibility is ripe for further investigation in additional taxa.

Moreover, do particular gene arrangements have functional consequences which can serve as a basis for selection? Certainly the observation of "position effects" in transformation experiments (see chapter 13) show that the DNA environment surrounding a gene can influence phenotype—but it remains an open question whether such effects are large enough to fixation of structural rearrangements, or common enough to be considered an important evolutionary mechanism. Another possibility which has frequented the literature both in evolutionary genetics, and in plant breeding, is the hypothetical existence of "co-adapted gene complexes," groups of genes which function well together to confer an adaptive phenotype, and in which recombination is suppressed.

How will these questions be investigated? Clearly, more detailed comparative analysis of well-studied taxa, and extension of comparative analysis to additional taxa, will be essential. Development of high-resolution maps of common, conserved genes in many taxa, will be necessary. Practical implementation of such analysis may be accelerated by technological changes, as well as availability of more extensive repertoires of molecular tools. For example, extension of the concept of artificial chromosomes to plants would open new door to investigating consequences of particular gene orders. Using "plant artificial chromosomes," it might be possible to test variations in gene order, in controlled experiments which could offer more definitive results than comparisons of the present state of gene order in different taxa.

In regard to a study of chromosome evolution across the angiosperms nearly three decades ago, Stebbins[18] commented that "The opportunities for profitable investigations of this sort are by no means at an end, and new techniques may extend them to degrees of clarity and certainty which at present can hardly be imagined." Despite the tremendous progress of the past 30 years, his words remain appropriate today.

6.2. ASSOCIATION OF PHENOTYPES WITH GENES: THE CHANGING DEFINITION OF A "MODEL SYSTEM"

Since the demonstration that DNA was the hereditary material, the goal of most experimental genetics has come to be establishment of direct relationships between phenotypes and specific DNA elements. Genetics is the study of genes. Rapidly-improving documentation of the repertoire, organization, and function of genes places growing importance on genetics as a central discipline which unifies the life sciences. Genome mapping offers a means by which study of any traits which show genetic variation, even complex traits in relatively difficult study organisms such as behavioral traits in humans, can ultimately be related to their underlying DNA elements. Genetics, and more specifically genome analysis, affords opportunities to alter the hereditary information of organisms with unprecedented speed, and transcend the bounds of sexual hybridization.

Genome analysis, has accelerated progress in association of phenotypes with genes in two specific ways:

(1) Using genetic maps or insertion mutagenesis, genes responsible for specific traits can be located, and ultimately isolated, with a minimum of a priori information. (see chapter 12 and 13).

(2) Using present-day sequencing technology, genes of unknown function can be isolated and characterized in enormous numbers. By comparison of sequence data with established databases, much might be learned about the probable function of a gene, quickly and efficiently. (see chapter 14).

For reasons which have been discussed in ample detail, *Arabidopsis thaliana* is the preferred model system for isolating and characterization of large numbers of randomly-chosen plant genes (see chapter 13). The fact that functional genes comprise 30-40% of the DNA of *Arabidopsis* (a much larger portion than in other higher plants), and that many genes in *Arabidopsis* are represented only by a single copy, makes *Arabidopsis* an excellent system for isolation of genes (see chapters 12-14 and 16). Moreover, the prospect of a contig map of the *Arabidopsis* chromosomes, a database of most DNA sequences expressed in *Arabidopsis*, and ultimately a complete DNA sequence of the *Arabidopsis* genome, clearly represents the beginning of a new era in plant genetics (see chapter 4 for citations).

Although *Arabidopsis* will clearly be the keystone in future understanding of genetic regulation of plant growth and development, identifying genes responsible for evolution of novel plant organs and functions will require molecular analysis of a wide range of plant taxa. The preferred model system(s) for study of many phenotypes, will be determined by unique genetic variation which has accumulated over evolutionary time periods, mediated either by natural selection or by human cultivation.

For example, crop plants are cultivated because of extraordinary attributes which are simply not found in more facile experimental systems, such as the maize "ear" (inflorescence), tomato "fruit" (berry), or cotton "fiber" (epidermal hair). Many genes in *Arabidopsis* are known to influence (for example) the morphology of the inflorescence. However, it remains unclear which of these, if any, are responsible for evolution of the maize ear, and indirectly, for the productivity of one of the world's leading grain crops, valued at more than $50 billion per year worldwide (see chapter 20). Despite the relative complexity of the maize genome, maize is clearly the preferred model system for investigation of how an exposed terminal inflorescence is relocated, emasculated, and enveloped in leaves—and considerable progress has been made toward answering these questions.[19-22] Likewise, to study reallocation of photosynthate into a large "fruit" (berry), tomato is clearly a preferred model. To study the genetics of subterranean storage organs, one might choose potato (tubers), or sorghum (rhizomes). The tremendous epidermal hair

elongation of cotton, makes it an attractive experimental system.

As a result of comparative mapping, we are afforded a new level of information to use in identifying specific genes which are high priorities for cloning, and in identifying the "model system" in which the target gene(s) might be most readily cloned.

Excellent examples derive from the genetic analysis of complex traits. Opponents of Mendelian genetic (particulate) theory argued that many characteristics showed "blending inheritance," with progeny intermediate between parents. This precipitated one of the great debates in the history of genetics, which was eventually resolved by the realization that "blending inheritance" might be accounted for by the independent transmission of many different Mendelian factors, together with the modifying effects of environment. QTLs have remained refractory to molecular cloning,[23] largely because of the fact that the effects of an individual "quantitative trait locus (QTL)" on a single individual are difficult to discern with confidence.

The suggestion has been put forth that QTLs might be more-simply cloned by identifying allelic mutations with discrete effects[24]—however, only recently has this possibility come close to realization. For example, comparative analysis shows that a discrete mutation preventing disarticulation ("shattering") of the mature sorghum inflorescence corresponds closely to QTLs explaining 25.8% and 16.8% of phenotypic variation in disarticulation of maize (on duplicate sites of chromosomes 1 and 5), and a QTL explaining 14.5% of variation in disarticulation of rice.[7]

Based on discrete inheritance of the phenotype, the genome of sorghum appears to be the system of choice for cloning of the "shattering" gene. The physically smaller genome of rice, usually the system of choice among the Poaceae for map-based cloning, segregates for several genes which each have only small effects, thus are more difficult to discern from environmental noise. Consequently, the simple genetic basis of the phenotype makes sorghum the preferred system for cloning. Further, comparative mapping enables us to use DNA probes or megabase DNA clones from rice, to aid in cloning of the sorghum gene. The existence of a much larger set of tools for analysis of rice, *Arabidopsis*, or other models, can be used to benefit genome analysis of other taxa, and enables us to place greater weight on other considerations such as the genetic basis of a trait in different taxa.

Finally, how will the vast body of information emerging from *Arabidopsis* genetics, and the information we anticipate on specific novel traits from a wide range of plant taxa (see above), relate to applied issues such as agricultural productivity? One of the initial motivations of botanical research was the need to feed and clothe humanity more efficiently. Although it is clear that evolution of common traits during domestication of different crops has involved mutations in at least some of the same genes,[7] it remains unclear whether genes which account for gross morphological variations between crop plants and their wild ancestors (for example, the size of the tomato fruit, or the size of the sorghum seed) are also related to variation among elite cultivars. In other words, do these genes remain important in the modern crop gene pool, or have they already been selected to fixation? In view of the role of contingency (chance) in evolution, it seems likely that isolation of genes associated with gross morphological variations specific to one taxon will afford opportunities to improve other taxa (Paterson et al).[7] The role of genome mapping in traditional plant improvement, is addressed in more detail in the following chapter.

Isolation of genes fundamental to a particular trait in many different taxa, may provide the means to confer the trait to additional taxa, with unprecedented speed and efficiency. As world populations continue to grow, and resources such as water and fertilizer become ever more precious, it may be of increasing importance to be able to conduct agriculture in non-traditional environments (e.g., hot, cold, wet, dry). Plant evolution has proceeded for about 200 million years or more, and yielded a wide range of genotypes adapted to remarkably diverse environments. This need may be more effectively met by domestication of native plants which have evolved in these environments, rather than trying to further expand the range of the small number of crops which presently sustain humanity. By isolating genes associated with domestication of existing crops, we may be able to dramatically accelerate the process of domesticating new crops.

There can be little doubt that models such as *Arabidopsis* and rice will continue to be central tools in cataloging plant genes and genomes, and associating botanical traits with specific DNA elements. By use of comparative information from facile models to guide analysis of more complex genomes, the issue of genome size may come to be less of an obstacle. Model systems research may come to emphasize criteria based on the phenotype, over criteria based on the genome.

REFERENCES

1. Dorit RL, L Schoenbach, W Gilbert. How big is the universe of exons? Science 1990; 250:1377-1382.
2. Gibbons A. Calculating the original family—of exons. Science 1990; 250:1342.
3. Bennetzen JL, and M Freeling. Grasses as a single genetic system: genome composition, collinearity and compatibility. Trends Genet 1993; 9:259-261.
4. Helentjaris T. Implications for conserved genomic structure among plant species. Proc Natl Acad Sci USA 1993; 90:8308-8309.
5. Shields R. Pastoral synteny. Nature 1993; 365:297-298.
6. Paterson AH, S Damon, JD Hewitt, D Zamir, HD Rabinowitch, SE Lincoln, ES Lander and SD Tanksley. Mendelian factors underlying quantitative traits in tomato: comparison across species, generations, and environments. Genetics 1991; 127:181-197.
7. Paterson AH, YR Lin, Z Li, KF Schertz, JF Doebley, SRM Pinson, SC Liu, JW Stansel, JE Irvine. Convergent domestication of ce-

real crops by independent mutations at corresponding genetic loci. Science, 1995; 269:1714-1718.

8. Pereira MG, M Lee, and PJ Rayapati. Comparative RFLP and QTL mapping in sorghum and maize. Poster 169 in the Second Internal Conference on the Plant Genome, New York: Scherago International, Inc., 1994.

9. Lin YR, KF Schertz and AH Paterson. Comparative analysis of QTLs affecting plant height and maturity across the Poaceae, in reference to an interspecific sorghum population. Genetics 1995; 141:391-411.

10. Fatokun CA, DI Menacio-Hautea, D Danesh, ND Young. Evidence for orthologous seed weight genes in cowpea and mungbean, based upon RFLP mapping. Genetics 1992; 132:841-846.

11. Kowalski SP, Lan T-H, Feldmann KA, et al. Comparative mapping of *Arabidopsis thaliana* and *Brassica oleracea* chromosomes reveals islands of conserved organization. Genetics 1994; 138:1-12.

12. Tugendreich S, MS Bogulski, MS Seldin, P Hieter. Linking yeast genetics to mammalian genomes: Identification and mapping of the human homolog of CDC27 via the expressed sequence tag (EST) data base. Proc Natl Acad Sci USA 1993; 90:10031-10035.

13. Dobzhansky T. Genetics and the Evolutionary Process. New York: Columbia University Press, 1970.

14. Cleland RE. Oenothera: Cytogenetics and Evolution. New York: Academic Press, 1972.

15. Wiens D, and BA Barlow. Permanent translocation heterozygosity and sex determination in East African Mistletoes. Science 1975; 187:1208-1209.

16. Nadeau JH and BA Taylor. Lengths of chromosomal segments conserved since divergence of man and mouse. Proc Natl Acad Sci USA 1984; 81:814-818.

17. Rieseberg LH, C Van Fossen, AM Desrochers. Hybrid speciation accompanied by genomic reorganization in wild sunflowers. Nature 1995; 375:313-316.

18. Stebbins GL. Chromosomal variation and evolution; polyploidy and chromosome size and number shed light on evolutionary processes in higher plants. Science 1966; 152:1463-1469.

19. Doebley J, A Stec, J Wendel, and M Edwards. Genetic and morphological analysis of a maize-teosinte F2 population: implications for the origin of maize. Proc Natl Acad Sci USA 1990; 87:9888.

20. Doebley J and A Stec. Inheritance of the morphological differences between maize and teosinte: comparisons of results for two F2 populations. Genetics 1993; 134:559-570.

21. Dorweiler J, A Stec, J Kermicle, J Doebley. *Teosinte glume architecture-1* : a genetic locus controlling a key step in maize evolution. Science 1993; 262:233.

22. Doebley J, A Stec, C Gustus. *teosinte branched 1* and the origin of maize: evidence for epistasis and the evolution of dominance. Genetics 1995; 141:333-346.

23. Tanksley SD, Ganal MW, Martin GB. Chromosome landing: A paradigm for map-based gene cloning in plant species with large genomes. Trends Genet 1995; 11:63-68.

24. Robertson DS. A possible technique for isolating genic DNA for quantitative traits in plants. J Theor Biol 1985; 117:1-10.

DNA Marker-Assisted Crop Improvement

Andrew H. Paterson

"Molecular biology offers new tools ... combining genes in new ways to create improved crops." *

7.1. INTRODUCTION

Crop plants evolved first by incidental consequences of human gatherers, and more recently through sophisticated plant breeding programs. While changes in cultural practices and mechanization have had significant impact on agricultural productivity, yield gains in most crops have been due to genetic improvement more than by any other single factor.[1]

Despite the gains already achieved, further improvements of agricultural productivity and quality are demanded continuously, by factors such as population growth, the spiraling cost of inputs such as water, fertilizer, and energy, concerns about the effects of agrichemicals on the ecosystem, and rapidly-changing consumer preferences.

In the past decade, "biotechnology" has come to play an increasing role in crop improvement. Detailed genetic maps of DNA markers have been constructed for most major crop plants (see chapters 15-22). Genome analysis has resulted in mapping of many agriculturally significant genes, and contributed to molecular cloning of a growing number of these genes (see chapter 12 and 13).

Genome analysis first began to be utilized, as a generally-applicable tool for improvement of many crops, about a decade ago. How do the accomplishments of genome analysis to date relate to our objectives of a decade ago? What areas still need to be addressed, to continue progressing toward our goals? What new opportunities are on the horizon? How can we best exploit these opportunities with a shrinking resource base for agricultural research?

---

* Board on Agriculture, National Research Council. *Genetic Engineering of Plants. Washington, DC: National Academy Press, 1984.*

Genome Mapping in Plants, edited by Andrew H. Paterson. © 1996 R.G. Landes Company.

This chapter is divided into three sections. The first section summarizes the present status of DNA marker utilization in crop genome analysis, relative to goals proposed near the beginning of the "DNA marker era." The second section discusses current needs in DNA-marker assisted crop improvement. The third section highlights new opportunities to improve the rate and efficiency of crop improvement, as a result of genome analysis.

7.2. PROGRESS TOWARD THE HISTORICAL OBJECTIVES OF CROP GENOME ANALYSIS

Several key articles which stimulated early interest in crop genome analysis[2-5] highlight four objectives which were viewed as areas where DNA markers had potential impact on classical plant breeding:

(1) Strain identification and plant variety protection.
(2) Assessment of genetic diversity.
(3) Accelerated introgression or backcross "conversion."
(4) Mapping complex traits relevant to crop improvement.

Goals 1 and 2 are addressed in detail by subsequent chapters in this volume (chapters 8 and 9). This chapter will focus on goals 3 and 4.

7.2.1. ACCELERATED BACKCROSS "CONVERSION"

A common objective in plant breeding is to transfer a specific gene (trait) into an otherwise-desirable genotype which has an easily-remedied defect. The "backcross" breeding method[6] was devised to meet this need—retaining the superior performance of an elite genotype, but extending the useful life of the genotype by remedying defects such as susceptibility to a new strain of disease. Because the cultivated gene pool of many crops represents only a small portion of the total genetic diversity within a taxon (see chapter 8), many genes for disease resistance or other traits are found only in wild or feral plants—use of the backcross method minimizes the like-

lihood that undesirable traits will be transferred from these exotic relatives, along with the target gene.

The potential for accelerating backcross breeding was appreciated even prior to the advent of DNA markers. Tanksley and Rick[7] recognized that isozymes might be used to select backcross progeny which had unusually large portions of "recurrent" parent genotype, reducing the number of backcross generations necessary to re-constitute a recurrent parent genotype. Complete genetic maps of DNA markers afforded more complete genome coverage than isozyme maps. Detailed discussion of the numerical advantages of DNA marker-aided backcrossing have been published elsewhere.[8]

DNA marker-aided backcrossing has been very widely-used by commercial firms—savings of even 1-2 generations means that an improved cultivar reaches the marketplace sooner. In addition, increased speed means that backcross-derived cultivars are more likely to remain competitive in the marketplace. In the past, one drawback of the backcross approach has been that development of new, improved cultivars renders a recurrent parent obsolete by the time a backcrossing program is completed.

In a hybridization of "biotechnologies," DNA marker-aided backcrossing is often practiced in conjunction with transformation of crop plants with exogenous genes. In many crops, transformation has been optimized only for a small number of genotypes, which are especially amenable to transformation (for unknown reasons). A considerable amount of work and cost is invested in obtaining a single transgenic plant which properly expresses a new gene—many transgenic plants express "transgenes" at suboptimal levels or not at all, as a result of ill-defined "position effects" (see chapter 13). Moreover, even if transgenes can be directly introduced into the target genotype, several generations of backcrossing may still be needed. Somaclonal variation can arise during the process of regenerating transformed plants, and can alter the performance of the transgenic plants in unexpected ways, relative to the original cultivar. "Genotype-independent" transformation methods may be devised in the future. However, in the meantime, many practitioners are optimizing transformation for a single genotype, then using marker-aided backcrossing to accelerate transfer of new genes to other genotypes.

7.2.2. INTROGRESSION

In the literature of a decade ago, it was envisioned that DNA marker technology would reduce the obstacles associated with introgression of useful genes from exotic germplasm. The principles underlying this hypothesis have generally been borne out—as discussed above, DNA markers can accelerate recovery of a "recurrent" genotype. Further, by using DNA markers to identify recombinants, introgressed chromosome segments might be "trimmed" to minimal size, reducing the extent to which the recurrent genotype is disrupted by undesirable alleles closely-linked to the target trait.[7]

Numerous "wide" crosses have been made for the purpose of genetic mapping, using cultivars or advanced breeding lines as one parent, and feral germplasm or even different congeneric species as the other parent. Further, many near-isogenic strains derived from wide crosses have been analyzed, and previously-introgressed genes have been mapped. However, despite improved capabilities at the molecular level, classical breeding programs have generally not expanded their introgression efforts, perhaps partly due to the high cost and cumbersome technology still associated with DNA markers, but also due to continuing reluctance to use exotic germplasm. In several plant taxa, loss of key public-sector positions dedicated to exotic germplasm introduction and population improvement, threaten to reduce the flow of new genetic variation from exotic sources.

Recently, some different approaches to introgression are being implemented, which are providing new incentive to attempt introgression projects. Classical introgression programs have targeted obvious traits from an exotic source which were clearly advantageous, and were lacking from the cultivated gene pool. However, several recent studies have shown that by random introgression of exotic chromatin into cultivated genotypes, cryptic variation can be revealed, variation which is not apparent from the phenotype of the exotic source. For example, the slow-growing *Lycopersicon pennellii* has been shown to contain particular chromosome segments which increase the rate of dry weight accumulation of cultivated tomato, a trait which was completely unexpected based on the plant's phenotype. Likewise, it was unexpected that *Lycopersicon pennellii* could contribute very much to the "soluble solids" concentration of the tomato fruit—but random introgression revealed that some *Lycopersicon pennellii* chromosome segments do enhance soluble solids of tomato. The approach of random backcrossing of exotic chromatin into elite genotypes, in a search for cryptic variation, has been dubbed "advanced-backcross" (AB)-QTL analysis.[11] The success of early examples in tomato[9-12] and cotton (A.H. Paterson and C.W. Smith, unpublished) may motivate expanded efforts in other crop species.

7.2.3. MAPPING OF COMPLEX TRAITS RELEVANT TO CROP IMPROVEMENT

The economic importance of particular plant species, together with the many advantages that might accrue to crop improvement through DNA-marker assisted selection, has precipitated virtually explosive growth in molecular genetic analysis of agriculturally important traits. Genetic mapping of characteristics associated with many aspects of crop productivity is published or in progress. A sampling of the sorts of characteristics which have been mapped, or soon will be, includes:

(1) Basic growth parameters such as plant height[13-16] and flowering time;[14,16,17,18]
(2) Morphological features such as tillering, rhizomatousness,[19] and many other morphological variants;[20]
(3) Yield components such as the size,

number, and harvestability of seed,[19,21-27] biomass and/or growth rates;[28,29]

(4) Quality parameters such as composition of fruit or seed,[30-36] shape of tubers,[37] or specific gravity of wood;[38]

(5) Impact of adversities such as diseases,[39-43] insects,[44,45] and abiotic factors;[46,47]

(6) Evolution of novel organs;[48] and

(7) Basic genetic phenomena such as transgression (progeny which exceed parental genetic potential[28,32]), heterosis (hybrid vigor[22,49]), and epistasis (nonlinear interaction between different genetic loci[50,51]).

Of particular potential value to the plant breeder are DNA markers diagnostic of traits which are difficult to measure. One excellent example of such a trait is nematode resistance—which requires one to dig up the plant and examine the roots. Nematode resistance was one of the first traits associated with a molecular marker in any plant.[52] DNA markers diagnostic of nematode resistance have been identified in tomato (see chapter 21), peanut (Burow, Simpson, Paterson, and Starr, submitted), and several other crops.

7.2.4. APPLICATIONS OF GENETIC MAP INFORMATION IN PLANT BREEDING PROGRAMS: DNA MARKER-ASSISTED SELECTION

The rate at which genetic map information is generated has dramatically outstripped application of this information in crop improvement programs. It has been clearly established that genes underlying both simply-inherited and complex traits can readily be "diagnosed" with DNA markers. Further, the idea that DNA marker-assisted selection affords improved efficiency in the improvement of quantitative traits has been demonstrated both at the theoretical level[53] and experimentally.[34]

There are at least two obvious reasons why relatively few breeding programs have adopted large-scale DNA-marker assisted breeding approaches. One is the level of technology still needed to practice DNA

marker-based assays. Another is the cost of the assays. Each of these obstacles are likely to be obviated by technological improvements—as discussed in more detail below.

A third, less obvious factor may be even more important. An increasing number of young scientists are being trained both in classical and genome-based approaches to crop improvement, with an appreciation of specific aspects of breeding programs that are particularly amenable to DNA-marker acceleration, and with both the technical and theoretical expertise to identify diagnostic molecular tools at minimal cost. Plant breeding programs are designed around particular "breeding methods," partly as dictated by the biology of the crop, and partly at the preference of the individual breeder. Usually, the program "advances" genetic material from year to year, gradually altering the nature and intensity of selection, often taking a decade or more between the time a cross is made, and the time a new cultivar is identified.

Application of genetic mapping to the population structures found in classical plant breeding programs is often not straightforward. While some breeding methods such as the "backcross"[6] method, or "single-seed descent"[54] are readily amenable to genetic mapping, more complex populations such as random or selective intermating are just beginning to be studied by genetic mapping,[55] and use of DNA markers in many additional population structures remains to be investigated.

The merger of plant breeding and genome mapping will not be accomplished simply by development of high-throughput DNA markers, but will also involve rational modification of breeding methods to best exploit marker information. The availability of marker information is a new and valuable component of plant breeding programs, which has not previously been factored into design of breeding programs. "Marker-assisted breeding" versus classical approaches differ profoundly, in the same way that breeding for monogenic traits differs from breeding for polygenic traits.

Basic quantitative parameters such as heritability may be critical in identifying specific aspects of breeding programs which can benefit the most from marker information, and which aspects might be best addressed by classical means.[34] Although many contemporary plant breeders are enthusiastically pursuing opportunities to accelerate progress through genome mapping, it seems likely that the merger of plant breeding and genome mapping will be consummated by young scientists emerging from training programs which encompass both the experience of classical plant breeding, and the technology of molecular genetics.

7.3. NEEDS

The technology associated with DNA markers remains tedious and expensive. Perhaps the most apparent need in DNA-marker assisted crop improvement is the development and implementation of simpler technology for assaying DNA markers. A second important need is a more detailed description of the allelic composition of elite crop gene pools. By being able to track the transmission of particular chromosome segments through decades, or even centuries of crop breeding, we might gain clues about specific genes or genomic regions which are important to crop performance, in a manner more simple and efficient than these clues might have been gained by mapping experiments.

7.3.1. DNA MARKER TECHNOLOGY

Many different opinions have been expressed regarding what characteristics a DNA marker assay should have, to be "optimal" for use in applied breeding programs. This author's opinion, regarding the characteristics of an ideal DNA marker system for crop breeding programs, follow:
(1) Requires minimal DNA per assay (nanogram quantities, or less). This criterion is widely-embraced, and points to a PCR-based assay. The RFLP technique generally requires about 250 ng DNA per individual per assay (assuming that 5 μg of DNA per lane is blot-

ted, and that a blot is used 20 times, the latter a somewhat generous assumption in some ta). By contrast, most PCR-based assays use 10-20 ng template DNA. Several techniques have been described for very simply obtaining small amounts of DNA suitable for PCR. One of these which is particularly appealing involves chipping a small piece of tissue from an ungerminated seed—the seed can then be genotyped while still quiescent, and only those seeds containing a target genotype need be planted.[56]

(2) Allelic richness. This criterion, too is widely embraced. Most crop gene pools represent a rather small number of closely-related genotypes, and many DNA marker alleles are common to most leading cultivars. While the importance of allelic richness is generally-acknowledged, two approaches have been employed to overcome this problem:

2a) The "brute force" approach—simply characterize so many candidate DNA elements that even a small subset provide the number of DNA polymorphisms needed. This approach is employed in the RAPD, AP-PCR, DAF, and AFLP, techniques (see chapter 3).

2b) Enrichment for rapidly-evolving DNA sequences—this is the basis for the popularity of the STM (= SSR = SSLP) method based upon DNA microsatellites (see chapter 10). While many tandemly-repeated sequences are rapidly-evolving, these short elements are more-easily detected by PCR.

(3) Technological simplicity. The importance of this criterion is sometimes debated—in the view of this author, technological simplicity empowers a greater number of individuals to participate in the activity. This is particularly important in plant breeding, because of the need to study the behavior of genes and genotypes in a wide range of environments, and because of differences in plant breeding methodology and "philosophy." High-throughput

and low cost (including fixed cost of equipment) are needed for on-site application of DNA marker information in breeding programs. Particular constraints include:

3a) Radioactivity—as of this writing, non-radioactive techniques remain more expensive than radio-isotopic procedures for labeling of DNA molecules, in most situations. (One noteworthy exception is remote locations such as crop research centers in developing countries, which cannot obtain radio-isotopes quickly enough to maintain high specific activities, and where shipping costs are prohibitive).

This may be largely a matter of time—as non-radioactive techniques are streamlined and simplified, costs may come down. Most existing DNA marker assays, and most new types of assays anticipated can, in principle, use non-radioactive assays.

3b) Electrophoresis—repeated handling of individual samples is prohibitively time-consuming. Moreover, assays which rely on both electrophoresis and fluorescence require very expensive machinery for detection of signals. Most assays devised to date rely upon electrophoresis for separation of different alleles from one another. Only one assay, "allele-specific oligonucleotides,"[57] bypasses electrophoresis, requiring only good temperature control.

(4) Universality. The utility of comparative mapping in cross-utilizing genetic information, suggests that identification of DNA reagents which can be applied equally well to divergent taxa would find greater utility than reagents specific to particular taxa. Such marker systems would meet the criterion previously described for markers useful in comparative mapping (chapter 3), of orthology (correspondence) between taxa, and conservation of DNA sequence. Many cDNAs show appreciable DNA sequence conservation over evolutionary distances of 100 million years or more, and partially-degenerate PCR primers can be used to assay orthologous loci.[58]

(5) Direct links to existing genetic maps. In view of the enormous amount of information which has already accumulated regarding the genetic control of agriculturally important traits, it is important that new techniques be easily tied to existing genetic maps. Preferably, such new techniques would be derived from previously-mapped DNA segments. However, because ease of application is a criterion of the new techniques sought, if necessary, additional markers can be applied to existing genetic maps, as well as to new populations and applications.

In the opinion of this author, the CAPS system represents the most logical starting point for development of advanced-generation assays which meet these criteria. CAPS are PCR-based markers which can easily use expressed sequences conserved across taxa (see chapter 3), thereby meeting both the criteria of "DNA efficiency," and "universality." The CAPS system itself does require electrophoresis and is not particularly allele-rich.[59-60] However, by using CAPS as a starting point, introns or 3' flanking sequences might provide a genomic domain which is more rapidly-evolving (hence more allele-rich). By combining PCR-based amplification with detection of alleles specific to a rapidly-evolving region of an intron, one might develop an assay which meets most of the above criteria.[61-62] Rapid progress in generation of DNA sequence information for large numbers of expressed sequences in many taxa (see chapter 14) provides a starting point for identifying PCR primers suitable for amplifying CAPS. Sequence analysis of genomic DNA amplified using CAPS primers could reveal allelic variation in introns suitable for ASO-based detection,[57] needed to design such a system.

7.3.2. FINGERPRINTING CROP GENE POOLS

A second important need is the comprehensive DNA fingerprinting of crop gene pools, including as many cultivars, breeding lines, and progeny as possible. Moreover, these data need to be integrated with both phenotypic information and pedigree information. The needs for this information are at least three-fold.

(1) A database of DNA marker alleles for the elite gene pool of a crop provides the information on specific DNA polymorphisms which is needed to design, execute, and analyze genetic mapping experiments, targeted at specific traits or specific crosses. By "fingerprinting" a large number of genotypes simultaneously, one gains economies of scale relative to what such an exercise might have required if done independently for many different genetic mapping experiments.

(2) The same database serves as a classification tool, describing the overall levels and patterns of variation within the crop gene pool, and illustrating subdivisions within gene pool such as "heterotic groups." Such information might be useful in making predictions about the performance of new germplasm, or selecting parents for crosses which are likely to yield new gene combinations, or afford an optimal degree of heterosis.[63-67]

(3) Genes associated with agriculturally important phenotypes might be tracked through pedigrees and gene pools. The gene pools of most crops trace back to a rather small number of "founder" genotypes, in some cases quite recently. While occasional ventures outside this group of founders have been employed to introgress genes for disease resistance (or other traits), the bulk of chromatin in elite crop varieties is comprised of chromosome segments which can be traced back to these "founders." By first identifying specific phenotypes which are shared among particular elite crop cultivars, then identifying chromosome segments which are identical by descent in these cultivars, one might "map" agricultur-

ally important genes or gene complexes more efficiently than by present approaches (such as described in chapter 4). This pedigree-based approach to genetic mapping has the advantages that many traits can be studied simultaneously, and that genes are studied within the context of elite germplasm rather than in exotic crosses which are genetically very distant. Early efforts at such pedigree-based analysis are being pursued in several crops (for example, see ref. 68), and it is likely that the future will hold much more of this sort of work.

7.4. OPPORTUNITIES

Despite the long list of crops and traits which have now been subjected to genetic mapping, we are only just beginning to truly implement DNA marker information in crop breeding programs. In the field, every growing season points to new phenotypes which are priorities for genetic analysis—and every germplasm expedition brings back new alleles worthy of our attention. In the lab, every passing year finds us with a more detailed understanding of the organization and composition of crop genomes, better understanding of the relationships among different crop genomes, a larger repertoire of tools for gene manipulation, and more efficient technologies for utilizing, and further expanding, our knowledge.

Two aspects of genome research have come into prominence only in the past few years, absent from the seminal literature of a decade ago.[2-5] "Comparative mapping," the alignment of chromosomes from different taxa based on common DNA probes, has been demonstrated in a few taxa, and is likely to be of increasing importance in the coming years. Molecular cloning of agriculturally important plant genes has been accomplished in a rapidly-growing number of cases (see chapters 12 and 13), and some comments on the potential of molecular cloning in crop improvement are warranted.

7.4.1. COMPARATIVE MAPPING

Perhaps the single greatest opportunity which has emerged in the past few years is the opportunity to exploit comparative information in crop improvement. Vavilov's "law of homologous series in variation"[69] was an early recognition of the fundamental similarity between different cultivated species. Over the past two decades, molecular genetics has revealed fundamental similarity in gene repertoire between taxa that have been reproductively-isolated for millions of years. Most plant breeders now recognize that similarities between their target species and related taxa transcend diversity in breeding objectives, however, there has previously been no conduit for exchange of genetic information between taxa which could not be hybridized.

Comparative maps provide such a conduit for communication—permitting information gathered during study of one species to be quickly and efficiently applied to related species. By using common DNA probes as reference loci to align the chromosomes of taxa which cannot be hybridized, it has been made clear that a relatively modest number of chromosomal rearrangements (inversions and/or translocations) distinguish the chromosomes of many major crops and model systems (chapters 15-22). Moreover, mutations which explain variation in agriculturally-important traits such as plant height, flowering, shattering, seed size, and other traits map to corresponding chromosomal locations in different taxa, far more often than can be explained by chance (see chapter 4)—suggesting that the mutations may be in orthologous genes.

Comparative genetic mapping can be used to provide a more comprehensive picture of the repertoire of genes which potentially influence a trait, and directly link individual genes to DNA markers. A "comparative phenotypic map" finds potential application both in crop improvement and in molecular cloning.[16] In breeding programs, a comparative phenotypic map might predict for example the locations of new reduced height mutants, in which allelic variation has not yet been found. DNA markers might guide a search for such new variants. Through establishing correspondence among the genes influencing a complex trait in different taxa, genes of interest in a large, highly duplicated genome such as maize might be more easily cloned based upon allelic variants in smaller genomes such as sorghum or rice.

With the advent of plant genome databases (for example, see ref. 70) it becomes ever-more possible, and ever-more fruitful for the breeder/geneticist to utilize information from distantly-related taxa. An "evolutionary" approach to improving the efficiency of crop improvement is rapidly becoming feasible. The value of comparative information makes it worthwhile for the geneticist not only to map a gene or QTL, but also to determine how that gene/ QTL fits into the repertoire that have already been identified for that taxon, and related taxa. At present, this involves a comprehensive search of literature, and perhaps mapping of a few additional markers as a basis for more precise comparison. However, as databases move toward greater integration of genotypic and phenotypic data, as well as qualitative and quantitative data,[71] this exercise will become increasingly simple.

7.4.2. MOLECULAR CLONING OF AGRICULTURALLY-IMPORTANT PLANT GENES

The search for DNA elements which specifically encode functions critical to plant productivity and quality, enjoys an array of tools and information seldom envisioned even a few short years ago. Nonetheless, cloning of agriculturally-important plant genes remains challenging. Many agriculturally important traits are influenced by multiple genes because the effects of individual genes can be obscured by the vagaries of nongenetic factors such as microenvironmental variation, and because plant genomes are large and complex, requiring a search through tens of thousands of protein-encoding sequences together with a much larger quantity of non-coding

DNA to find the single element that directs a particular biochemical step. Often, this search is done in the absence of knowledge about exactly what the step is, biochemically speaking—the step is often known only from its phenotypic consequences. Technologies such as insertion mutagenesis (see chapter 13), and map-based cloning (see chapter 12), are amenable to isolation of genes known only from phenotype. Moreover, development of physical maps for several plant genomes (see chapter 5), together with comprehensive programs to identify and catalog the majority of expressed sequences in several plant taxa (see chapter 14), provide an ever-more powerful database for searching complex genomes for specific genes.

A number of genes have been cloned from crop plants and model systems, mostly by use of naturally-occurring or induced mutant alleles with discrete phenotypic effects. In principle, the path to cloning of relatively discrete mutant alleles is clear in many plant taxa. In practice, however, the path often turns out to be tortuous, due to hidden obstacles such as imperfect penetrance or expressivity, gaps in genetic maps, repetitive DNA elements, transient or low levels of expression of target genes, lack of efficient transformation techniques, or other headaches. "Quantitative trait loci (QTLs)," the class of genetic loci which account for the inheritance of many complex measures of plant quality or productivity, remain refractory to molecular cloning. In principle, the information needed for map-based cloning of QTLs can be obtained,[33] however, in no case has a QTL yet been cloned. In practice, the first QTLs to be cloned might well take advantage of discrete mutant alleles at loci which correspond to QTLs. This approach was suggested several years ago,[72] and shows increasing promise as comparative maps enable us to broaden the search for discrete mutants to include divergent taxa.[27]

Molecular cloning is frequently only loosely-related to crop improvement—first of all, most of the genes which have been cloned to date come from *Arabidopsis*, which is not a crop. Secondly, it is much faster and easier to manipulate a plant gene by making a cross between different genotypes, than it is to clone the gene and introduce it extra-genetically into a different genotype. The merits of molecular cloning remain a topic which is sometimes hotly debated by basic and applied scientists respectively. The high cost, and high level of risk associated with cloning plant genes is implicitly acknowledged, in that only a small fraction of proposed cloning projects are considered worthy of public funding.

In the view of this author, molecular cloning will gain increasing acceptance as a crop improvement technique. In many instances, molecular cloning is motivated by gaining a greater understanding of plant development, revealing processes, pathways, and regulatory controls which are relevant to study of a wide range of plant taxa. Model systems such as *Arabidopsis* will be related to crop genomes by several means, including comparative mapping, isolation and study of heterologous genes, or direct analysis of *Arabidopsis* genes transformed into other taxa. The complexity of cloning plant genes will continue to require "prioritization," choosing the few genes which appear most valuable and most tractable to pursue first. Comparative mapping affords a means to prioritize genes for cloning, identifying those genes which appear to be important in a wide range of taxa.

7.5. SUMMARY

Plant breeding is itself evolving, as it directs the evolution of crop gene pools. A widely-held fear of a few years ago, that the classical plant breeder would be replaced by a "biotechnologist," seems to be fading, as many plant breeders find themselves the key player in determining which biotechnology products are competitive in the marketplace, and which are not. The newfound ability to study the plant genome as a set of components, rather than as a single phenotype, offers a new level of resolution to plant breeding endeavors.

Even within the boundaries of sexual hybridization, cryptic variation is being found which is clearly useful in breeding programs. The possibility of reaching outside the boundaries of sexual hybridization to seek such variation adds further opportunity. It is truly an exciting time to be a plant breeder!

The promise of DNA markers for crop improvement, based upon goals and expectations of a decade ago, is well on its way to being realized. In many areas of crop improvement, molecular markers are now commonplace, and accepted as a valuable adjunct to classical breeding methods. The major obstacles to seamless integration of molecular marker information with classical plant breeding are technological, and the advent of the polymerase chain reaction (PCR) has offered valuable new approaches. Recent documentation of the extent of similarity of gene order, along the chromosomes of different crop species, provides a conduit for communication of information between scientists working on different taxa. Such information exchange holds the promise of greater efficiency in genetic analysis of crops, more w.despread use of existing information, and more rapid accumulation of new information about the genetic control of agriculturally important traits.

REFERENCES

1. Fehr W. Genetic contributions to yield gains of five major crop plants. Spec. Publ. No. 7. Madison, WI: Crop Sci Soc America, 1984.
2. Burr B, SV Evola, FA Burr, JS Beckmann. The application of restriction fragment length polymorphism to plant breeding. pp. 45-59 In: JK Setlow and A Hollaender (eds.), Genetic Engineering: Principles and Methods. NY: Plenum Publishing Co, 1983.
3. Soller M and JS Beckmann. Genetic polymorphism in varietal identification and genetic improvement. Theor Appl Genet 1983; 67:25-33.
4. Tanksley SD. Molecular markers in plant breeding. Plt Mol Biol Rptr 1983; 1:3-8.
5. Helentjaris T, G King, M Slocum, C Siedenstrang, S Wegman. Restriction fragment polymorphisms as probes for plant diversity and their development as tools for applied plant breeding. Plant Mol Biol 1985; 5:109-118.
6. Briggs FN. The use of the backcross in crop improvement. Amer Nat 1938; 72:285-292.
7. Tanksley SD and CM Rick. Isozyme gene linkage map of the tomato: applications in genetics and breeding. Theor Appl Genet 1980; 57:161-170.
8. Tanksley SD, ND Young, AH Paterson, MW Bonierbale. RFLP mapping in plant breeding: New tools for an old science. Biotechnology 1989; 7:257-264.
9. DeVicente MC, SD Tanksley. QTL analysis of transgressive segregation in an interspecific tomato cross. Genetics 1993; 134: 585-596.
10. Eshed Y, D Zamir. Introgressions from *Lycopersicon pennellii* can improve the soluble solids yield of tomato hybrids. Theor Appl Genet 1994; 88:891-897.
11. Tanksley SD and Nelson JC. Advanced backcross QTL analysis: A method for the simultaneous discovery and transfer of valuable QTLs from unadapted germplasm into elite breeding lines. Theor Appl Genet 1995; in press.
12. Tanksley SD, Grandillo S, Fulton TM et al. Advanced backcross QTL analysis in a cross between an elite processing line of tomato and its wild relative *L. pimpinellifolium.* Theor Appl Genet 1995; in press.
13. Beavis W, D Grant, M Albertsen, and R Fincher. Quantitative trait loci for plant height in four maize populations and their associations with qualitative loci. Theor Appl Genet 1991; 83:141-145.
14. Koester RP, PH Sisco, and CW Stuber. Identification of quantitative trait loci controlling days to flowering and plant height in two near isogenic lines of maize. Crop Sci 1993; 33:1209-1216.
15. Pereira MG, M Lee, and PJ Rayapati. Comparative RFLP and QTL mapping in sorghum and maize. Poster 169 in the Second Internal Conference on the Plant Genome, New York: Scherago Internal, Inc., 1994.
16. Lin YR, KF Schertz and AH Paterson.

Comparative analysis of QTLs affecting plant height and maturity across the Poaceae, in reference to an interspecific sorghum population. Genetics 1995; 141:391-411.

17. Kowalski SD, T-H Lan, KA Feldmann, AH Paterson. QTL mapping of naturally-occurring variation in flowering time of Arabidopsis thaliana. Mol Gen Genet 1994; 245:548-555.

18. Li Z, SRM. Pinson, JW Stansel, WD Park. Identification of quantitative trait loci for heading date and plant height in cultivated rice. Theor Appl Genet 1995; 91:374-381.

19. Paterson AH, KF Schertz, YR Lin, SC Liu, YL Chang. The weediness of wild plants: Molecular analysis of genes influencing dispersal and persistence of johnsongrass, *Sorghum halepense* (L.) Pers. Proc Natl Acad Sci USA 1995; 92:6127-6131.

20. Kennard WC, Slocum MK,Figdore SS, et al. Genetic analysis of morphological variation in *Brassica oleracea* using molecular markers. Theor Appl Genet 1994; 87:721-32.

21. Stuber CW, MD Edwards, JF Wendel. Molecular-marker-facilitated investigations of quantitative-trait loci in maize. II. Factors influencing yield and its component traits. Crop Sci 1987; 27:639-648.

22. Stuber CW, SE Lincoln, DW Wolff, T Helentjaris, ES Lander. Identification of genetic factors contributing to heterosis in a hybrid from two elite inbred lines using molecular markers. Genetics 1992; 132:823-839.

23. Abler BSB, MD Edwards, CW Stuber. Isozymatic identification of quantitative trait loci in crosses of elite maize inbreds. Crop Sci 1991; 31:267-274.

24. Fatokun CA, DI Menacio-Hautea, D Danesh, ND Young. Evidence for orthologous seed weight genes in cowpea and mungbean, based upon RFLP mapping. Genetics 1992; 132:841-846.

25. Doebley J, A Bacigalupo, A Stec. Inheritance of kernel weight in two maize-teosinte hybrid populations: Implications for crop evolution. J Hered 1994; 85:191-195.

26. Schon CC, AE Melchinger, J Boppenmaier, E Brunklaus-Jung, RG Herrmann et al RFLP mapping in maize: Quantitative trait

loci affecting testcross performance of elite European flint lines. Crop Sci 1994; 34:378-389.

27. Paterson AH, YR Lin, Z Li, KF Schertz, JF Doebley, SRM Pinson, SC Liu, JW Stansel, JE Irvine. Convergent domestication of cereal crops by independent mutations at corresponding genetic loci. Science 1995; 269:1714-1718.

28. DeVicente MC and SD Tanksley. QTL analysis of transgressive segregation in an intraspecific tomato cross. Genetics 1993; 134:585-596.

29. Bradshaw HD, and RF Stettler. Molecular genetics of growth and development in Populus. IV. Mapping QTLs with large effects on growth, form, and phenology of traits. Genetics 1995; 139:963-973.

30. Osborn TC, Alexander DC, Fobes JF. Identification of restriction fragment length polymorphisms linked to genes controlling soluble solids content in tomato fruit. Theor Appl Genet 1987; 73:350-356.

31. Tanksley SD and JD Hewitt. Use of molecular markers in breeding for soluble solids in tomato: a re-examination. Theor Appl Genet 1988; 75:811-823.

32. Paterson AH, ES Lander, JD Hewitt, S Peterson, SE Lincoln and SD Tanksley. Resolution of quantitative traits into Mendelian factors by using a complete map of restriction fragment length polymorphisms. Nature 1988; 335:721-726.

33. Paterson AH, JW Deverna, B Lanini and SD Tanksley. Fine mapping of quantitative trait loci using selected overlapping recombinant chromosomes, in an interspecies cross of tomato. Genetics 1990; 124:735-742.

34. Paterson AH, S Damon, JD Hewitt, D Zamir, HD Rabinowitch, SE Lincoln, ES Lander and SD Tanksley. Mendelian factors underlying quantitative traits in tomato: comparison across species, generations, and environments. Genetics 1991; 127:181-197.

35. Weller JI, M Soller, T Brody. Linkage analysis of quantitative traits in an interspecific cross of tomato, *Lycopersicon esculentum* x *Lycopersicon pimpinellifolium*, by means of genetic markers. Genetics 1988; 118:329-340.

36. Teutonico RA, and TC Osborn. Mapping

of RFLP and qualitative trait loci in *Brassica rapa* and comparison to the linkage maps of *B. napus*, *B. oleracea*, and *Arabidopsis thaliana*. Theor Appl Genet 1994; 89: 885-894.

37. Van Eck HJ, JME Jacobs, P Stam, J Ton, WJ Stiekema, E Jacobsen. Multiple alleles for tuber shape in diploid potato detected by qualitative and quantitative genetic analysis using RFLPs. Genetics 1994; 137:303-309.

38. Groover A, M Devey, T Fiddler, J Lee, R Megraw, T Mitchell-Olds, B Sherman, S Vujcic, C Williams, D Neale. Identification of quantitative trait loci influencing wood specific gravity in an outbred pedigree of loblolly pine. Genetics 1994; 138:1293-1300.

39. Bubeck DM, MM Goodman, WD Beavis, D Grant. Quantitative trait loci controlling resistance to gray leaf spot in maize. Crop Sci 1993; 33:838-847.

40. Leonards-Schippers C, W Gieffers, R Schaefer-Pregl, E Ritter, SJ Knapp, F Salamini, C Gebhardt. Quantitative resistance to Phytophthora infestans in potato: A case study for QTL mapping in an allogamous plant species. Genetics 1994; 137:68-77.

41. Wang G, DJ MacKill, JM Bonman, SR McCouch, MC Champoux, RJ Nelson. RFLP mapping of genes conferring complete and partial resistance to blast in a durably resistant rice cultivar. Genetics 1994; 136:1421-1434.

42. Li Z, SRM Pinson, MA Marchetti, JW Stansel, WD Park. Characterization of quantitative trait loci in cultivated rice contributing to field resistance to sheath blight (*Rhizoctonia solani*). Theor Appl Genet 1995; 91:374-381.

43. Jung M, T Weldekidan, D Schaff, A Paterson, S Tingey, J Hawk. Generation means analysis and genetic mapping of anthracnose stalk rot resistance in maize. Theor Appl Genet 1995; in press.

44. Nienhuis J, T Helentjaris, M Slocum, B Ruggero and A Schaefer. Restriction fragment length polymorphism analysis of loci associated with insect resistance in tomato. Crop Sci 1987; 27:797-803.

45. Bonierbale MW, RL Plaisted, O Pineda, SD Tanksley. QTL analysis of trichome-mediated insect resistance in potato. Theor Appl Genet 1994; 87:973-987.

46. Martin B, J Nienhuis, G King, A Schaefer. Restriction fragment length polymorphisms associated with water use efficiency in tomato. Science 1989; 243:1725-1728.

47. Reiter RS, JG Coors, MR Sussman and WH Gabelman. Genetic analysis of tolerance to low-phosphorus stress in maize using restriction fragment length polymorphisms. Theor Appl Genet 1991; 82:561-568.

48. Doebley J, A Stec, J Wendel, and M Edwards. Genetic and morphological analysis of a maize-teosinte F2 population: implications for the origin of maize. Proc Nat Acad Sci USA 1990; 87:9888.

49. Xiao J, J Li, L Yuan, SD Tanksley. Dominance is the major genetic basis of heterosis in rice as revealed by QTL analysis using molecular markers. Genetics 1995; 140:745-754.

50. Doebley J, A Stec, C Gustus. *Teosinte branched 1* and the origin of maize: evidence for epistasis and the evolution of dominance. Genetics 1995; 141:333-346.

51. Lark KG, K Chase, F Adler, LM Mansur, JH Orf. Interactions between quantitative trait loci in soybean in which trait variation at one locus is conditional upon a specific allele at another. Proc Nat Acad Sci USA 1995; 92:4656-4660.

52. Rick CM, JF Fobes. Association of an allozyme with nematode resistance. Rep Tomato Genet Coop 1974; 24:25.

53. Lande R, and R Thompson. Efficiency of marker-assisted selection in the improvement of quantitative traits. Genetics 1990; 124:743-746.

54. Brim CA. A modified pedigree method of selection in soybeans. Crop Sci 1966; 6:220.

55. S-C Liu, SP Kowalski, T-H Lan, KA Feldmann, AH Paterson. Genome-wide High-Resolution Mapping by Recurrent Intermating, Using *Arabidopsis thaliana* as a Model. Genetics 1996; 142:247-258.

56. Wang G, R Wing, and AH Paterson. PCR amplification of DNA extracted from single seeds, facilitating DNA-marker assisted selection. Nucl Acids Res 1993; 21:2527.

57. Conner BJ, AA Reyes, C Morin, K Itakura, RL Teplitz, RB Wallace. Detection of sickle-cell β^s—globin allele by hybridization with synthetic oligonucleotides. Proc Nat Acad Sci USA 1983; 80:278-282.

58. Mazzarella R, V Montanaro, J Kere, R Reinbold, A Ciccodicola, M D'Urso, D Schlessinger. Conserved sequence-tagged sites: A phylogenetic approach to genome mapping. Proc Natl Acad Sci USA 1992; 89:3681-3685.

59. Konieczny A, and FM Ausubel. A procedure for mapping Arabidopsis mutations using co-dominant ecotype-specific PCR-based markers. Plant J 1993; 4:403-410.

60. Jarvis P, C Lister, V Szabo, C Dean. Integration of CAPS markers into the RFLP map generated using recombinant inbred lines of *Arabidopsis thaliana*. Plt Mol Biol 1994; 24:685-687.

61. Nickerson DA, Kaiser R, Lappin S et al. Automated DNA diagnostics using an ELISA-based oligonucleotide ligation assay. Proc Natl Acad Sci USA 1990; 87:8923-27.

62. Barany F. Genetic disease detection and DNA amplification using cloned thermostable ligase. Proc Natl Acad Sci USA 1991; 88:189-193.

63. Lee M. Molecular genetic diversity among maize inbred lines: taxonomic and plant breeding implications. Plant Sci 1989; 68:69-72

64. Godshalk EB, M Lee, KR Lamkey. Relationship of restriction fragment length polymorphisms to single-cross hybrid performance of maize. Theor Appl Genet 1990; 80:273-280.

65. Melchinger AE, M Lee, KR Lamkey, W Woodman. Genetic diversity for restriction fragment length polymorphisms - relation to estimated genetic effects in maize inbreds. Crop Sci 1990; 30:1033-1040.

66. Melchinger AE, M Lee, KR Lamkey, AR Hallauer, W Woodman. Genetic diversity for restriction fragment length polymorphisms and heterosis for 2 diallel sets of maize inbreds. Theor Appl Genet 1990; 80:488-496.

67. Melchinger AE, M Messmer, M Lee, W Woodman, KR Lamkey. Diversity and relationships among U.S. maize inbreds revealed by restriction fragment length polymorphisms. Crop Sci 1991; 31:669-678.

68. Wang G, J Dong, AH Paterson. Genome composition of cultivated *Gossypium barbadense* reveals both historical and recent introgressions from *G. hirsutum*. Theor Appl Genet 1995; 91:1153-1161.

69. Vavilov NI. The law of homologous series in variation. J Genet 1922; 12:1.

70. Altenbach S, et al. USDA plant genome research program. Adv Agron 1996; 55:113-166.

71. Byrne PF, MB Berlyn, EH Coe. Accessing QTL data in the maize genome database. San Diego, CA: Proc Plant Genome III, 1995:15.

72. Robertson DS. A possible technique for isolating genic DNA for quantitative traits in plants. J Theor Biol 1985; 117:1-10.

CHAPTER 8

MOLECULAR CHARACTERIZATION FOR PLANT GENETIC RESOURCES CONSERVATION

Stuart M. Brown and Stephen Kresovich

"A logical investigator must therefore approach the problem of the origin and the evolution of the cultivated plants to the fullest, while applying new and realistic methods."*

8.1. INTRODUCTION

Plant genetic resources provide the foundation for the maintenance and improvement of crop agriculture. Throughout the course of history, plant genetic resources have been acquired, exchanged, selected, improved, and preserved. As the 21st century approaches, society at large and the scientific community in particular have become keenly aware of the 'value' of ready access to genetic resources. Ex situ conservation of plant genetic resources in seed banks and repositories has evolved to serve its user community, i.e., breeders and evaluators, scientists desiring particular characters or ideotypes for study, and the broader agricultural sector globally. In complement with in situ management strategies for conservation of plants within their native environments, ex situ maintenance will be expected to play a greater role in the future for conservation of agricultural biodiversity.

> "Without a continued source of variability, the ability of plant breeders to improve agronomic performance that is based on complex genetic combinations could decline. Diversity can allow

---

* Academician Nikolai Invanovich Vavilov (1887-1943) Director, V.I. Lenin All-Union Institute of Plant Industry

Genome Mapping in Plants, edited by Andrew H. Paterson. © 1996 R.G. Landes Company.

farmers the opportunity to hedge genetically based risks with respect to annually unforeseen difficulties that accrue from environmental, pest, and disease pressures."[1]

The fundamental mission of plant genetic resources managers is to preserve as broad a sample of the extant genetic diversity of target species as is scientifically and economically feasible. This broad swath must also include currently recognized genes, traits, genotypes, and phenotypes. The justification for this effort lies in the future utility of as yet unidentified genetic information for agriculture, medicine, and other potential uses of plants. In addition to the direct economic utility of genetic resources, the decline of global biodiversity due to human impact presents ecological and philosophical arguments for preservation of unique plant ideotypes that may otherwise become extinct. Because it is impossible to directly survey plant populations for all possible gene and genotypic combinations that may contribute to valuable traits, data from molecular markers can be used as a substitute with the assumption that the diversity of marker loci directly reflects the diversity of useful genes.[2]

Across the world, most genetic resource collections are being expanded to incorporate wild and weedy relatives of crop species. Genes introduced from wild relatives constitute an important source of variation for the improvement of domesticated species. Conservation of these wild and weedy accessions is, and will be, particularly challenging because these entries are often highly heterogeneous and heterozygous. The application of molecular marker-based strategies for characterization can help to maximize the genetic diversity of wild and weedy material selected for inclusion in collections and in core subsets of collections.[2] Financial support for genetic resources conservation has historically been constrained (despite the concurrent interest and value); therefore, maximizing the genetic diversity and useful traits and types of collections allows for the allocation of limited funds among the largest number of species.

The primary tasks of curation can be divided into four major categories including: (1) acquisition, (2) maintenance, (3) characterization, and (4) utilization.[3] As will be demonstrated in this chapter, data from molecular markers can positively impact operations in all of these categories, for example:

(1) acquisition: Data on the diversity of existing collections can be used to plan acquisition strategies. In particular, calculations of genetic distances can be used to identify particularly divergent subpopulations that might harbor valuable genetic variation that is underrepresented in current holdings.

(2) maintenance: Molecular markers can be used to confirm duplicate accessions. Markers can be applied to monitor changes in heterogeneity and heterozygosity (genetic drift) as accessions are regenerated. Molecular data on diversity may provide essential information for the development of core collections that accurately represent the entire collection.

(3) characterization: The genetic diversity within collections must be assessed in the context of the total available genetic diversity for each species. When available, existing passport data documents the geographic location where each accession was acquired. However, many records are missing or incorrect. Molecular markers may allow for characterization based on gene, genotype, and genome, which provide more accurate and detailed information than classical phenotypic or passport data.

(4) utilization: Users of collections benefit from genetic information that allows them to quickly identify valuable traits. On a more fundamental level, molecular marker information may lead to the identification of useful genes contained in collections and transferal of these genes into well-adapted cultivars.

8.2. CURATORIAL NEEDS FOR EFFECTIVE, LONG-TERM CONSERVATION

The goals of effective ex situ curation can be quite complex. This complexity is founded on the need to resolve numerous operational, logistical, and biological questions. Curatorial needs may be summarized as four specific issues:[4]

(1) identity: the determination that an accession or cultivar is catalogued correctly, is true to type, and maintained properly;

(2) relationship: the degree of relatedness among individuals in an accession or groups of accessions within a collection;

(3) structure: the partitioning of variation among individuals, accessions, populations, and species (genetic structure is influenced by in situ demographic factors such as population size, mating patterns, pollination biology, and migration); and

(4) location: the presence of a desired gene or gene complex in a specific accession, as well as the mapped site of a desired DNA sequence on a particular chromosome in an individual, or on a cloned DNA segment.

The application of molecular markers for the resolution of problems of genetic resources conservation is a nascent, yet constructive undertaking, requiring collaboration among curators and molecular biologists. The challenge of long-term, ex situ maintenance of crop species compels conservators and users of genetic resources to articulate what constitutes quality in collections. Molecular markers will serve as tools in this endeavor.

8.3. DESCRIPTION OF A DESIRABLE MOLECULAR MARKER

When considering the application of molecular markers to questions of plant genetic resources conservation, both technical and operational issues must be considered. For example, technical issues relevant to marker characteristics include discriminatory ability, sensitivity, reproducibility, and the ability to be used for further genetic analysis or diagnostics.

Operational issues include protocol characteristics, time, and cost. The ideal molecular marker must be easy to employ, timely, cost effective, highly informative, and reliable (accurate with the desired level of precision). In order to be easy to employ, timely, and cost effective, sample preparation must be simple and the assay (including data generation, collection, organization, and analysis) should be suitable for increased throughput via automation. A high information content necessitates a marker assay that detects high heterozygosity which provides discriminatory ability among closely related individuals, as well as the generation of data from multiple genomic sites, using a single assay. Reliability implies reproducibility of results from assay to assay both within and across laboratories, as well as unambiguous data analysis. In addition, a codominant marker would be advantageous for investigations involving mapping, pedigree analysis, and other plant breeding applications.

In the following sections, the advantages and disadvantages of several types of PCR (polymerase chain reaction)-based molecular markers are evaluated for the measurement of diversity within the context of genetic resources conservation. PCR-based markers, rather than the more traditional markers such as isozymes and RFLPs (restriction fragment length polymorphisms) are the focus of this chapter because these newer markers have the potential for widespread, low-cost, large-scale application that is suitable for the needs of genetic resources conservation as well as for plant breeding and intellectual property rights protection. While no single type of marker satisfies all criteria, each of the four curatorial needs may be addressed using one or more of these markers.

Particular attention has been paid to the appropriate levels of taxonomic distinctions that can be made using each marker type. Markers must detect sufficient polymorphism among samples to address questions of both identity and relatedness. Too much variability, however, can preclude an accurate analysis. In highly diverged lineages, if the genetic distance $(D)^5$ is too great, then back/parallel mutations to identical alleles can occur at high frequencies. Therefore, the presence of identical alleles does not necessarily imply identity by descent. Examples are cited which highlight investigations that have made good use of the characteristics of each marker to address practical questions of genetic resources conservation.

8.4. EXAMPLES OF MOLECULAR MARKERS USEFUL IN GENETIC RESOURCES CHARACTERIZATION

8.4.1. RANDOM AMPLIFIED POLYMORPHIC DNA (RAPD)

A family of molecular marker techniques, based on the ability of short (8-12 bp) primers to amplify at random segments of genomic DNA, has proven to be of value for genetic characterization in a broad range of crop species. The most widely used of these techniques is known as Random Amplified Polymorphic DNA (RAPD), which employs a single 10-base primer of arbitrary sequence (of at least 60% GC content) in each PCR reaction.[6] The amplification yields many DNA fragments ranging in size from less than 100 bp to greater than 2 kb. These fragments are anonymous in the sense that their genomic origins are not known. Differences in the fragment patterns amplified from each genomic DNA sample are generally attributed to mutations at primer binding sites, preventing the annealing of a primer. Because multiple anonymous fragments are produced by a single amplification, data are recorded as the presence or absence of each fragment in the pattern. These fragments are scored as dominant Mendelian elements in diploid organisms.

The primary advantages of random primer PCR techniques are: (1) the lack of a requirement for pre-existing knowledge about DNA sequences in the organisms to be studied, and (2) the simplicity of PCR. PCR-based assays require only small amounts of crude genomic DNA preparations from each sample, results can be obtained in a single day, the necessary laboratory equipment is not expensive, and they are not technically challenging. In addition, cost per assay is low, the assay is readily scaled up to handle large numbers of samples, and many steps can be automated. Each primer provides data from many sites in the genome so that rare polymorphisms between closely related samples can be detected more quickly than with a single locus assay.

Because of the assay characteristics noted above, RAPD analysis has been widely used for the characterization of plant, animal, and microbial genetic resources (for an extensive review see ref. 7). Recently, Virk et al[8] evaluated a number of rice (*Oryza sativa* L.) accessions in an attempt to resolve the question of unwanted duplication within the collection. Based on a selected test array, 100 random primers provided sufficient data to discriminate duplicates. In general, the RAPD assay can provide a good 'first cut' at molecular characterization, independent of particulars regarding the research question and organism. RAPD data is most appropriate for resolving questions of genetic identity rather than relationship, structure, or location.

The primary disadvantage of random primer PCR is the lack of lab-to-lab reproducibility of results. Amplifications are influenced by the purity of genomic DNA, composition of the PCR reactions (particularly the concentration of sample DNA and Mg^+, amplification profile, and thermal cycler. This inconsistency is often manifested as uneven intensity of fragments; a fragment that appears to be missing from one sample may be too lightly stained with ethidium bromide. Another problem affecting the interpretation of RAPD fragment patterns is the uncertainty that fragments of similar size from two different DNA samples are actually produced from the same site in the genome. In response, considerable effort has been devoted to improving the reproducibility and reliability of RAPD assays, resulting in a proliferation of acronym-bearing techniques. The Sequence Characterized Amplified Regions (SCAR) technique, developed by Paran and Michelmore,[9] uses the sequence of polymorphic fragments, detected initially by RAPD analysis, to generate longer (15-30 bp) PCR primers that are specific for a single site in the genome. Fragments produced with SCAR primers are much less subject to inconsistent amplification due to variation in PCR conditions.

8.4.2. AMPLIFIED FRAGMENT LENGTH POLYMORPHISM (AFLP)

Another recently developed modification of RAPD is known as Amplified Fragment Length Polymorphism (AFLP). This technique shares some characteristics with both RFLP and RAPD analyses.[10] AFLP uses restriction enzyme-digested genomic DNA as the template for a PCR reaction with primers that contain the restriction enzyme recognition site as well as a number of additional 'arbitrary' nucleotides. The amplified products are then resolved by polyacrylamide gel electrophoresis. This approach would allow the simultaneous screening of a large number of anonymous markers, randomly distributed throughout the genome. Polymorphisms could be detected between samples while retaining some of the specificity and reproducibility of RFLP assays.

The primary drawback of the AFLP assay may be the inherent generation of great quantities of information, which may require some form of automated data collation prior to analysis. Nonetheless, this drawback could be easily resolved by improvements in computer technology. In addition, the markers produced by the AFLP assay must initially be treated in the same analytical manner as RAPD-type markers, i.e., as dominant Mendelian elements. Therefore, additional analyses are required prior to the use of a fragment as a genetic allele.

8.4.3. VARIABLE NUMBER TANDEM REPEATS (VNTRs)

All eukaryotic genomes contain some regions of highly repetitive DNA, which have higher mutation rates than single copy DNA. These repetitive regions have been categorized into three classes. The first class of repeated DNA sequence elements was discovered by density gradient centrifugation of sheared total genomic DNA from various eukaryotic organisms. This revealed small 'satellite' bands above and/or below the main mass of DNA. This satellite DNA was found to be composed of long tandem repeats of 100 to 5000 bp core sequences clustered at chromosome ends (telomeres) and centromeres. Jeffreys et al[11] described another class of repetitive DNA as 'minisatellites', with 10-100 bp core repeats occurring in tandem arrays of up to 1000 units, distributed throughout the genome. An even more ubiquitous class of repetitive DNA, known as 'microsatellites',[12] is composed of 1-6 bp core units that are tandemly repeated throughout the genome in arrays of a few to many thousands of copies. From these latter two classes of repetitive DNA, molecular markers are being developed for use in genetic resources characterization.

Variations in the number of tandem repeats of minisatellite DNA (VNTRs) have been used as molecular markers to detect high levels of polymorphism, even between closely related individuals and populations of a single species. Mutation rates at minisatellite loci have been estimated as high as 2×10^{-3} per meiosis. VNTRs are generally assayed by a multilocus RFLP approach that involves digesting genomic DNA with a restriction endonuclease, blotting to a membrane, and hybridizing with a labeled probe. The probe can be a copy of any minisatellite repeat unit, but a set of probes developed by Jeffreys et al[11] gave good results across widely divergent groups of species. The multiple, 'anonymous' fragments generated with a VNTR probe are very useful for determining parentage or identity (genetic fingerprinting), but do not allow for the identification of alleles or the determination of genotypes. Thus these markers are difficult to use for measurements of relatedness among individuals or for phylogenetic analyses.[13] Ramakrishna et al[14] used VNTR probes to generate DNA fingerprints of rice cultivars. They identified sufficiently high levels of polymorphism to distinguish between cultivars, but no variation was found between individuals of the same cultivar.

8.4.4. Simple Sequence Repeats (SSRs)

SSRs (also known as microsatellites) are less polymorphic than minisatellites, but occur more frequently in the genome. The variability of SSR loci in humans has been estimated at one mutation per 10^4-10^5 meiosis events.[15] Allelic variation occurs at an SSR locus as a result of changes in the number of times the core unit is tandemly repeated, altering the length of the region. Differences in length at an SSR locus are detected by DNA amplification via PCR, using a pair of oligonucleotide primers that complement unique sequences flanking the SSR. The sizes of the amplified products are precisely determined by acrylamide gel electrophoresis. This type of assay generates single-locus data that provide alleles and genotypes. Evolutionary distances can be calculated between the various alleles at a locus based on the assumption of a stepwise mutation process that results in the alleles closest in size being most closely related by descent from a common progenitor.

SSRs exhibit high levels of polymorphism in most plant species, even among closely related genotypes, with large numbers of alleles found at each locus and heterozygosity values often greater than 0.7.[16] Codominant Mendelian inheritance has been observed for these alleles in many different species.[17-19] Because SSRs can be assayed by PCR, only small quantities of sample DNA are required; the reactions are inexpensive, and can be completed (from sample preparation to genotype analysis) in just two days. Primers for up to 12 loci have been combined in a single-tube 'multiplex' PCR reaction (Kresovich et al, unpublished data). This allows for the simultaneous scoring of multilocus genotypes, providing significant savings in both time and reagents. The primary drawback of SSR-based markers is the effort and cost required to develop specific primer pairs for each polymorphic locus. This involves the cloning and sequencing of large numbers of genomic DNA fragments containing SSRs. However, much progress is occurring in this area, so costs associated with marker isolation and characterization are likely to diminish.

To date, only a few molecular characterizations of plant genetic resources have been conducted using SSR markers in crop species, but preliminary results are very promising. Rongwen et al[17] have surveyed polymorphism at seven loci across 96 diverse soybean (*Glycine max* L.) genotypes, finding an average gene diversity of 0.87. Thomas and Scott[20] found heterozygosity values of 0.69-0.88 within individual grape (*Vitis vinifera* L.) cultivars and an average of about eight alleles for five different SSR loci across 26 cultivars, as well as additional alleles in other *Vitis* species. Diversity among approximately 50 ecotypes of seashore paspalum turf grass (*Paspalum vaginatum* Swartz) was assayed with five SSR markers.[21] The SSR markers were highly polymorphic, with an average of 14 alleles per locus. A phenetic analysis of paspalum ecotypes, based on SSR data, agreed with relationships determined previously by RAPD analysis. Saghai-Maroof et al[16] screened 207 accessions of wild and cultivated barley (*Hordeum vulgare* L.) with four microsatellite markers. Two of these markers identified three alleles each, but the other two markers revealed 28 and 37 alleles, potentially the largest numbers of alleles ever observed at a single locus in plants.

Weising et al[22] assayed polymorphism at SSR loci in plants and fungi by an RFLP approach, using labeled SSR oligonucleotides as probes against filters containing restriction enzyme-digested genomic DNA. This procedure generates highly polymorphic multilocus fingerprints much like VNTRs, but it shares some of the drawbacks in data analysis of scoring anonymous fragments, as well as the technical limitations inherent in a filter hybridization-based assay. Bhat et al[23] used SSR oligonucleotide hybridization to identify unique fingerprint patterns for 16 genotypes of banana and plantain (*Musa* spp. L.), highlighting the value of this marker for cultivar identification.

DNA Sequencing

DNA sequencing provides the ultimate fine-scale measurement of diversity, since all markers are derived (directly or indirectly) from sequence polymorphisms. Evolutionary distances between alleles at a single locus can be directly calculated from the number of nucleotide changes. Until recently, the determination of DNA sequences has been technically very demanding and time consuming. New technology, based on direct sequencing of PCR products using fluorescent dye-terminator chemistry,[24] now allows the semi-automated collection of sequence data from many samples. Direct PCR sequencing can be targeted to any genomic location, either in a highly variable region or adjacent to a particular gene of interest.

However, one major drawback of using DNA sequence information for phylogenetic or diversity measurements is that different genes evolve at different rates, so that gene diversity is not necessarily equivalent to species diversity.[25] Since the determination of DNA sequence is very laborious and costly, most studies have relied on sequences of one or a few loci from each accession, thus potentially generating biased results. This inconsistency can be observed most clearly when phylogenies that are determined using sequence data from different loci do not agree. The only way to obtain an unbiased phylogeny is to use sequence data from multiple loci across the genome that are subject to different selection pressures, but this requires a tremendous amount of laboratory work. As a result, in the near term, a sequencing-based approach is generally not practical for molecular characterization of genetic resources.

8.5. CONCLUSIONS AND FUTURE DIRECTIONS

Because of its technical objectives of high genetic resolution and high throughput, the Human Genome Project now serves as a major source of new technologies for plant biology in general and for genetic analysis in particular. In complement, the Human Genome Diversity Project will provide additional technical and conceptual breakthroughs relevant to plant genetic resources conservation. Continued emphasis on linking resolution, throughput, and assay cost will lead to the development of analytical methods independent of gel-based electrophoretic techniques. These methods will ultimately be of use in plant biology and crop improvement. For example, colorimetric genetic analyses for specific alleles and/or DNA sequences will become routine.[26,27]

Plant genetic resources conservation also will benefit conceptually from an improved understanding of plant genome evolution. The insights provided through comparative mapping efforts within plant families will cause curators to reconsider what and how genetic resources will be maintained in the future. In addition, comparative mapping will facilitate our collective ability to identify useful genes and co-adapted gene complexes for use in crop improvement.

One direct outgrowth of the collaboration between genetic resources curators and molecular biologists is an ever increasing amount of molecular marker data associated with accessions in various collections. As investigators utilize markers developed by others, the value of those markers increases. Additional data about a group of markers may shed new light on earlier studies conducted with those same markers. A need is developing for a central clearinghouse for plant molecular marker data so that investigators can access primer and probe DNA sequences, map locations, diversity indices, and genotypes of particular accessions. Existing computer technology could make such a cross-referenced database publicly accessible over the Internet.

As noted previously, financial support for plant genetic resources conservation will always be limited in light of the multiple challenges presented. However, an increase in our fundamental understanding of plant genome organization, evolution, and diversity will provide curators with the theories and techniques to improve the quality of their ever more critical efforts.

REFERENCES

1. Smith JSC, Smith OS. Fingerprinting crop varieties. Adv Agron 1992; 47:85-140.
2. Schoen DJ, Brown AHD. Conservation of allelic richness in wild crop relatives is aided by assessment of genetic markers. Proc Natl Acad Sci USA 1993; 90:10623-10627.
3. Committee on Managing Global Genetic Resources: Agricultural Imperatives, Nat Acad Sci. Managing Global Genetic Resources: The U.S. National Plant Germplasm System. Wash DC: National Academy Press, 1991:171.
4. Kresovich S, Williams JGK, McFerson JR et al. Characterization of genetic identities and relationships of *Brassica oleracea* L. via random polymorphic DNA assay. Theor Appl Genet 1992; 85:190-196.
5. Nei M. Genetic distance between populations. Am Natur 1972; 106:283-292.
6. Williams JGK, Kubelik AR, Livak KJ et al. DNA polymorphisms amplified by arbitrary primers are useful as genetic markers. Nucleic Acids Res 1990; 18:6531-6535.
7. Caetano-Anolles G. MAAP - A versatile and universal tool for genome analysis. Pl Molec Biol 1994; 25:1011-1026.
8. Virk PS, Newbury HJ, Jackson MT et al. The identification of duplicate accessions within a rice germplasm collection using RAPD analysis. Theor Appl Genet 1995; 90:1049-1055.
9. Paran I, Michelmore RW. Development of reliable PCR-based markers linked to downy mildew resistance genes in lettuce. Theor Appl Genet 1993; 85:985-993.
10. Zabeau M, Vos P. Selective restriction fragment amplification: a general method for DNA fingerprinting. EPO Patent No. 0534858A1. 1993.
11. Jeffreys AJ, Wilson V, Thein SL. Hypervariable 'minisatellite' regions in human DNA. Nature 1985; 314:67-73.
12. Tautz D, Trick M, Dover, GA. Cryptic simplicity in DNA is a major source of genetic variation. Nature 1986; 322:652-656.
13. Lynch M. Estimation of relatedness by DNA fingerprinting. Mol Biol Evol 1988; 5:584-599.
14. Ramakrishna W, Chowdari KV, Lagu MD et al. DNA fingerprinting to detect genetic variation in rice using hypervariable DNA sequences. Theor Appl Genet 1995; 90:1000-1006.
15. Oudet C, Heilig R, Mandel JL. An informative polymorphism detectable by polymerase chain reaction at the 3' end of the dystrophin gene. Am J Human Genet 1990; 84:283-285.
16. Saghai-Maroof MA, Biyashev RM, Yang GP, et al. Extraordinarily polymorphic microsatellite DNA in barley: species diversity, chromosomal locations, and population dynamics. Proc Natl Acad Sci USA 1994; 91:5466-5470
17. Rongwen J, Akkaya MS, Bhagwat AA et al. The use of microsatellite DNA markers for soybean genotype identification. Theor Appl Genet 1995; 90:43-48.
18. Bell CJ, Ecker J. Assignment of 30 microsatellite loci to the linkage map of Arabidopsis. Genomics 1994; 19:137-144.
19. Dow BD, Ashley MV, Howe, HF. Characterization of highly variable $(GA/CT)_n$ microsatellites in the bur oak *Quercus macrocarpa*. Theor Appl Genet 1995; 91:137-141.
20. Thomas MR, Scott NS. Microsatellite repeats in grapevine reveal DNA polymorphisms when analyzed as sequence-tagged sites. Theor Appl Genet 1993; 86:985-990.
21. Liu ZW, Jarret RL, Kresovich S et al. Characterization and analysis of simple sequence repeat (SSR) loci in seashore paspalum (*Paspalum vaginatum* Swartz). Theor Appl

Genet 1995; 91:47-52.

22. Weising K, Ramser J, Kaemmer D et al. Oligonucleotide fingerprinting in plants and fungi. In: Burke T, Dolf G, Jeffreys AJ et al., eds. DNA Fingerprinting: Approaches and Applications. Basel: Birkhäuser, 1991:312-329.

23. Bhat KV, Bhat SR, Chandel KPS et al. DNA fingerprinting of *Musa* cultivars with oligodeoxyribonucleotide probes specific for simple sequence repeat motifs. Genet Anal: Biomol Engr 1995; 12:45-51.

24. Kwok P-Y, Carlson C, Yager RD et al. Comparative analysis of human DNA varia-

tions by fluorescence-based sequencing of PCR products. Genomics 1994; 23:138-44.

25. Doyle JJ. Gene trees and species trees: molecular systematics as one character taxonomy. Syst Bot 1991; 17:144-163.

26. Nickerson DA, Kaiser R, Lappin S et al. Automated DNA diagnostics using an ELISA-based oligonucleotide ligation assay. Proc Natl Acad Sci USA 1990; 87:8923-27.

27. Barany F. Genetic disease detection and DNA amplification using cloned thermostable ligase. Proc Natl Acad Sci USA 1991; 88:189-193.

CHAPTER 9

DNA Fingerprinting and Plant Variety Protection

Stephen Smith and Tim Helentjaris

9.1. INTRODUCTION

Much of the recent genetic gains in crop productivity have come from efforts of the private sector. Significant investment is dependent upon strong intellectual property protection (IPP). For example, privately funded plant breeding played major roles during the birth and later growth of U.S. hybrid corn agriculture. Commercial investments were due to the inherent protection of inbred lines afforded by hybrids, annual purchases by farmers and the consequent abilities of enterprising financial underwriters to gain returns from their investments. However, success in these endeavors was, is and can only be due to effective plant breeding.

In contrast, IPP for self-pollinated crops can only be provided through *sui generis* forms of variety protection or by utility patents. Nor can breeders of hybrid crops now depend upon trade secrets for effective protection. Inbred lines are vulnerable to misappropriation and molecular marker technologies greatly facilitate determination of pedigrees, reverse engineering, pedigree reconstruction, and the creation of varieties that may differ only cosmetically from a previously released variety. More sophisticated methods of IPP such as recently revised Plant Variety Protection in the U.S., European Union and Australia or Utility Patents are urgently required to encourage the creation of newly productive varieties from the combined efforts of traditional breeding and biotechnologies. On the other hand, detailed plant variety profiling methodologies are instrumental to strengthen IPP through their abilities to provide positive identification of proprietary germplasm and to check veracity of pedigrees. Therefore, plant variety profiling and IPP exist in a dynamic

Genome Mapping in Plants, edited by Andrew H. Paterson. © 1996 R.G. Landes Company.

relationship and it must be determined how data from molecular profiling technologies should be used with regard to IPP. Matters are further complicated since not all of the questions can be answered with complete objectivity and there is rapid technological change. Consequently, plant breeders, registration authorities, molecular biologists, and legal experts are re-examining variety description in relation to IPP at new levels of detail.

9.2. EMERGENT TRENDS THAT HAVE RECENTLY SPURRED THE INCREASED USE OF DNA VARIETY PROFILING TO SUPPORT IPP

Three recently emergent trends highlighted the increased need to characterize and protect plant varieties. First, a reduction in germplasm security because of misappropriation and reverse engineering was noted.[1] These suspicions were subsequently confirmed.[2] Second, advances in molecular marker and transformation capabilities have increased the speed and precision by which breeders can manage germplasm. Transformation and marker assisted backcrossing are becoming more commonplace in the industry. Some genetic changes might only be cosmetic and represent attempts to plagiarize or counterfeit protected varieties. Molecular marker systems also provide numerous opportunities to reveal differences between varieties, although many differences might be unimportant in terms of pedigrees or agronomic performance differences. Molecular markers provide an increasingly detailed understanding of pedigrees and breeding strategies. For example, RFLP profiles of maize hybrids can facilitate the identification, preferential extraction and rapid introgression of germplasm across proprietary breeding programs. Third, there has been an increase in competitiveness as new corporations have entered into plant breeding due in large part to new technological opportunities and effective IPP. Consequently, plant breeders have sought to improve both the technical and legal bases of protection.

9.3. OUTLINE OF PROCEDURES AND REQUIREMENTS FOR OBTAINING IPP FOR PLANT VARIETIES

The United States was first among nations to invoke special laws granting IPP for plant varieties. The Plant Patent Act of 1930 provided protection for non-tuberous asexually propagated varieties. Subsequently, the United States extended IPP to sexually reproduced varieties in the Plant Variety Protection (PVP) Act of 1970. IPP became available to breeders in many more countries following their enactment of PVP legislation based upon the 1978 Convention of the Union pour le Protection des Obtentions Végétale (UPOV). In April of 1995, new PVP laws came into effect in both the United States and the European Union (EU) that reflect the revised 1991 UPOV Convention. Following a 1980 decision by the U.S. Supreme Court and a ruling in 1985 from the Patent Office Board of Appeals and Interference, it has been possible to obtain patent protection for biological processes, genes, DNA sequences, microorganisms and plant varieties in the U.S. The situation regarding patents granted for plant variety protection in the E.U. remains confused.

In order to be eligible for protection by PVP or by patent, a variety must be shown to be distinct, uniform, and stable (DUS). In addition, to qualify for utility patent protection, there must be evidence of usefulness and nonobviousness. An enabling description is provided by a seed deposit. Evaluation of DUS criteria mandates that varieties are described and compared in detail. Distinctness is determined by officials in PVP offices or other agencies acting for the official body that grants PVP certificates.

In the 1991 UPOV Convention, there is an important new provision that describes the concept of an Essentially Derived Variety (EDV). A variety is an EDV when:

1) "it is predominantly derived from the initial variety or from a variety that is

itself predominantly derived from the initial variety, while retaining the expression of the essential characteristics that result from the genotype or combination of genotypes;

2) it is clearly distinguishable from the initial variety; and

3) except for the differences which result from the act of derivation, it conforms to the variety in the expression of the essential characteristics that result from the genotypes or combination of genotypes of the initial variety."[3]

Determination of EDV is not usually in the domain of PVP officials; it is the responsibility of breeders. (However, the Australian PVP Act prescribes an important role for the PVP examiner to decide EDV status.) The EDV concept is important for two main reasons. First, it can help promote agricultural productivity. Genetic engineering can allow exotic genes to be rapidly inserted into varieties and it will be to the maximum benefit of agriculture that such traits are introduced into well adapted varieties. Therefore, there must be a continual succession of varieties of improved performance for a multitude of quantitatively inherited characteristics that affect yield and stress resistance. The EDV concept protects plant breeding by preventing a skewing of support and effort toward biotechnologies. Varieties modified by genetic engineering are distinct from previously released varieties. Without a functional EDV concept, a genetic engineer could effectively pirate germplasm that took ten years or more to develop. Yet it is this very germplasm upon which the transgene has its dependence as a useful and salable commodity. Such a potential inequity in IPP is possible through the potency of genetic engineering to effect rapid, small, but distinctive genomic change. However, the support of agriculture by breeding would be damaged by such inequity. Without EDVs, there could be a shift of effort away from the improvement of quantitative traits using a relatively broad base of germplasm. Instead, resources might focus on a narrow germ-

plasm base with effort concentrated toward improvement of relatively few traits using backcrossing procedures or genetic engineering. The EDV concept can stimulate productivity by providing a framework of symbiosis within which classical breeders and genetic engineers can be equitably rewarded and encouraged. Second, the EDV concept can promote genetic distance between varieties and thereby guard against a narrowing of the germplasm base. Intense selection by pathogens might then be avoided and genetic resources would be less prone to erosion. The EDV concept has been criticized as vague and complicated. However, EDVs are inevitable because of technological change. The demands for increased agricultural productivity mandate that the challenges and opportunities afforded by new technologies be realized through a practically usable framework to identify EDVs and determine their legal, commercial and royalty status. More complete reviews of the technical and legal aspects of IPP for plants are available.[4]

9.4. DATA THAT ARE USED TO CHARACTERIZE PLANT VARIETIES

Variety descriptions are prerequisites for the granting of PVP and Utility Patents and are necessary to allow breeders to identify germplasm and monitor protection. In order to fulfill these requirements, variety descriptions must be discriminative, free from environmental effects, interpretable in genetic terms and reflective of pedigree and genetic constitution.

9.4.1. MORPHOLOGICAL AND BIOCHEMICAL DATA

Morphological data provide the basis of descriptions for variety registration and identification. To many, these data are irreplaceable because they represent the products of expressed genes and their omnipresence makes them convenient indicators. However, G x E effects, increasing difficulties in showing discrimination (for some crops), increasing costs in data collection, the length of time required to collect data and a lack of knowledge of genetic control, are leading PVP examiners and breeders towards the use of molecular data. Morphological data do not usually allow genetic distances to be calculated; consequently, these data are of little to no use in determining EDV status.

The use of biochemical methods for variety identification has been thoroughly reviewed.[5-8] Isozymic data are of good taxonomic quality because they are unaffected by G x E interactions and genetic control is known. These characters are used as supplementary evidence of distinctness. The practice in France is to include isozyme loci within a class of simply inherited traits; differences for at least three simply inherited characters are needed to show distinctness. Isozymic data can place lines on a one year "fast track" for registration in France. Isozymic data are also used to verify pedigrees and identities of seedlots. Electrophoretic data from seed storage proteins are also used to improve the organizational efficiency and analytical precision of field growout trials through the pregrouping of varieties (Ghijsen, personal communication). Isozymic data are supporting evidence for utility patents of maize inbred lines.[9-13]

However, even at their most discriminative, biochemical data alone cannot provide a level of unique varietal identification equivalent to that already available from morphological data. Genomic coverage afforded by biochemical markers is also incomplete. This lack of discriminative detail from biochemical data has led to failures in detecting some instances of misappropriation or "cosmetic" breeding. Biochemical markers do not provide genetic distance measures that are sufficiently precise or reflective of pedigree distances or heterosis for them alone to be determinative of EDV status.

9.4.2. DNA BASED METHODS OF VARIETAL DESCRIPTION

Varietal profiling methods that directly utilize DNA could potentially address all of the limitations associated with morphological and biochemical data. DNA based profiles are not affected by G x E effects and the basis of inheritance can be understood. Profiles can be interpreted in terms of the presence or absence of specific alleles. DNA based markers can be found in great abundance through the whole genome. Powers of discrimination can be very high, even between very closely related varieties. Methods can be standardized. Cost and speed of data acquisition are acceptable for variety identification.

DNA fingerprinting is a term that was first coined by Jeffreys et al[14] to describe a multilocus RFLP assay for hypervariable regions of human minisatellite sequences where polymorphisms are from variable numbers of tandem repeats (VNTRs). The first reference to DNA fingerprinting in plants was by Ryskov et al[15] who reported the use of the M13 repeat phage probe. Human minisatellite probes[16] and synthetic simple repeat oligonucleotide probes[17] have also been utilized. Reviews of DNA fingerprinting with these methods have been provided by Nybom[18,19] and by Weising et al.[20] Reviews of the use of single and double loci probes to discriminate among plant varieties are available.[6,21]

A rapid expansion of identification abilities came from a profusion of new methods that utilize the polymerase chain reaction (PCR) process. One class of methods employs "classical" PCR methods in which two directionally opposed and relatively long (15-25 b) primers compliment DNA sequences that flank the amplifiable region. Another class of methods utilizes single relatively short (5-15 b) primers that are of arbitrary sequence with respect to target DNA. Greater throughput capacities and efficiencies combined with detection precisions down to the level of a single base are hallmarks of newly emergent profiling technologies.[22-26] Among these new technologies are oligomer hybridization dot-blot assays[27,28] and genome scanning methods such as Genome Mismatch Scanning (GMS),[29] Representational Difference Analysis (RDA)[30] and techniques to spread DNA in two-dimensional arrays.[31] Further developments using sequence hybridization assays performed as multiple arrays on silicon chips or as microlaboratories etched on glass will soon be available.[32-34]

9.4.3. DNA Methods in Current Use

Three broad categories of DNA based molecular marker technologies are in current use for the purposes of varietal identification; these are:

1) Hybridization based assays such as Restriction Fragment Length Polymorphisms (RFLPs) usually allow one or two loci per probe to be evaluated through differences in either the detected fragment length between two defined digestion sites or sequence polymorphisms within the restriction sites themselves. Alternatively, multilocus approaches based on minisatellites or Variable Number Tandem Repeats (VNTRs) provide data from 2-20 loci per probe.

2) DNA Polymerase Chain Reaction (PCR) methods that utilize primers of sequence that are arbitrary to that of the target genome. For example, Randomly Amplified Polymorphic DNA (RAPD),[35] Arbitrarily Primed Polymerase Chain Reaction (AP-PCR),[36] DNA Amplification Fingerprinting (DAF),[37] and Amplified Fragment Length Polymorphisms (AFLPs), (also termed Selective Fragment Length Amplification).[38] These methods allow 3-30 genomic sites to be assayed simultaneously.

3) DNA PCR methods that use primers synthesized to specific, complementary sequences in the target DNA of varieties. Sequence Tagged Sites (STS),[39] Sequence Characterized Amplified Regions (SCARs) and Microsatellites, also termed Simple Sequence Repeats (SSRs)[25,26,40-43] are examples of this category of methods.

All current widely used methods utilize size separation of DNA fragments. Discrimination power increases as the number of polymorphic bands increase; how-ever, reliability and the ability to score fragments as alleles usually become more complicated as profile complexity increases. VNTR and arbitrary primer methods, in particular, often result in complex multibanded profiles. Allelic scoring is complicated for arbitrary primer methods where relatively little sequence homology is required to initiate DNA amplification, hence, DNA fragments of similar length can be amplified from different chromosome locations by the same primer. Although SSR technology requires the added effort to make sequence data available, the ability to then repeatedly, accurately and efficiently array data in the form of discrete alleles at 3-10 mapped loci simultaneously,[25,26,43] makes this methodology especially appropriate and attractive as a means to support IPP. The U.S. PVP Office has recently requested soybean breeders to submit seed for SSR profiling (Grace, personal communication); maize SSRs are publicly available and procedures for efficient variety profiling are being developed.[42]

Abilities of marker technologies to reveal associations that reflect pedigrees among related inbreds or varieties, are dependent upon the number of polymorphic loci assayed, linkage among loci, degree of genomic coverage and number of alleles/locus that can be recognized. Associations among related maize inbred lines as shown from RFLP, AP-PCR, RAPD and AFLP data (Fig. 9.1) are similar and reflective of pedigree because polymorphisms and degree of genomic coverage were not limiting. RFLP data were from 160 probes that revealed 1205 polymorphic bands.[44] The number of bands scored for other methods were AP-PCR (258), RAPD (393) and AFLP (347). In contrast, isozymic data from 25 loci could not distinguish two inbreds and associations of lines were less reflective of pedigree relationships (Fig. 9.1). For crops such as soybean, tomato and wheat, polymorphisms among elite varieties have been more difficult to find other than at regions of introgressed exotic germplasm. Nonetheless, over 95% of elite

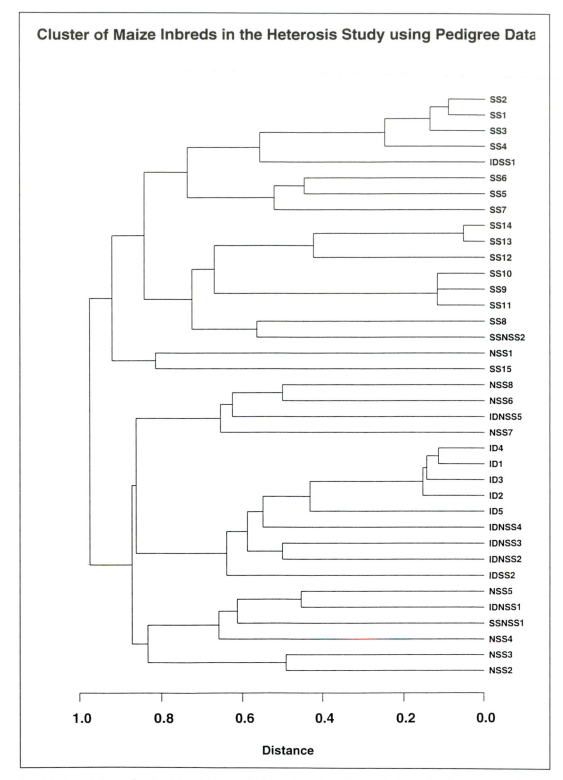

Fig. 9.1. *Associations of maize inbred lines on the basis of pedigree, isozymic, RFLP, RAPD, AP-PCR and AFLP data. Inbred lines are coded to reflect pedigree backgrounds. ID is Iodent, SS is Iowa Stiff Stalk Synthetic and NSS is non-Stiff Stalk.*

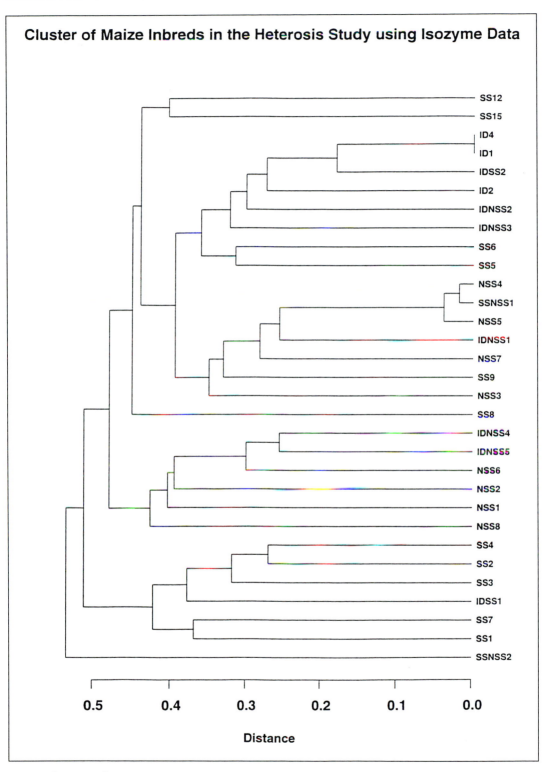

Fig. 9.1. (continued)

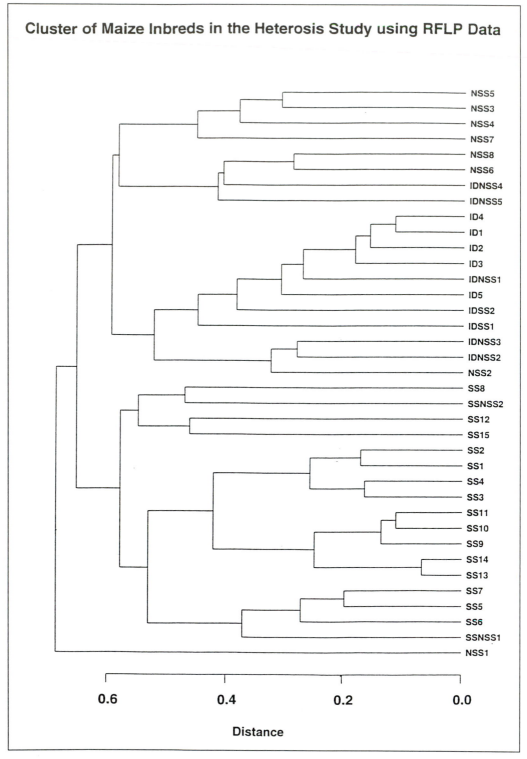

Fig. 9.1. (continued)

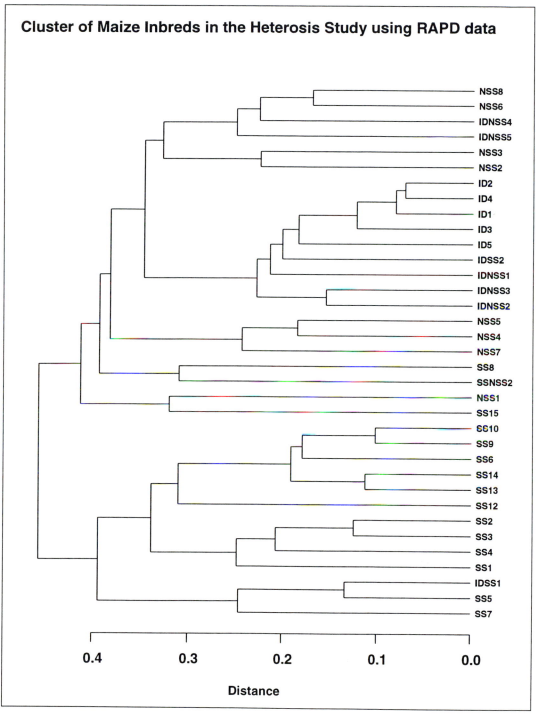

Fig. 9.1. (continued)

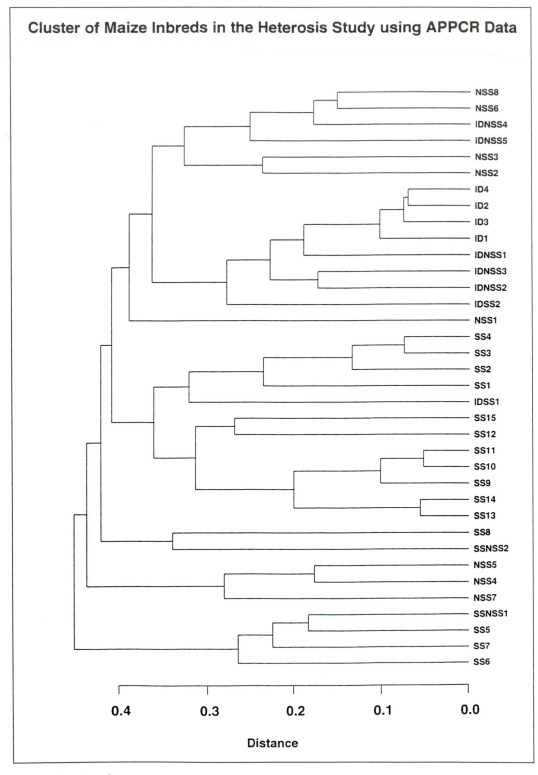

Fig. 9.1. (continued)

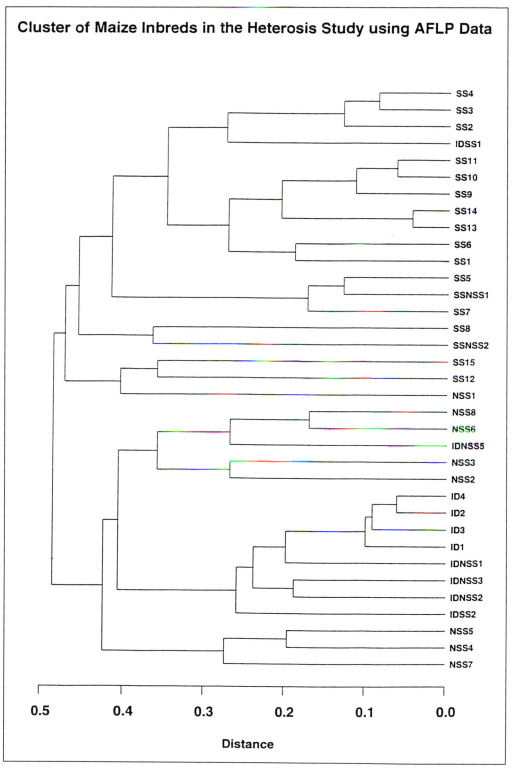

Cluster of Maize Inbreds in the Heterosis Study using AFLP Data

Fig. 9.1. (continued)

soybean varieties can be distinguished and associations reflective of pedigree can be shown by using just 33 RFLPs (Webb, personal communication). SSRs could potentially remove most, if not all, of the limitations in revealing polymorphisms and in obtaining more complete genomic coverage for plants[40,41,45] as has already been achieved for the human genome.[46]

9.5. FUTURE ISSUES AND CHALLENGES IN THE USE OF DNA VARIETY PROFILES TO HELP SUPPORT IPP

9.5.1. THE REGISTRATION PROCESS; EVALUATION OF DUS

Plant breeders and PVP examiners alike would like to conduct the registration process more quickly and less expensively. Varietal profiling methods that utilize DNA have the potential to meet these goals. However, there are many concerns to be addressed and challenges to be met before complete advantage can be taken of DNA profiling methods. First, there are major concerns that the use of these traits as descriptors in the registration process would devalue distinctness and IPP would be powerfully undermined unless a threshold of minimum distance could be defined for distinctness. The hierarchical system of classifying traits with respect to showing distinctness that is used in France as a model of minimum distance might be developed to meet this challenge. Second, under current regulations, trait differences between varieties can only qualify as indicators of distinctness after their expression has been shown to be uniform within a variety. DNA based assays of variety profiles, would necessitate the collection and analysis of data from many tens of individual marker loci from one or several laboratories. These data would be additional to current practice and perhaps superfluous for most needs of varietal discrimination. Furthermore, breeders might have to conduct additional assays because residual heterozygosity would remain without direct selection. Use of currently available DNA pro-

filing technologies for these assays could require very significant additional efforts by breeders and registration authorities. Current profiling technologies are not practically feasible for assays of uniformity for more than a very few loci due to cost and limited throughput. Third, there is a basic and fundamental debate on the qualification of data from DNA marker technologies for use in the establishment of *de novo* varietal descriptions and for the determination of distinctness. There is an argument that data from most DNA technologies do not conform to the requirements of UPOV[3] where in Article 1 of the 1991 UPOV Convention a variety is stated to be:

i) "defined by the expression of the characteristics resulting from a given genotype of combination of genotypes;" and

ii) "distinguished from any other plant grouping by the expression of at least one of the said characteristics."

Therefore, DNA sequence data from non-coding regions of the genome are eliminated from the acts of defining and distinguishing varieties if "expression" is interpreted specifically and solely to mean gene products. For example, the components of most RFLP, arbitrary primer, and microsatellite marker loci are not gene products. However, these data are visible expressions of genotype and they are consequently evidence of varietal differences. Also, DNA in non-coding regions could play important roles in gene regulation and thereby affect expression. From this standpoint marker data could be included within the scope of characteristics that define varieties for registration. It seems incongruous to many that human DNA profiles can be accepted as credible evidence in cases of capital offense but that data from molecularly equivalent sources should not be acceptable as evidence of identity and distinctness in cultivated varieties of plants. RFLP data have provided evidence of the unique identity of inbred lines in Utility Patents.[10,11]

Any future widespread use of DNA profiles in variety registration will proceed cautiously and depend upon further technological development. First, there is the definition of "expression" that is yet unresolved. Second, the development of new DNA technologies with ten to one hundred fold increases in throughput efficiencies and cost effectiveness is required to provide assays of uniformity and stability. Third, development of profiling procedures that are less onerous and cumbersome than RFLP technology is desirable. Fourth, standard and repeatable methods of data generation, acquisition, and comparison for technologies should be agreed upon. Fifth, agreement of a boundary of minimum distance to define distinctness is needed. Technological development, accumulation of experience and the needs for international harmonization mandate much concerted effort. Consequently, the routine widespread use of DNA based profiles in variety registration might be unlikely before the second decade of the 21st century.

9.5.2. DETERMINATION OF EDV STATUS

UPOV[47] stresses the importance of using data from characteristics that are free from environmental effects and which can be interpreted genetically to determine EDV status. Implementation of the EDV provision mandates that genetic conformity between varieties is measured. DNA based molecular methods are excellent means for establishing genetic conformity since they can provide precise measures of genetic distance between varieties that reflect pedigree. Determination of where an EDV boundary threshold lies involves an element of subjectivity in its resolution. However, objectivity can be brought to bear by providing crop specific data that allow pedigree, molecular marker, and agronomic performance (including heterosis) data to be directly compared. Determination of a threshold boundary can then be made upon a more formal and informed basis.

In the United States, discussions within ASTA are resulting in an increased awareness and capability to determine EDV status. Sets of guidelines to help resolve

EDV status are being established.[48] Boundary levels for EDVs are emerging and will be further examined before official pronouncements can be made (Geadelmann, personal communication). ASSINSEL is examining additional studies in tomato and in maize. Studies of similarly relevant data should be made as soon as possible for all major cultivated species. Results from these studies will form an important basis to promote the smooth release of increasingly productive and new genetic diversity into agriculture.

9.5.3. OVERSIGHT AND MONITORING OF GERMPLASM USAGE BY BREEDING ORGANIZATIONS

Privately funded plant breeders cannot rely solely on the existence of legal instruments and the registration process to guarantee germplasm security. Unauthorized use of a protected inbred line as a parent of a hybrid might still occur despite the great likelihood that such an occurrence would be detected by DNA methods of profiling. The responsibility to detect misappropriation rests with the owner of protected germplasm. Consequently, many privately funded breeders routinely profile patented inbred lines and other varieties that are commercially released by competitors. Breeders are vigilant, for instances, where protected inbred lines have been used directly either as parents of hybrids or as initial varieties to develop EDVs.

9.6. CONCLUSIONS

Highly discriminative "fingerprinting" technologies are available to all breeding and biotechnology organizations regardless of their size and resources. Technological developments continually increase the discriminative power and cost effectiveness of profiling. Detailed and discriminative genetic profiles provide extremely powerful and effective procedures to allow meaningful and valid comparisons among inbred lines, varieties and hybrids to be made with respect to germplasm identification and ownership. These profiles when used in conjunction with pedigree and performance data provide a complete source of information to protect intellectual property rights that relate to plant varieties. Thereby molecular marker data help promote investment, productivity and use of genetic diversity in agriculture.

REFERENCES

1. Brown WL. The coming debate over ownership of plant germplasm. In: American Seed Trade Association, ed. Proc. 39th Ann. Corn and Sorghum Ind Res Conf., Chicago, IL, 1984:44-51.

2. U.S. District Court. Pioneer Hi-Bred International, Inc., an Iowa Corporation Plaintiff, Holden Foundation Seeds, Inc., et al, Defendants. 1987:Case No. 81-60-E.

3. UPOV. UPOV International Convention for the Protection of New Varieties of Plants, UPOV Publication No. 221(E). International Union for the Protection of New Varieties of Plants, Geneva, 1992:31.

4. Baenziger PS, Kleese R, Barnes RF, ed. Intellectual Property Rights: Protection of Plant Materials. Crop Sci Soc Amer 1993:187.

5. Cooke RJ. Electrophoresis in plant testing and breeding. Adv Electrophor 1988; 2:171-261.

6. Smith JSC, Smith OS. Fingerprinting crop varieties. Adv Agron 1992; 47:85-140.

7. Cooke RJ. Gel electrophoresis for the identification of plant varieties. J Chromatog A, 1995; 698:281-299.

8. Cooke RJ. The characterisation and identification of crop cultivars by electrophoresis. Electrophor 1984; 5:59-72.

9. Troyer AF. Inbred corn line HBA1. United States Patent No. 4,594,810. United States Patent Office, Washington, DC; 1986.

10. Noble SW. Inbred Corn Line PHHB4. United States Patent No. 5,444,178. United States Patent Office, Washington, DC; 1995.

11. Chapman MA. Inbred Corn Line PHTE4. United States Patent No. 5,453,564. United States Patent Office, Washington, DC; 1995.

12. Jensen SD, Williams NE. Inbred Corn Line PHK29. United States Patent No. 4,812,600. United States Patent Office, Washington, DC; 1989.

13. Cavanah JA. Inbred Corn Line PHN37. United States Patent No. 5,082,991. United States Patent Office, Washington, DC; 1992.

14. Jeffreys AJ, Wilson V, Thein SL. Hypervariable 'minisatellite' regions in human DNA. Nature 1985; 314:67-73.

15. Ryskov AP, Jincharadze AG, Prosnyak MI, Ivanov PL, Limborska SA. M13 phage DNA as a universal marker for DNA fingerprinting of animals, plants and microorganisms. FEBS Letters; 1988:388-392.

16. Dallas JF. Detection of DNA "fingerprints" of cultivated rice by hybridization with a human minisatellite probe. Proc Natl Acad Sci USA 1988; 85:6831-6835.

17. Weising K, Weigand F, Driesel AJ, Kahl G, Zischler H, Epplen JT. Polymorphic simple GATA/GACA repeats in plant genomes. Nucleic Acids Res 1989; 17:10128.

18. Nybom H. Applications of DNA fingerprinting in plant breeding. In: Burke T, Dolf G, Jeffreys AJ, Wolff R, ed. DNA Fingerprinting Approaches and Applications. Birkhauser Verlag, Basel, Switzerland, 1991:294-311.

19. Nybom H. DNA fingerprinting - A useful tool in fruit breeding. Euphytica 1994; 77:59-64.

20. Weising K, Beyermann B, Ramser J, Kahl G. Plant DNA fingerprinting with radioactive and digoxigenated oligonucleotide probes complementary to simple repetitive DNA sequences. Electrophor 1991; 12: 159-69.

21. Smith JSC. Identification of cultivated varieties by nucleotide analysis. In: Wrigley CW, ed. Identification of food-grain varieties. Amer Assoc Cereal Chem, St. Paul, MN, 1995:131-150.

22. Barany F. Genetic disease detection and DNA amplification using cloned thermostable ligase. Proc Natl Acad Sci USA, 1991; 88:189-193.

23. Nickerson DA, Kaiser R, Lappin S, Stewart J, Hood L, Landegren U. Automated DNA diagnostics using an ELISA-based oligonucleotide ligation assay. Proc Natl Acad Sci USA 1990; 87:8923-8927.

24. Nickerson DA, Whitehurst C. Boysen C,

Charmley P, Kaiser R, Hood L. Identification of clusters of biallelic polymorphic sequence-tagged sites (pSTSs) that generate highly informative and automatable markers for genetic linkage mapping. Genomics 1992; 12:377-387.

25. Ziegle JS, Su Y, Corcoran KP, Nie L, Mayrand E, Hoff LB, McBride LJ, Kronick MN, Diehl SR. Application of automated sizing technology for genotyping microsatellite loci. Genomics 1992; 14:1026-31.

26. Schwengel DA, Jedlicka AE, Nanthakumar EJ, Weber JL, Levitt RC. Comparison of fluorescence-based semi-automated genotyping of multiple microsatellite loci with autoradiographic techniques. Genomics 1994; 22:46-54.

27. Guo A, Guilfoyle RA, Thiel AJ, Wang R, Smith LM. Direct Fluorescence analysis of genetic polymorphisms by hybridization with oligonucleotide arrays on glass supports. Nucl Acids Res 1994; 22:5456-5465.

28. Nikiforov TT, Rendle RB, Goelet P, Rogers Y-H, Kotewicz ML, Anderson S, Trainor GL, Knapp MR. Genetic bit analysis: a solid phase method for typing single nucleotide polymorphisms. Nucl Acids Res 1994; 22:4167-4175.

29. Nelson SF, McCusker JH, Sander MA, Kee Y, Modrich P, Brown PO. Genomic mismatch scanning: a new approach to genetic linkage mapping. Nature Genetics 1993; 4:11-18.

30. Lisitsyn N, Lisitsyn N, Wigler M. Cloning the differences between two complex genomes. Science 1993; 259:946-951.

31. Uitterlinden AG, Vijg J. Denaturing gradient gel electrophoretic analysis of minisatellite alleles. Electrophor 1991; 12:12-16.

32. Grossman PD, Bloch W. Brinson E. Chang CC, Eggerding FA, Fung S, Iovannisci DA, Woo S, Winn-Deen ES. High-density multiplex detection of nucleic acid sequences: oligonucleotide ligation assay and sequence coded separation. Nucleic Acids Res 1994; 22:4527-4534.

33. Jacobson SC, Ramsey JM. Microchip electrophoresis with simple stacking. Electrophor 1995; 16:581-486.

34. Chetverin AB, Kramer FR. Oligonucleotide arrays: New concepts and possibilities. Bio/Technol 1994; 12:1093-1099.

35. Williams JGK, Kabelik AR, Livak KJ, Rafalski JA, Tingey SV. DNA polymorphisms amplified by arbitrary primers are useful as genetic markers. Nucl Acids Res 1990; 18:6531-6535.

36. Welsh J, McClelland M. Fingerprinting genomes using PCR with arbitrary primers. Nucl Acids Res 1990; 18:7213-7218.

37. Caetano-Annolles G, Bassam BJ, Gresshoff PM. DNA amplification fingerprinting using very short arbitrary oligonucleotide primers. Bio/Technol 1991; 9:553-557.

38. Zabeau M. Vos P. Selective Restriction fragment amplification: A general method for DNA fingerprinting. European Patent Application No. 0534858A1, European Patent Office, Paris, 1993.

39. Talbert LE, Blake NK, Chee PW, Blake TK, Magyar GM. Evaluation of "sequence-tagged-site" PCR products as molecular markers in wheat. Theor Appl Genet 1994; 87:789-794.

40. Akkaya MS, Bhagwat AA, Cregan PB. Length polymorphisms of simple sequence repeat DNA in soybean. Genetics 1992; 132:1131-1139.

41. Senior ML, Heun M. Mapping maize microsatellites and polymerase chain reaction confirmation of the targeted repeats using a CT primer. Genome 1993; 36:884-89.

42. Senior ML. Chin ECL, Kresovich S, Lee M, Smith JSC. The development and mapping of microsatellite-based markers in maize - a continuing effort. In: Amer Soc Agron, ed. Agron Abstr 1995; p. 169.

43. Glowatzki-Mullis M-L, Gaillard C, Wigger G, Fries R. Microsatellite-based parentage control in cattle. Animal Genetics 1995; 26:7-12.

44. Smith JSC, Smith OS, Bowen SL, Tenborg RA, Wall SJ. The description and assessment of distances between inbred lines of maize. III A revised scheme for the testing of distinctiveness between inbred lines utilizing DNA RFLPs. Maydica 1991; 36: 213-226.

45. Akkaya MS, Shoemaker RC, Specht JA, Bhagwat AA, Cregan PB. Integration of

simple sequence repeat DNA markers into a soybean linkage map. Crop Sci 1995; 35:1439-1445.

46. Reed PW, Davies JL, Copeman JB, Bennett ST, Palmer SM, Pritchard LE, Gough SCL, Kawagucki Y, Cordell HJ, Balfour KM, Jenkins SC, Powell EE, Vignal A, Todd JA. Chromosome-specific microsatellite sets for fluorescence-based, Semi-automated genome mapping. Nature Genetics 1994; 7:390-94.

47. UPOV. Essentially derived varieties. Sixth meeting with International Organizations, Geneva, Oct. 30, 1992. Union Pour le Protections d'Obtentions Végétales Mimeo IOM/6/2, Geneva, Switzerland.

48. Hunter RB, Science based identification plant material. In: Baenziger PS, Kleese R, Barnes RF, ed. Intellectual Property Rights: Protection of Plant Materials. Crop Sci Soc Amer, 1993:93-100.

CHAPTER 10

APPLICATIONS OF REPETITIVE DNA SEQUENCES IN PLANT GENOME ANALYSIS

Xinping Zhao and Martin W. Ganal

10.1. INTRODUCTION

Higher plant genomes consist of unique (or "single copy") and repetitive DNA sequences. The nuclear DNA content can vary widely among plant species, even within the same taxonomic family, and repetitive DNA accounts for the majority of this variation.[1-3] Therefore, information about the structure and organization of repetitive DNA sequences is important to our understanding of plant genomes.

Several lines of evidence suggest that repetitive sequences may have functional significance in eukaryotic genomes.[4,5] In plants, it has been found that cell and nuclear size as well as duration of mitosis and meiosis are correlated with the nuclear DNA content. Tandemly repetitive sequences are frequently associated with chromosomal landmarks such as centromeres and telomeres.[6,7] This suggests that repetitive sequences might play a role in chromosome pairing, meiotic recombination, or speciation.[3,4,8-10] However, some repetitive elements appear to have no functional significance, and simply accumulate in the genome because they do not produce any phenotypic disadvantage.

Repetitive DNA sequences were the first DNA components to be studied at the molecular level, and are very useful for many aspects of plant genome analysis. Specifically, repetitive DNA sequences have been used for understanding genome evolution, analysis of phylogenetic relationships among related taxa, genetic mapping, DNA fingerprinting, physical mapping and gene cloning. This chapter discusses our current understanding of repetitive DNA sequences in the plant genome. Emphasis will be put on methodology of studying repetitive DNA sequences and their applications in plant genome analysis, and specific examples are presented. Readers are referred to recent reviews for general description

Genome Mapping in Plants, edited by Andrew H. Paterson. © 1996 R.G. Landes Company.

of repetitive DNA, general discussion of genome evolution, and novel features of selected plant species.[11-13]

10.2. GENERAL METHODS FOR STUDYING REPETITIVE DNA SEQUENCES

10.2.1. DNA REASSOCIATION STUDIES

DNA reassociation (renaturation) kinetics, also known as DNA:DNA hybridization or Cot analysis, has proven to be valuable for studying overall organization of a given plant genome, and relationships between closely related plant genomes.[1,14-17] Repetitive DNA sequences are fast annealing, while low copy sequences are slowly annealing. Such studies provide an overall picture of the relationship between these two classes of sequences. Estimates of number, length, and arrangement of repetitive DNA sequences are also generated by these analyses. Most higher plant genomes contain a substantial amount of repetitive DNA and the differences in genome sizes of closely related species are due primarily to the repetitive DNA content.[17-19] For example, DNA reassociation studies suggest that 50-65% of the tetraploid cotton genome is composed of low copy DNA and 35-50% are repetitive DNA, and differences in genome size between A and D genome species are mainly due to the variation in repetitive DNA.[19-21] While these studies have generated a considerable amount of information about genome organization and evolution of plants, they do not allow detailed molecular characterization of individual repetitive DNA sequences.

10.2.2. ISOLATION AND CHARACTERIZATION OF INDIVIDUAL REPETITIVE DNA SEQUENCES

10.2.2.1. Isolation of Repetitive DNA Sequences

Tandemly repeated DNA sequences which occur in uninterrupted arrays one after the other are called satellites. They were first isolated as satellite bands in density gradient centrifugation because the base composition of some prominent tandem repeats is different from the remainder of the nuclear DNA. Today, the term satellite DNA has become synonymous with any tandemly repeated DNA even if it does not form a peak on cesium chloride density gradients. An efficient method to isolate satellites is to clone prominent DNA fragments from an agarose gel containing genomic DNA digested with restriction enzymes. This method is based on the fact that most plant satellites have a high copy number and high homology, and often have repeating units large enough to contain restriction enzyme sites. A number of plant satellites have been isolated using this method.[22-25]

However, the majority of plant repetitive DNA sequences can not be identified so simply, and must be isolated from genomic libraries. Libraries are screened with radioactively labeled genomic DNA, then with organellar (chloroplast and mitochondria) DNA sequentially, to distinguish between nuclear repetitive sequences and organellar DNA clones. The rationale for such screening strategies is that repetitive sequences are present in multiple copies in the plant genome and therefore hybridize to genomic DNA much stronger than low copy or single copy sequences. Since many organellar sequences are contained in each plant cell, a selection against these sequences is necessary.

Plasmid libraries are used for this purpose in most cases, because the small size of DNA elements cloned in plasmids means that no additional subcloning is usually required to isolate individual repetitive elements, e.g., each plasmid usually contains only one repetitive DNA element. Such libraries are often constructed by cloning DNA digested with a single frequent-cutting restriction enzyme such as Sau3A I or Taq I. Repetitive sequences which do not contain the appropriate recognition sequences cannot be isolated from such a library. To isolate repetitive DNA with a better genome representation, two methods are used. The first method takes advantage of randomness of DNA fragments generated by physical shearing.[26] The second method uses multiple restriction enzymes to generate nearly random DNA fragments.[27]

10.2.2.2. Organization of repetitive sequences

Southern blot hybridization to cloned repetitive sequences is the most commonly used method to study organization of repetitive DNA in the genome. Two basic organization patterns of repetitive DNA are recognized: tandem and interspersed. A typical hybridization pattern of a tandemly repeated DNA sequence is the "ladder-pattern." This pattern is characterized by a series of hybridization bands that are the monomer and respective multimers of a basic repeating unit. The appearance of ladders is caused by sequence heterogeneity, resulting from single nucleotide sequence changes in the restriction enzyme recognition sites in some members of the repeat family.[28,29]

Interspersed repeats, whose members are scattered throughout genomes, produce a smear pattern with fragments of all sizes and not of defined size classes. To get a clearer picture of the organization of a repeat sequence, many restriction enzymes are needed in Southern blot analysis.[27]

If a restriction enzyme does not cut in the basic repeat unit of a large tandemly repeated sequence, the resulting fragments are usually too large for separation by conventional gel electrophoresis. Pulsed field gel electrophoresis (PFGE) allows the analysis of entire arrays of satellites, and the genomic organization of interspersed repeats, because this technique has the capacity to separate DNA fragments of up to several million base pairs. In plants, it has demonstrated that PFGE combined with Southern blot hybridization is very useful for studying the organization of repetitive DNA sequences.[23,30-33]

Characterization of individual clones containing large genomic fragments has provided detailed information about the organization of repetitive DNA elements, and the interspersion of repeats among single copy elements. Restriction mapping and cross-hybridization analysis are normally used for this purpose. For example, different restriction maps are generated from randomly selected lambda clones containing (GGC) microsatellite, suggesting that arrays of this repeat type are located in different sequence environments.[34] Sometimes, such analyses reveal clusters of different repetitive sequences.[35] In maize, up to 37 different repetitive DNA elements have been detected in a single 280 kb YAC. In this case, individual repetitive elements are interspersed with each other, and spatially separate from single copy elements.[36]

10.2.2.3. Identification of the repeated unit

It is relatively easy to identify the repeat unit for tandemly repeated DNA by analyzing the ladder-pattern as described in the previous section. However, the repeat unit size for interspersed repetitive sequences can not be determined so simply, because in many cases, the interspersed repetitive sequences are isolated as short DNA fragments that do not contain the entire unit. To determine the repeat unit of an interspersed repeat, a genomic library with longer inserts, such as a lambda library, is screened with the cloned fragment. Positive clones with large inserts are selected for restriction mapping, cross-hybridization studies, and DNA sequencing.[35,37] Based on these data, the repeat unit size can be deduced in many cases.

10.2.2.4. Determination of the number of copies of the repeated unit found in a genome

Although arbitrary terms have often been used to describe abundance of repetitive sequences in eukaryotic genomes such as highly repetitive and moderately repetitive, a detailed characterization of repetitive sequences will require more accurate estimations of their copy numbers in the genome.

Quantitative dot blot or slot blot hybridization provides the most reliable estimates of copy numbers of repetitive sequences.[38] Serial dilutions of genomic DNA and cloned DNA are loaded on filters and hybridized to the cloned sequence under saturated hybridization conditions. After hybridization and autoradiography, the signals are analyzed densitometrically and the copy number of a repetitive sequence can be estimated by comparing hybridization signals. Genome size and repeat unit size have to be known as a prerequisite. Genome sizes for many important plant species can be found in the published literature.[39,40] Copy number estimations for interspersed repeats are especially sensitive to the hybridization conditions because such families are frequently more heterogeneous than tandem repeat families.

As an alternative, genomic libraries are used to estimate the copy number of a repetitive sequence by screening with the cloned repetitive sequence, but such an estimation is much less accurate than quantitative slot blot hybrdization analysis because this method is dependent on organization of repetitive sequences and randomness of the genomic libraries used. For example, tandemly repeated DNA is located in one or a few regions of the genome, therefore, fewer clones are likely to hybridize, resulting in an underestimation of copy number in the genome. The cloning bias of the libraries is another source of variation.

10.2.2.5. Genome/species specificity of the repeated unit

Distribution of a repetitive DNA in a plant genus or family is an important piece of information regarding the evolution of repetitive sequences. Southern blot and slot blot hybridization are often used to investigate the copy number and distribution of a repetitive element. As a consequence of rapid changes of repetitive DNA sequences in plant genomes, repetitive sequences are often species- and genome-specific in a wide variety of plant taxa and require the analysis under different hybridization stringency conditions.

10.2.2.6. Chromosomal locations of repeated units

Two general methods are used to locate repetitive DNA sequences onto chromosomes: RFLP mapping and in situ hybridization. It is relatively simple to map tandem repeats into chromosomes by RFLP mapping as long as there are polymorphisms present between mapping parents (see chapter 3). However, it is much more difficult to map interspersed repeats by RFLP analysis because of the complex hybridization patterns generated by such repeats. Increasing hybridization and washing stringency conditions is one way to reduce the number of detected fragments. Alternatively, flanking single copy sequences from large insert clones have to be used to localize interspersed repeats on genetic maps.[34,35]

More commonly, localization of repetitive DNA sequences is accomplished by in situ hybridization (see chapter 11). This technique allows the cytological localization of repetitive DNA sequences onto the chromosomes of a plant. Currently, this is accomplished by the labeling of cloned repetitive DNA sequences with biotin or digoxygenin and the use of this probe for direct hybridization onto denatured metaphase chromosomes. After removing excess hybridization probe by washing, the signal is detected by the use of fluorescent dye-linked antibodies specific for the label on the DNA probe. This technique, called fluorescence in situ hybridization (FISH), provides an efficient way to find the precise physical localization of a given repetitive DNA sequence at the cytological level. Because multiple copies of a repetitive DNA sequence are present in a plant genome, it is much easier to assign a chromosomal location to a repetitive DNA than to single copy sequences using in situ hybridization.

10.2.2.7 DNA sequence analysis

Sequence analysis provides direct information on sequence organization and evolution of repeat families. It also allows the study of homogeneity or heterogeneity of different members of a repeat family. By sequencing several members of the repeat family, a consensus sequence for the family can be obtained.[29,41] Information from DNA sequences of repeats derived from related species is frequently used to analyze evolution of both the repeat family and the species involved.[42]

10.3. IMPORTANT REPETITIVE DNA SEQUENCES AND THEIR APPLICATIONS IN PLANT GENOME ANALYSIS

10.3.1. SATELLITES

Variation in the number of repeat units among individuals is a universal feature of satellite DNA, regardless of the type and length of repeats. This is because tandemly repeated DNA sequences evolve by unequal crossing over during DNA replication, in combination with gene conversion and alternating cycles of mutation, amplification and deletion.[3,12,43] Unequal crossing-over between individual repeat units occurs much more frequently than, for example, point mutations, and thus causes the copy number of tandemly-repeated sequences to vary markedly even in closely related organisms.[7]

Two types of satellite sequences are present in all eukaryotic genomes. These are the ribosomal RNA genes (also called rDNA) and the 5S rRNA genes. Each of these two RNA gene repeats consist of a transcribed unit and intergenic spacer sequences. Due to the conservation of coding sequences, combined with rapid evolution of nearby intergenic spacers, the rDNAs have been very useful for phylogenetic studies in a wide range of organisms, including plants. In higher plants, the genes for 16-18S, 5.8S and 25S rRNA occur in large tandemly repeated units (mostly approximately 10-20 kb) at one or several "nucleolus organizer regions"

(NOR) which are characterized cytologically as "secondary constrictions." NORs are some of the most prominent cytogenetic landmarks, and can constitute 5-10% of the total nuclear DNA.[44] The genetic mapping of ribosomal RNA genes and other macrosatellites by means of RFLPs along with the localization of these genes onto metaphase chromosomes by in situ hybridization provide important landmarks for the alignment of genetic maps and physical maps.[45] In tomato, the rRNA genes occupy almost the entire short arm of chromosome 2 and are associated with a region of highly reduced recombination.[46] In *Arabidopsis*, there are two NOR regions on chromosome 2 and 4 that comprise approximately 6% of the genome. The NORs segregate as Mendelian factors with no detectable recombination within the individual arrays.[47]

In plants, the rRNA genes at the NORs are physically separated from the 5S rRNA genes. The repeating units of the 5S rRNA genes are in the range of a few hundred base pairs to 1 kb with a similar copy number as the other rRNA genes. Therefore, the 5S rRNA genes constitute a much smaller proportion of the genome. In many organisms, they are found at a single location, such as in tomato, where they are located near the centromere of chromosome 1 in a single continuous array of approximately 500 kb.[48] In pea, they are found at three loci probably in one continuous array at each locus.[49] For barley and wheat, the 5S rRNA genes are also present in large tandemly repeated arrays on several chromosomes, and these arrays are extremely variable so that PFGE analysis of the arrays exhibits an extremely high level of variation even between closely related cultivars.[50,51]

Besides these functionally defined rDNA sequences, most higher eukaryotic genomes contain other tandemly repeated sequences that are called macrosatellites. These macrosatellites contain large numbers of tandem repeats (up to more than hundred thousand copies of a five to several hundred base pair repeating unit) and occur in large clusters in the size range of tens to thousands of kilobases. Most macrosatellites share two other general features: (1) they are not transcribed to any significant extent, and (2) they are replicated late in S-phase of the cell cycle. Due to their prominence in many plant genomes, a large number of satellites have been isolated and characterized in various plant species (for review see ref. 11). Macrosatellites are present in a limited number of large arrays on most chromosomes of a plant. Generally, macrosatellites are associated with heterochromatin and thus provide good cytogenetic landmarks for the alignment of genetic and physical maps in the same way as the rRNA genes. In situ hybridization with such repeats allow the precise karyotyping even of organisms whose chromosomes can not be discriminated by other means.[52] In *Arabidopsis* and the Brassicaceae, the most prominent macrosatellites are tightly associated with the centromeric regions of the chromosomes,[53-55] while in many other plants, macrosatellites are associated with the telomeric regions of the chromosomes.[6,26] In tomato, 20 of the 24 chromosome ends are associated with the macrosatellite repeat TGR I and only a few interstitial loci are found. PFGE and Southern blot analyses show that the length of the individual arrays in tomato range between 25 and more than 1000 kb.[52]

10.3.2. TELOMERES

Telomeric sequences are a special kind of tandemly repeated DNA sequences, which constitute the most terminal fragments of eukaryotic chromosomes and are responsible for maintaining the integrity of chromosomes.[56] Telomeres have been characterized for many plants, and are usually heptameric repeats with the consensus sequence CCCTAAA or derivatives thereof.[57-59] The length of the telomeric arrays in plants ranges from a few kilobases to tens of kilobases. In several higher plants, the telomeres are tightly associated with macrosatellites or other repetitive se-

quences. In barley, a 118 bp satellite is found directly adjacent to many telomeres.[32,60] In tomato, the TGR I repeat is found to be a few kb to 100 kb away from the telomeres.[58] In maize, moderately repeated sequences are directly associated with the telomere, while in *Arabidopsis* the telomere-associated sequences are single copy.[61]

Telomeres are some of the most highly variable sequences in plants and show extreme length variation between cultivars.[62] This variation is heritable.[61] Interestingly, sequences similar to the telomere are also associated with centromeric regions of several organisms.[61,63] In maize, *Arabidopsis*, and tomato, such sequences could be mapped to the presumed position of centromeres on several chromosomes by RFLP mapping. Furthermore, the isolation of a centromere-associated repeat from maize B chromosomes revealed that this sequence had significant homology to the telomeric repeat.[64]

The genetic mapping of telomeres is an extremely important point in genome mapping because it gives information about the completeness and genome coverage of a genetic map (chapter 3). Telomeres or telomere-associated macrosatellites have been mapped in tomato,[31] barley,[32] corn,[61] and rice.[65] These genetic mapping experiments show that the genetic maps for these plants are basically complete.

10.3.3. MINISATELLITES

Minisatellites occur interspersed throughout the nuclear DNA of almost all eukaryotes and are defined by repeating units of 10-40 bp that occur in tens to hundreds of copies at each locus. Their contribution to a given genome in terms of kilobases is usually very small although they occur at many locations in the genome. However, they have attracted interest because of the fact that they evolve via unequal crossing over and this results in variable numbers of copies in the respective arrays between individuals.[66] Because of this, they are often called VNTR (variable number tandem repeat) loci.[67] Minisatellites have been found in many plant species by cross-hybridization to known sequences from, for example, isolated human minisatellite sequences.[68] They constitute good probes for fingerprinting of plant varieties or germplasm because they detect multiple loci in the genome (see chapters 8 and 9).

10.3.4. MICROSATELLITES

Microsatellites are characterized by a very short repeat unit of 1 to 5 bases. They occur in short arrays which are widely-distributed in the DNA of all eukaryotic genomes, including plants.[69] Microsatellites are also called simple sequences, simple sequence repeats (SSRs), simple repetitive DNA sequences, short tandem repeats (STRs), and simple sequence motifs (SSMs), or called dinucleotide repeats, trimeric repeats, and tetrameric repeats based on the length of repeat unit. The major source of variation is replication slippage resulting in different array length.[70] Microsatellites currently gain tremendous importance as markers for genetic mapping and fingerprinting.

Microsatellite oligonucleotides are excellent DNA fingerprinting probes (see chapter 9). One of the problems using this approach is the difficulty in defining allele relationships with complex DNA fingerprints. Another problem is that microsatellites with four base-repeats might not be randomly distributed in a genome but clustered at the centromeres or telomeres. This limits the usefulness of such multilocus probes in genetic linkage analysis and population studies.

For use as DNA markers in genetic mapping, locus-specific microsatellite markers have to be isolated. This can not be done by using the microsatellite sequences as probes (since they are interspersed at many locations in the genome) but requires isolation of flanking sequences specific to each microsatellite locus, followed by use of the polymerase chain reaction (PCR). This involves the following steps (see also Fig. 10.1):

(1) Isolation of microsatellite-containing clones from genomic libraries by hybridization with synthetic oligonucleotides. This can be performed with any genomic library but for convenience regarding sequencing, small insert libraries (< 1 kb) are preferred.

(2) Determination of the DNA sequences of microsatellite-containing clones.

(3) Design of locus-specific oligonucleotide primers in the sequences that flank the region containing a microsatellite.

(4) Amplification of the respective genomic region by PCR using the oligonucleotide primers.

(5) Analysis of the size of the amplified product on high resolution gels such as polyacrylamide gels, because the size differences are only one or a few bases depending on the repeat size and number.

Because of these steps, the generation of microsatellite markers requires much more work than other marker systems and is relatively expensive. However, because microsatellite markers are locus-specific, multiallelic, and highly informative, they are extremely useful for organisms that display limited sequence variation.[71] Microsatellite markers are also sequence tagged sites (STSs) useful for combining genetic linkage maps and physical maps.

Because of their high level of DNA polymorphism, microsatellites have become the principal molecular markers in genome analysis for human and other animal systems.[72,73] In the last few years, initial studies on the frequencies of microsatellite repeats and the isolation of a small number of sequences have been reported for plants such as rice, wheat, *Arabidopsis* and others.[74-81] A major difference between plant and animal systems is the relative abundance of various kinds of microsatellite motifs. While microsatellites with CA/GT repeat are the most abundant in animals, they are less abundant in many plants, where AT/TA repeats are much more prominent, followed by GA/CT repeats. In plants, microsatellites with repeat unit of 3-4 bases are generally less abundant, and

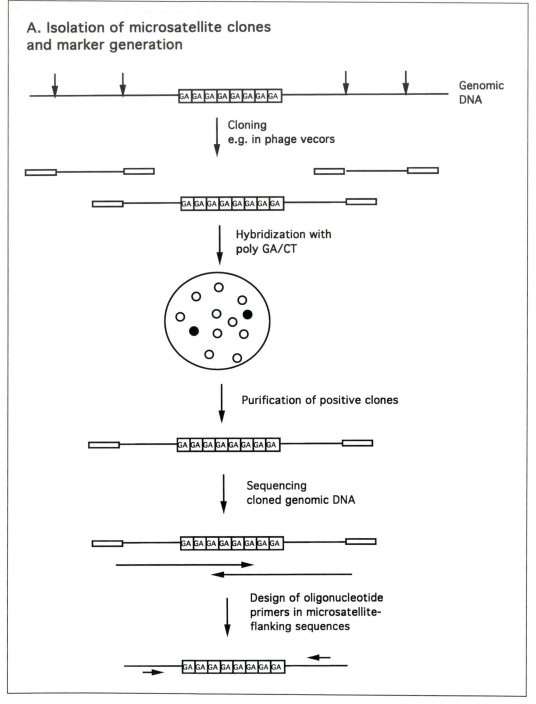

Fig. 10.1 (A). Principles of the isolation and analysis of microsatellite markers in plants. (continued on facing page)

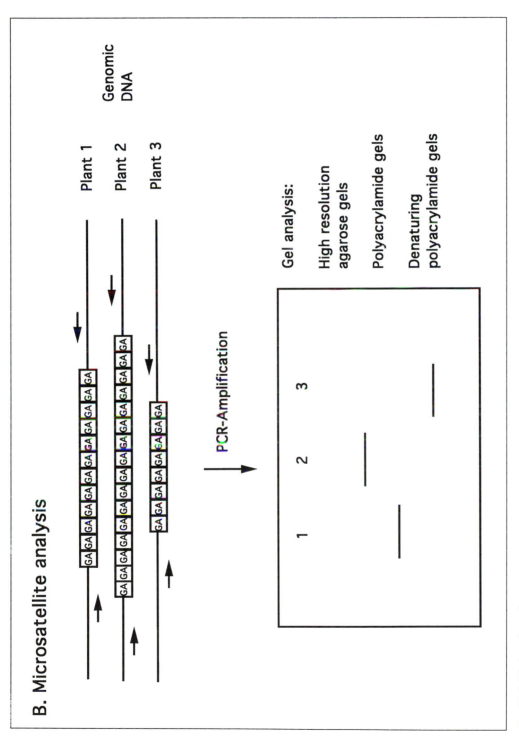

Fig. 10.1 (B). (continued)

appear to be clustered in specific regions of the genome such as centromeres and telomeres.

Most microsatellite loci in plants exhibit a higher level of polymorphism than other marker systems. Besides being excellent molecular markers for genetic mapping, microsatellite markers are found very useful for population genetics, variety identification and protection (chapter 9), monitoring of seed purity and hybrid quality, gene tagging, germplasm evaluation, and phylogenetic studies (chapter 8). In some allopolyploid plants (e.g., wheat), microsatellite markers are often genome-specific and do not amplify loci from homoeologous chromosomes, while in others (e.g., cotton) they amplify loci from all genomes simultaneously.

The analysis of microsatellite markers requires only small amounts of DNA because the products are amplified by PCR. This allows an almost complete automation of the steps required for microsatellite analysis (DNA isolation, PCR, product separation, and data analysis). Microsatellite analysis has been automated so that up to several hundred thousand data points can be generated in a very short time by the use of automated DNA sequencers.[82] In this way, microsatellite analysis requires no radioactive material or any other complex handling steps and, because of this, is of much interest for plant breeding applications which require high throughput.[71,83]

10.3.5. INTERSPERSED REPETITIVE SEQUENCES

Interspersed repetitive sequences comprise the largest proportion of repetitive sequences in most plant genomes.[18] They are usually divided into two classes based on their interspersion pattern.[84] Long interspersed nuclear elements (LINEs) have a repeat length of several kilobases and the average distance between the repeating units is in the range of hundreds of kilobases. As in other eukaryotes, many plant LINEs have features that are typical of mobile genetic elements such as terminal repeats and/or similarity to reverse transcribed sequences (retrotransposons).[85,86] In fact, most transposons are members of this class of genomic elements.[87] The features and use of transposable elements are described in detail in chapter 13 (for review see ref. 88).

Short interspersed nuclear elements (SINEs) have a repeat length of 100 to several hundred base pairs, occur in large numbers of copies and the average distance between the repeats in genomic DNA is in the range of 1-20 kb. The *Alu*-repeats of the human genome are the prototype of this repeat class.[89] Only a few plant species are known to contain short interspersed repeats (sugar beet, rye), while most contain long interspersed repeats.[18,37,90] Nevertheless, unlike animal genomes, plant genomes usually lack interspersed repetitive sequences with abundance as high as the *Alu*-repeats of human (ca. 10^6 copies). Interspersed repetitive sequences with a high abundance in plant genomes should be very useful for physical mapping (chapter 5), in a similar way as human *Alu*-repeats.[91] Such interspersed repeats have been isolated from several plants. For example, in tomato, TGR II has a copy number of 77,000,[26] and in cotton, a few repeats are estimated to have approximately 100,000 copies.[27]

Similar to macrosatellites, interspersed repeats are often specific for a given genus, genome, or even species because they evolve much more rapidly than single copy or low copy elements, and this might be associated with speciation.[4,11,18] These repetitive sequences have provided very useful phylogenetic tools to study evolutionary relationships among species.[22,92-94] For example, the major interspersed repeat (TGR II) of tomato is specific to the genus *Lycopersicon*, and is absent from the closely related genus *Solanum*, of which potato is a member. In contrast, more than 99% of the single copy markers used in RFLP mapping show cross-hybridization.[26,95] Likewise, major repeated elements of the potato genome are not present in the tomato genome.[96] Thus, these elements can be used for rapid identification of somatic hybrids between tomato and potato generated by protoplast fusion.[96] In cotton, the most prominent interspersed repeats are either genome-specific or genome-enriched, and provide an important tool for the discrimination of the two genomes in the allotetraploid cotton for the purpose of chromosome walking (Zhao and Paterson, in preparation).

A short period interspersed repetitive element of sugar beet has been identified that is specific to a wild beet species (*Beta procumbens*) and is not present in the cultivated sugar beet (*Beta vulgaris*). This isolated interspersed repeat can be used to monitor the presence of chromosomes or chromosome fragments in cultivated beet that carry a nematode resistance gene derived from this wild beet species.[90] Similarly, species-specific repetitive elements have been used in other plants as hybridization probes to identify introgressed chromosome segments and to identify parental chromosomes in interspecific hybrids.[96-98]

Several advantages have been recognized for application of repetitive DNA in phylogenetic studies. Repetitive elements cover a large proportion of a plant genome, therefore, more information can be generated by using repetitive elements as hybridization probes. Both RFLPs and abundance variation can be scored. Conclusions generated from repetitive sequences are therefore more reliable especially for the analysis of allopolyploid species such as wheat, cotton, and oats.[92,93,99]

10.4. GENERAL CONCLUSIONS AND FUTURE DIRECTIONS

The analysis of repetitive sequences is extremely important for plant genome analysis because repetitive sequences represent the majority of the genomic DNA in most plants. Future directions in the analysis of repetitive sequences in plants will likely include the following:

(a) Repetitive sequences are found at key positions of eukaryotic chromosomes, such as centromeres, secondary constrictions and telomeres. In this context, repetitive sequences have to be analyzed and mapped for the alignment of genetic and physical maps in plants. A thorough understanding of the function of repetitive sequences at chromosomal landmarks are important for the understanding of chromosome function. In the long term, this might lead to the defining of a minimal set of sequences for chromosome function, and the construction of plant artificial chromosomes.

(b) Macro-, mini- and microsatellites will gain much importance for DNA fingerprinting. They will be used for the characterization of breeding lines, variety protection, monitoring of seed purity and patenting. Interspersed repeated sequences can be used for the rapid identification of somatic hybrids and introgression from wild plant species during crop improvement.

(c) More studies are likely to be centered on using repetitive sequences in the study of plant genome evolution and phylogenetic relationship among closely related species.

(d) Microsatellite sequences will be used as genetic markers much more broadly in the future once sets of markers have been developed for important crop plants. Their high level of polymorphism and potential for automated analysis is currently being shown in human and animal genome analysis. These analytical techniques will be transferred to plants in the near future.

(e) Much knowledge has to be gained in the future to understand the structure, function and evolution of interspersed repetitive sequences during plant evolution and speciation, since they are the largest class of repetitive DNA sequences in plant genomes and they will be very useful for plant genome analysis in many aspects.

REFERENCES

1. Flavell RB, Bennett MD, Smith JB, Smith DB. Genome size and the proportion of repeated nucleotide sequence DNA in plants. Biochem Genet 1974; 12:257-69.
2. Bennett MD, Smith JB. Nuclear DNA amounts in angiosperms. Phil Trans R Soc London B 1976; 274:227-274.
3. Flavell RB. Sequence amplification, deletion, and rearrangement - major sources of

variation during species divergence. In: Dover GA, Flavell RB, eds. Genome Evolution. London: Academic Press, 1982: 301-323.

4. Flavell RB. Repetitive DNA and chromosome evolution in plants. Phil Trans R Soc Lond B Biol Sci 1986; 312:227-42.

5. Ting SJ. A binary model of repetitive DNA sequence in *Caenorhabditis elegans*. DNA Cell Biol 1995; 14:83-5.

6. Bedbrook JR, Jones D, O'Dell M et al. A molecular description of telomeric heterochromatin in *Secale* species. Cell 1980; 19:545-560.

7. Miklos GLG. Localized highly repetitive DNA sequences in vertebrate and invertebrate genomes. In: MacIntyre RJ, ed. Molecular Evolutionary Genetics. New York: Plenum, 1985:241-321.

8. John B, Miklos GL. Functional aspects of satellite DNA and heterochromatin. Int Rev Cytol 1979; 58:1-114.

9. Brutlag DL. Molecular arrangement and evolution of heterochromatic DNA. Ann Rev Genet 1980; 4:121-144.

10. Flavell RB. Repetitive sequences and genome architecture. In: Ciferri O, Dure L, eds. Structure and Function of Plant Genomes. New York: Plenum Press, 1982: 1-14.

11. Lapitan NLV. Organization and evolution of higher plant nuclear genomes. Genome 1992; 35:171-181.

12. Charlesworth B, Sniegowski P, Stephan W. The evolutionary dynamics of repetitive DNA in eukaryotes. Nature 1994; 371: 215-20.

13. Dean C, Schmidt R. Plant genomes: a current molecular description. Annu Rev Plant Physiol Plant Mol Biol 1995; 46:395-418.

14. Bendich AJ, Anderson RS. Characterization of families of repeated DNA sequences from four vascular plants. Biochemistry 1977; 16:4655-63.

15. Murray MG, Cuellar RE, Thompson WF. DNA sequence organization in the pea genome. Biochemistry 1978; 17:5781-5790.

16. Murray MG, Palmer JD, Cuellar RE et al. Deoxyribonucleic acid sequence organization in the mung bean genome. Biochemistry 1979; 18:5259-5266.

17. Evans IJ, James AM, Barnes SR. Organization and evolution of repeated DNA sequences in closely related plant genomes. J Mol Biol 1983; 170:803-26.

18. Flavell R. The molecular characterization and organization of plant chromosomal DNA sequences. Ann Rev Plant Physiol 1980; 31:569-596.

19. Geever RF, Katterman F, Endrizzi JE. DNA hybridization analyses of *Gossypium* allotetraploid and two closely related diploid species. Theor Appl Genet 1989; 77:553-559.

20. Walbot V, and Dure LS. Developmental biochemistry of cotton seed embryogenesis and germination. VII. Characterization of the cotton genome. J Mol Biol 1976; 101:503-536.

21. Wilson JT, Katterman FR, Endrizzi JE. Analysis of repetitive DNA in three species of *Gossypium*. Biochem Genet 1976; 14: 1071-5.

22. Zhao X, Wu T, Xie Y et al. Genome-specific repetitive sequences in the genus *Oryza*. Theor Appl Genet 1989; 78:201-209.

23. Wu T, Wang Y, Wu R. Transcribed repetitive DNA sequences in telomeric regions of rice (*Oryza sativa*). Plant Mol Biol 1994; 26:363-75.

24. Harrison GE, Heslop-Harrison JS. Centromeric repetitive DNA sequences in the genus *Brassica*. Theor Appl Genet 1995; 90:157-165.

25. Grebenstein B, Grebenstein O, Sauer W et al. Characterization of a highly repeated DNA component of perennial oats (*Helictotrichon*, Poaceae) with sequence similarity to a A-genome-specific satellite DNA of rice (*Oryza*). Theor Appl Genet 1995; 90:1101-1105.

26. Ganal MW, Lapitan NLV, Tanksley SD. A molecular and cytogenetic survey of major repetitive DNA sequences in tomato (*Lycopersicon esculentum*). Mol Gen Genet 1988; 213:262-268.

27. Zhao X, Wing RA, Paterson AH. Cloning and characterization of the majority of repetitive DNA in cotton (*Gossypium* L.). Genome 1995:in press.

28. Grellet F, Delcasso D, Panabieres F et al. Organization and evolution of a higher plant alphoid-like satellite DNA sequence. J Mol

Biol 1986; 187:495-507.

29. Wu TY, Wu R. A new rice repetitive DNA shows sequence homology to both 5S RNA and tRNA. Nucleic Acids Res 1987; 15:5913-23.

30. Moore G, Cheung W, Schwarzacher T et al. BIS 1, a major component of the cereal genome and a tool for studying genomic organization. Genomics 1991; 10:469-476.

31. Ganal MW, Broun P, Tanksley SD. Genetic mapping of tandemly repeated telomeric DNA sequences in tomato (*Lycopersicon esculentum*). Genomics 1992; 14:444-8.

32. Röder MS, Lapitan NL, Sorrells ME et al. Genetic and physical mapping of barley telomeres. Mol Gen Genet 1993; 238: 294-303.

33. Kolchinsky A, Gresshoff PM. A major satellite DNA of soybean is a 92-base pairs tandem repeat. Theor Appl Genet 1995; 90:621-625.

34. Zhao X, Kochert G. Characterization and genetic mapping of a short, highly repeated, interspersed DNA sequence from rice (*Oryza sativa* L.). Mol Gen Genet 1992; 231:353-9.

35. Zhao X, Kochert G. Clusters of interspersed repeated DNA sequences in the rice genome (*Oryza*). Genome 1993; 36:944-53.

36. Springer PS, Edwards KJ, Bennetzen JL. DNA class organization on maize *Adh1* yeast artificial chromosomes. Proc Natl Acad Sci USA 1994; 91:863-867.

37. Aledo R, Raz R, Monfort A et al. Chromosome localization and characterization of a family of long interspersed repetitive DNA elements from the genus *Zea*. Theor Appl Genet 1995; 90:1049-1100.

38. Rivin CJ, Cullis CA, Walbot V. Evaluating quantitative variation in the genome of *Zea mays*. Genetics 1986; 113:1009-1019.

39. Arumuganathan K, Earle ED. Nuclear DNA content of some important plant species. Plant Mol Biol Rep 1991; 9:208-218.

40. Bennett MD, Smith JB. Nuclear DNA amounts in angiosperms. Philos Trans R Soc London B 1991; 334:309-345.

41. Xia X, Rocha PS, Selvaraj G et al. Genomic organization of the canrep repetitive DNA in *Brassica juncea*. Plant Mol Biol 1994; 26:817-32.

42. Ingham LD, Hanna WW, Baier JW et al. Origin of the main class of repetitive DNA within selected *Pennisetum* species. Mol Gen Genet 1993; 238:350-6.

43. Smith GP. Evolution of repeated DNA sequences by unequal crossover. Science 1976; 191:528-35.

44. Hemleben V, Ganal M, Gerstner J et al. Organization and length heterogeneity of plant ribosomal RNA genes. In: Kahl G, ed. Architecture of Eukaryotic Genes. VCH: Weinheim, 1988:371-383.

45. Werner JE, Endo TR, Gill BS. Toward a cytogenetically based physical map of the wheat genome. Proc Natl Acad Sci USA 1992; 89:11307-11311.

46. Tanksley SD, Ganal MW, Prince JP et al. High density molecular linkage maps of the tomato and potato genomes. Genetics 1992; 132:1141-60.

47. Copenhaver GP, Doelling JH, Gens JS et al. Use of RFLPs larger than 100 kbp to map the position and internal organization of the nucleolus organizer region on chromosome 2 in *Arabidopsis thaliana*. Plant J 1995; 7:273-286.

48. Lapitan NLV, Ganal MW, Tanksley SD. Organization of the 5S ribosomal RNA genes in the genome of tomato. Genome 1991; 34:509-514.

49. Ellis TN, Lee D, Thomas CM, et al. 5S rRNA genes in *Pisum*: sequence, long range and chromosomal organization. Mol Gen Genet 1988; 214:333-342.

50. Röder MS, Sorrells ME, Tanksley SD. 5S ribosomal gene clusters in wheat: pulsed field gel electrophoresis reveals a high degree of polymorphism. Mol Gen Genet 1992; 232:215-20.

51. Röder MS, Sorrells ME, Tanksley SD. Pulsed-field gel analysis of 5S and satellite DNA in barley. Genome 1995; 38:153-157.

52. Lapitan NLV, Ganal MW, Tanksley SD. Somatic chromosome karyotype of tomato based on in situ hybridization of the TGR I satellite repeat. Genome 1989; 32:992-98.

53. Martinez-Zapater JM, Estelle MA, Somerville CR. A highly repeated DNA sequence in *Arabidopsis thaliana*. Mol Gen Genet 1986; 204:417-423.

54. Maluszynska J, Heslop-Harrison JS. Local-

ization of tandemly repeated DNA sequences in *Arabidopsis thaliana*. Plant J 1991; 1:159-166.

55. Xia X, Selvaraj G, Bertrand H. Structure and evolution of a highly repetitive DNA sequence from *Brassica napus*. Plant Mol Biol 1993; 21:213-24.

56. Zakian VA. Structure and function of telomeres. Ann Rev Genet 1989; 23:579-604.

57. Richards EJ, Ausubel FM. Isolation of a higher eukaryotic telomere from *Arabidopsis thaliana*. Cell 1988; 53:127-136.

58. Ganal MW, Lapitan NL, Tanksley SD. Macrostructure of the tomato telomeres. Plant Cell 1991; 3:87-94.

59. Richards EJ, Chao S, Vongs A et al. Charaterization of *Arabidopsis thaliana* telomeres isolated in yeast. Nucleic Acids Res 1992; 20:4039-4046.

60. Kilian A, Kleinhofs A. Cloning and mapping of telomere-associated sequences from *Hordeum vulgare* L. Mol Gen Genet 1992; 235:153-156.

61. Burr B, Burr FA, Matz EC et al. Pinning down loose ends: mapping telomeres and factors affecting their length. Plant Cell 1992; 4:953-960.

62. Broun P, Ganal MW, Tanksley SD. Telomeric arrays display high levels of heritable polymorphism among closely related plant varieties. Proc Natl Acad Sci USA 1992; 89:1354-7.

63. Richards EJ, Goodman HM, Ausubel FM. The centromere region of *Arabidopsis thaliana* chromosome 1 contains telomere-similar sequences. Nucleic Acids Res 1991; 19:3351-3357.

64. Alfenito MR, Birchler JA. Molecular characterization of a maize B chromosome centric sequence. Genetics 1993; 135:589-597.

65. Wu KS, Tanksley SD. Genetic and physical mapping of telomeres and macrosatellites of rice. Plant Mol Biol 1993; 22:861-72.

66. Jeffreys AJ, Wilson V, Thein SL. Hypervariable 'minisatellite' regions in human DNA. Nature 1985; 314:67-73.

67. Nakamura Y, Leppert M, O'Connell P et al. Variable number of tandem repeat (VNTR) markers for human gene mapping. Science 1987; 235:1616-1622.

68. Dallas JF. Detection of DNA "fingerprints"

of cultivated rice by hybridization with a human minisatellite DNA probe. Proc Natl Acad Sci U S A 1988; 85:6831-5.

69. Tautz D, Renz M. Simple sequences are ubiquitous repetitive components of eukaryote genomes. Nucl Acids Res 1984; 12:4127-4138.

70. Tautz D, Trick M, Dover GA. Cryptic simplicity in DNA is a major source of genetic variation. Nature 1986; 322:652-656.

71. Mazur BJ, Tingey SV. Genetic mapping and introgression of genes of agronomic importance. Curr Opin Biotech 1995; 6:175-182.

72. Weissenbach J. A second generation linkage map of the human genome based on highly informative microsatellite loci. Gene 1993; 135:275-8.

73. Beattie CW. Livestock genome maps. Trends Genet 1994; 10:334-8.

74. Akkaya MS, Bhagwat AA, Cregan PB. Length polymorphisms of simple sequence repeat DNA in soybean. Genetics 1992; 132:1131-9.

75. Lagercrantz U, Ellegren H, Andersson L. The abundance of various polymorphic microsatellite motifs differs between plants and vertebrates. Nucleic Acids Res 1993; 21:1111-5.

76. Morgante M, Olivieri AM. PCR-amplified microsatellites as markers in plant genetics. Plant J 1993; 3:175-82.

77. Wu KS, Tanksley SD. Abundance, polymorphism and genetic mapping of microsatellites in rice. Mol Gen Genet 1993; 241:225-35.

78. Zhao X, Kochert G. Phylogenetic distribution and genetic mapping of a (GGC)n microsatellite from rice (*Oryza sativa* L.). Plant Mol Biol 1993; 21:607-14.

79. Saghai Maroof MA, Biyashev RM, Yang GP et al. Extraordinarily polymorphic microsatellite DNA in barley: species diversity, chromosomal locations, and population dynamics. Proc Natl Acad Sci USA 1994; 91:5466-70.

80. Becker J, Heun M. Barley microsatellites: allele variation and mapping. Plant Mol Biol 1995; 27:835-845.

81. Röder MS, Plaschke J, König SU et al. Abundance, variability and chromosomal location of microsatellites in wheat. Mol Gen

Genet 1995; 246:327-333.

82. Mansfield DC, Brown AF, Green DK et al. Automation of genetic linkage analysis using fluorescent microsatellite markers. Genomics 1994; 24:225-233.

83. Rafalski JA, Tingey SV. Genetic diagnostics in plant breeding: RAPDs, microsatellites and machines. Trends Genet 1993; 9:275-80.

84. Singer MF. Highly repeated sequences in mammalian genomes. Int Rev Cytol 1982; 76:67-112.

85. Sentry JW, Smyth DR. An element with long terminal repeats and its variant arrangements in the genome of *Lilium henryi*. Mol Gen Genet 1989; 215:349-354.

86. Moore G, Lucas H, Batty N et al. A family of retrotransposons and associated genomic variation in wheat. Genomics 1991; 10:461-68.

87. Smyth DR. Dispersed repeats in plant genomes. Chromosoma 1991; 100:355-359.

88. McDonald JF. Evolution and consequences of transposable elements. Curr Opin Genet Dev 1993; 3:855-864.

89. Deininger PL. SINES: Short interspersed repeated DNA elements in higher eukaryotes. In: Berg DE, Howe MM, eds. Mobile DNA. Washington, DC: Amer Soc Microbiol, 1989:619-636.

90. Schmidt T, Junghans H, Metzlaff M. Construction of *Beta procumbens*-specific DNA probes and their application for the screening of *B. vulgaris* x *B. procumbens* (2n=19) addition lines. Theor Appl Genet 1990; 79:177-181.

91. Kass DH, Batzer MA. Inter-*Alu* polymerase chain reaction: advancements and applications. Anal Biochem 1995; 228:185-193.

92. Dvoràk J, Zhang H-B. Variation in repeated nucleotide sequences sheds light on the phylogeny of the wheat B and G genomes. Proc Natl Acad Sci USA 1990; 87:9640-44.

93. Fabijanski S, Fedak G, Armstrong K et al. A repetitive sequence probe for the C genome in *Avena* (Oats). Theor Appl Genet 1990; 79:1-7.

94. Crowhurst RN, Gardner RC. A genome-specific repeat sequence from kiwifruit (*Actinidia deliciosa* var. *deliciosa*). Theor Appl Genet 1991; 81:71-78.

95. Zamir D, Tanksley SD. Tomato genome is comprised largely of fast-evolving, low copy-number sequences. Mol Gen Genet 1988; 213:254-261.

96. Schweizer G, Ganal M, Ninnemann H et al. Species-specific DNA sequences for identification of somatic hybrids between *Lycopersicon esculentum* and *Solanum acaule*. Theor Appl Genet 1988; 75:679-684.

97. Itoh K, Iwabuchi M, Shimanoto K. In situ hybridization with species-specific DNA probe gives evidence for asymmetric nature of *Brassica* hybrids obtained by X-ray fusion. Theor Appl Genet 1991; 81:356-362.

98. Zhang H-B, Dvoràk J. Isolation of repeated DNA sequences from *Lophophyrum elongatum* for detection of *Lophophyrum* chromatin in wheat genomes. genome 1990; 33:283-294.

99. Rayburn AL, Gill BS. Isolation of a D-genome specific repeated DNA sequence from *Aegilops squarrosa*. Plant Mol Biol Rep 1986; 4:102-109.

CHAPTER 11

CURRENT STATUS AND POTENTIAL OF FLUORESCENCE IN SITU HYBRIDIZATION IN PLANT GENOME MAPPING

Jiming Jiang and Bikram S. Gill

11.1. INTRODUCTION

The fluorescence in situ hybridization (FISH) technique, developed by Langer-Safer et al,[1] is an important genome mapping tool in human and other mammalian species. In addition to the localization of single probes on specific chromosome regions, multi-color FISH has given rise to a new class of applications where the relative positions of two or more probes can be visualized directly. FISH to human metaphase spreads allows rapid chromosomal localization of probes with a 1-3 megabase (Mb) resolution,[2,3] whereas FISH to interphase nuclei allows the ordering of probes in a range of 50-1000 kilobases (kb).[4,5] The most recent development of FISH using highly extended chromatin has improved the resolving power for ordering probes to as little as a few kb.[6,7] The application of FISH in plant species has been limited mainly because of the low sensitivity of the technique in plant chromosome preparations. The low sensitivity has hindered the wide application of this technique as a routine mapping tool in plants. However, significant progress has been achieved recently. The recent advances and future potential of this technique in plant genome mapping will be reviewed in this chapter.

Genome Mapping in Plants, edited by Andrew H. Paterson. © 1996 R.G. Landes Company.

11.2. REFINEMENT OF THE FISH TECHNIQUE

A major limitation of the FISH technique is its low sensitivity when it is applied in plant chromosome preparations. It has been technically difficult to detect DNA probes containing only a few kb of DNA sequences, whereas DNA probes as small as a few hundred basepairs can be visualized on human metaphase chromosomes.[8] Recently this problem was overcome, at least for some plant genomes if not all, by the development a FISH technique using large insert genomic DNA clones as probes. In this method, a genomic DNA clone can be labeled as a probe. The signals from the repetitive DNA sequences in the genomic DNA clone, which may interfere with the detection of the unique sequences in the clone, can be suppressed by preannealing the probe with Cot-1 DNA prepared from the same plant species as the cloned DNA. Jiang et al[9] successfully mapped both anonymous and specific bacterial artificial chromosome (BAC) clones on rice metaphase and interphase chromosomes (see Fig. 11.1A). A similar procedure was also demonstrated in sorghum[10] and cotton.[11]

FISH using genomic DNA clones provides an alternative method to map small DNA probes. A large genomic DNA clone can be isolated by screening a genomic DNA library using the small DNA probe. This large genomic DNA clone can then be mapped using FISH. Thus, any DNA markers can be mapped by FISH by this strategy as long as the corresponding genomic DNA clone can generate a detectable signal. However, this method is technically dependent on the amount of the repetitive DNA sequences in the genomic DNA clones. We have mapped more than 30 different rice BAC clones and all the clones generated excellent FISH signals (Jiang et al, unpublished data). But we had limited success in blocking the repetitive DNA sequences in mapping BAC clones of wheat. A similar problem was also reported in cotton.[11] Apparently, the difficulty in wheat and cotton is due to a much higher percentage of repetitive DNA sequences in the BAC clones compared to that in rice. Therefore, further improvement of technique, especially the efficient blocking of repetitive DNA sequences, is necessary before this method can be widely used in various plant species.

A significant step forward of the FISH technology during the last few years is the application of highly sensitive CCD (charge coupled device) cameras and computer digital imaging analysis.[3,12,13] The advantages of using these new technologies include:

1. The CCD cameras are highly sensitive to very weak fluorescent signals. Such signals can be documented and reproduced more readily. It is difficult to analyze such signals using conventional photographic systems;

2. It becomes possible to map a large number of clones in a relatively short time with computer-assisted imaging analysis; conventional photographic systems are too expensive and time-consuming for large scale FISH mapping; and

3. Quantitative data on measurements of inter- and intrasignal distances can be easily and accurately assessed using computer software. Application of these new technologies in plant FISH mapping has been reported from several laboratories.[9,14,15]

11.3. GENOME STRUCTURE IN ALLOPOLYPLOID SPECIES

Polyploidy plays an important role in the evolution of plants. At least a third of all Angiosperm species and about 70% of the grasses are polyploids. Most polyploids in nature are probably allopolyploids that were derived from two or more progenitor species. Genome composition and repatterning after polyploidization is one of the most important areas of research to polyploid species. The conventional "genome analysis" strategy, involving production of a sexual hybrid of the polyploid species with one of its presumed progenitor species and analysis of the chromosome pairing behavior in the hybrid, has been the most popular method. However, many

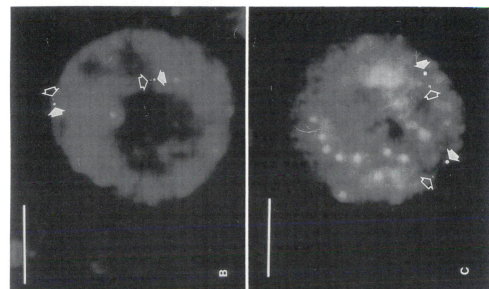

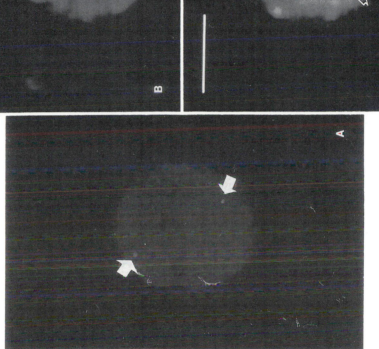

Fig. 11.1 (A) FISH mapping of a rice bacterial artificial chromosome clone, UCD6 provided by Dr. P.C. Ronald at the University of California-Davis, on an rice interphase nucleus. The clone was labeled by digoxigenin-11-dUTP and detected by a rhodamine-conjugated anti-digoxigenin antibody (arrows). The nucleus was stained by DAPI. (B) Interphase FISH mapping of pA1-Lc and pSh2.5 • SstISalI. Probes pA1-Lc (solid arrows) and pSh2.5 • SstISalI (open arrows) are separated by an average distance of 0.50 μm in interphase nuclei. (C) Interphase FISH mapping of KSU3/4 (solid arrows) and KSU16 (open arrows). Probes KSU3/4 and KSU16 are separated by an average distance of 2.32 μm in interphase nuclei. Figure 11.1A was taken directly under an Olympus BX60 microscope. FISH signals in Figures 11.1B, 11.1C were captured by a CCD camera and the photos were taken from merged images on a computer screen. Bars represent 10 μm.

factors can influence the patterns of chromosomal pairing. The inability to distinguish inter- and intra-genomic chromosomal pairing may also significantly complicate the pairing data. Erroneous conclusions can be drawn because of the different interpretation of the chromosome pairing results.

Development of the so called "genomic in situ hybridization" or "GISH" technique added a powerful method to examine the genome structure in allopolyploid species. In the GISH technique, genomic DNA from one of the progenitor of the allopolyploid species is labeled as a probe, and either a differently labeled or unlabeled genomic DNA from the second progenitor is added in the hybridization mixture during the in situ hybridization. This will result in differentiation of chromosomes from different genomes.[16,17] If the probe preferentially hybridizes to a specific set of chromosomes in the polyploid species, it is a good indication that the diploid species from which the probe DNA was isolated is one of the progenitors of the polyploid species. Using this strategy, Bennett et al[18] demonstrated that *Milium vernal* (2n=2x=8) is the donor of the eight large chromosomes in *Milium montianum* (2n=4x=22). Similar work was also carried in several polyploid species, including tobacco (*Nicotiana tabacum*, 2n=4x=48, genomically SSTT),[19] tetraploid wheat (*Triticum timopheevii*, 2n=4x=28, A^{t}A^{t}GG),[20] hexaploid wheat (*Triticum aestivum*, 2n=6x=42, AABBDD),[21] oat (*Avena sativa*, 2n=6x=42, AACCDD),[22,23] and onion (*Allium wakegi*).[24] The GISH technique is especially powerful to detect translocations involving chromosomes from different genomes. Up to nine S/T intergenomic chromosomal translocations were detected in tobacco.[19] Similarly, a number of A/C or D/C intergenomic translocations were reported in oat.[22,23] Detection of such intergenomic translocations provided a direct visualization of genome repatterning after polyploidization. Such translocations may also play an important role during the evolution of polyploid species. As a good ex-

ample, a cyclic translocation involving chromosomes 6A^t, 1G and 4G in *Triticum timopheevii* (A^{t}A^{t}GG) was found in all lines of this species.[20,25] Such translocation was named as "species-specific chromosomal translocation." A similar species-specific translocation involving chromosomes 4A, 5A, and 7B was reported in a different tetraploid wheat *Triticum turgidum* (2n=4x=28, AABB) and hexaploid wheat *Triticum aestivum* (AABBDD).[26,27] Detection of different species-specific chromosomal translocations in the two different polyploid wheat groups provides strong evidence to support the diphyletic hypothesis of the evolution of polyploid wheats.[20]

The GISH technique can also be applied to meiotic chromosome preparations. Application of GISH in chromosome pairing analysis gives a new life to the conventional "genome analysis" strategy. Using this technique, the homology between different genomes can be more accurately assessed because of the differentiation between interspecific and intraspecific chromosomal pairing in meiotic cells.[28-31]

11.4. POTENTIAL IN LARGE SCALE PHYSICAL MAPPING

Development of RFLP (restriction fragment length polymorphism) technology has revolutionized genetic linkage mapping. RFLP marker-based genetic maps have been developed in many plant species. In contrast to genetic mapping, very little has been accomplished in physical mapping during the last decade mainly due to the lack of general and economical techniques. For most of the major crop species, scientists are unable to correlate the genetic mapping data to the physical map.

There are three possible methods for large scale physical mapping. First, DNA contigs can be assembled using large insert DNA clones, such as cosmid, YAC (yeast artificial chromosome), or BAC clones. The first such contig covering chromosome 4 of *Arabidopsis thaliana* has been reported recently.[32] Although physical maps generated by this strategy have tremendous utility, this method is dependent on a saturated STS (sequence-tagged site) map and high-quality large insert DNA libraries. So far, it has been applied only in plant species with small genomes, including rice (400 Mb)[33] and *Arabidopsis thaliana* (100 Mb).[32] For many crop species with much larger genomes, physical mapping by contig assembly is not practical. In addition, contig assembly is technically difficult for polyploid species, such as wheat and oats, or species with extensive duplications in the genome, such as maize and soybean.

The second method for large scale physical mapping is to use cytogenetic stocks such as deletion or translocation lines. This strategy has been successful in wheat. Potentially an unlimited number of wheat deletion stocks can be generated by using a "gametocidal" chromosome from a wild *Aegilops* species.[34,35] RFLP markers can be mapped to a specific chromosomal region using deletion stocks.[36,37] Unfortunately, there are no systems available in other plant species to produce and characterize a large number of deletion or translocation stocks. The resolution of a physical map based on cytogenetic stocks is completely dependent on the number of such stocks available.

FISH is potentially the third method to do large scale physical mapping. Species with relatively small genomes which are amenable to pachytene analysis, such as rice and tomato, are ideal for such a strategy. The basic steps of the strategy include:

1. to develop a genomic DNA library, such as a BAC library;
2. to select genomic DNA clones using RFLP markers that have been used in linkage mapping; and
3. to map the genomic DNA clones on meiotic pachytene chromosomes. For example, the haploid DNA content of tomato is estimated to be approximately 950 Mb,[38] and the total length of the 12 tomato pachytene chromosomes is about 380 microns.[39] On average, one micron of tomato pachytene chromosome contains about 2.5 Mb of

DNA. The fluorescence signal produced by a typical cosmid probe is usually less than 0.3 micron in diameter.[4] Thus, two genomic DNA clones separated by 2 Mb can possibly be resolved on tomato pachytene chromosomes by FISH. This resolution is comparable to that by FISH on human metaphase chromosomes. The FISH mapping data generated from such a strategy can directly be correlated to the RFLP linkage mapping data. Even though the resolution of the physical map based on FISH will be relatively low compared to that based on contig assembly, the map can be produced in a much more efficient and affordable way.

11.5. POTENTIAL IN FINE PHYSICAL MAPPING

In plants, estimation of physical distance between closely linked DNA markers has been exclusively dependent on 'long range restriction mapping' using pulsed-field gel electrophoresis (PFGE). However, PFGE is technically difficult for plant species with large complex genomes. The resolution of PFGE is usually within about one megabase, and depends on the presence and methylation status of rare restriction sites. Because of the discrepancy between genetic and physical distances, sometimes even genetically very-closely linked DNA markers cannot be resolved by PFGE.[40] In polyploids, or species such as maize in which a large part of the genome is duplicated, PFGE analysis will be very difficult because most probes identify multiple bands that significantly complicate the hybridization results.

Interphase FISH mapping is a potential alternative for ordering DNA sequences, especially for plant species in which PFGE can hardly be applied. In the human genome, the minimum DNA distance for resolving two probes on interphase nuclei is about 50 kb.[4] FISH analysis using a set of DNA probes separated by 100 to 2000 kb revealed that the folding of chromatin in interphase nuclei of human was according to a random walk model.[5] At genomic (physical) distances < 2000 kb, the interphase distance between probes is linearly related to the genomic distance. Thus, within this limit, the genomic distance can be estimated based on the interphase distance. The relationship between genomic and interphase distance does not follow the random walk model with genomic distance > 2000 kb, suggesting that some constraining higher order structure exists at these distances.[5] Interphase distance measurements from different regions of the human and other mammalian genomes suggested that the relationship is linear, from 50 kb up to 500-800 kb.[2,4,41-43] Above 500 to 800 kb, the calibration curve had a decreasing gradient, and the mean values of independent experiments were distributed over a wider area.

At present, we do not know whether the relationship between interphase and genomic distance in plant species is similar to those in mammalian species. Recently, we have initiated an interphase FISH mapping experiment in maize.[44] Two probes were applied in the experiment. Probe pA1-Lc contains the *a1* (anthocyaninless) gene and probe pSh2.5•SstISalI contains the *sh2* (shrunken kernel) gene. The physical distance between these two probes is 140 kb.[45] FISH results showed that the two probes completely overlapped on metaphase chromosomes, but were separated by a distance of about 0.50 μm on interphase nuclei (Fig. 11.1B). This result suggests that the resolving power of interphase FISH in maize can be as little as at least 140 kb. Interphase mapping was also applied using probes without a known physical distance. RFLP probes KSU3/4 and KSU16 are 1 centiMorgan (cM) proximal and 2-3 cM distal to the *Rp1* (a rust resistance gene) locus on the genetic linkage map.[46,47] These two probes are separated with an average distance of 2.32 μm on interphase nuclei (Fig. 11.1C). The physical distance between KSU3/4 and KSU16 can be estimated as approximately 650 kb if the *a1-sh2* interval is used as a

direct calibration distance (140kb/0.62μm).

The most recent advances in high-resolution FISH mapping in mammalian species is the application of highly extended linear chromatin.[6,48,49] Although the chromatin in interphase nuclei is relatively decondensed, it is not extended and the intact nuclear membrane limits the maximum resolution that can be achieved. Several techniques have been developed for reducing the condensation of interphase chromatin, including the use of human sperm pronuclei after fusion of human sperm and hamster eggs,[50] the treatment of cells with chemicals causing chromatin decondensation, and the disruption of nuclear lamins[51] and extreme lysis leading to release of histone-depleted DNA.[6,48,49,52] The resolution of FISH on extended chromatin was demonstrated to be close to the theoretical lower limit of ~1 kb.[7] FISH techniques using highly extended chromatin, interphase nuclei, and mitotic and meiotic (pachytene) chromosomes, potentially provide very important supplemental tools for fine physical mapping in plant species in the future.

ACKNOWLEDGMENTS

The demonstrated FISH work was partially supported by start-up funds from College of Agriculture and Life Science, University of Wisconsin-Madison to J.J. We thank Fenggao Dong for technical assistance. Contribution No. 96-304-B from the Kansas Agricultural Experiment Station, Kansas State University.

REFERENCES

1. Langer-Safer PR, Levine M, Ward DC. Immunological method for mapping genes on *Drosophila* polytene chromosomes. Proc Natl Acad Sci USA 1982; 79:4381-4385.

2. Lawrence JB, Singer RH, McNeil JA. Interphase and metaphase resolution of different distance within the human dystrophin gene. Science 1990; 249:928-932.

3. Lichter P, Tang CC, Call K, Hermanson G, Evans GA, Housman D, Ward DC. High-resolution mapping of human chromosome 11 by in situ hybridization with cosmid

clones. Science 1990; 247:64-69.

4. Trask BJ, Pinkel D, van den Engh G. The proximity of DNA sequences in interphase cell nuclei is correlated to genomic distance and permits ordering of cosmids spanning 250 kilobase pairs. Genomics 1989; 5:710-717.

5. van den Engh G, Sachs R, Trask BJ. Estimating genomic distance from DNA sequence location in cell nuclei by a random walk model. Science 1992; 257:1410-1412.

6. Parra I, Windle B. High resolution visual mapping of stretched DNA by fluorescent hybridization. Nature Genet 1993; 5:17-21.

7. Florijin RJ, Bonden LAJ, Vrolijk H, Wiegant J, Vaandrager J-W, Baas F, den Dunnen JT, Tanke HJ, van Ommen G-JB, Raap AK. High-resolution DNA fiber-FISH for genomic DNA mapping and color barcoding of large genes. Hum Mol Genet 1995; 4:831-836.

8. Viegas-Péquignot E, Berrard S, Brice A, Apiou F, Mallet J. Localization of a 900-bp-long fragment of the human choline acetyltransferase gene to 10q11.2 by nonradioactive in situ hybridization. Genomics 1991; 9:210-212.

9. Jiang J, Gill BS, Wang GL, Ronald PC, Ward DC. Metaphase and interphase fluorescence in situ hybridization mapping of the rice genome with bacterial artificial chromosomes. Proc Natl Acad Sci USA 1995; 92:4487-4491.

10. Hanson RE, Zwick MS, Choi S, Islam-Faridi MN, McKnight TD, Wing RA, Price HJ, Stelly DM. Fluorescent in situ hybridization of a bacterial artificial chromosome. Genome 1995; 38:646-651.

11. Woo S-S, Jiang J, Gill BS, Paterson AH, Wing RA. Construction and characterization of a bacterial artificial chromosome library of *Sorghum bicolor*. Nucleic Acids Res 1994; 22:4922-4931.

12. Reid T, Baldini A, Rand TC, Ward DC. Simultaneous visualization of seven different DNA probes by in situ hybridization using combinatorial fluorescence and digital imaging microscopy. Proc Natl Acad Sci USA 1992; 89:1388-1392.

13. Ballard SG, Ward DC. Fluorescence in situ hybridization using digital imaging micros-

copy. J Histochem & Cytochem 1993; 41:1755-1759.

14. Busch W, Herrmann RG, Martin R. Refined physical mapping of the *Sec-1* locus on the satellite of chromosome 1R of rye (*Secale cereale*). Genome 1995; 38:889-893.

15. Fuchs J, Schubert I. Localization of seed protein genes on metaphase chromosomes of *Vicia faba* via fluorescence in situ hybridization. Chromosome Res 1995; 3:94-100.

16. Le HT, Armstrong K, Miki B. Detection of rye DNA in wheat-rye hybrids and wheat translocation stocks using total genomic DNA as a probe. Plant Mol Biol Rep 1989; 7:150-158.

17. Schwarzacher T, Leitch AR, Bennett MD, and J.S. Heslop-Harrison JS. In situ localization of parental genomes in a wide hybrid. Ann Bot 1989; 64:315-324.

18. Bennett ST, Kenton AY, Bennett MD. Genomic in situ hybridization reveals the allopolyploid nature of *Milium montianum* (Gramineae). Chromosoma 1992; 101: 420-424.

19. Kenton A, Parokonny AS, Gleba YY, Bennett MD. Characterization of the *Nicotiana tabacum* L. genome by molecular cytogenetics. Mol Gen Genet 1993; 240: 159-169.

20. Jiang J, Gill BS. Different species-specific chromosome translocations in *Triticum timopheevii* and *T. turgidum* support the diphyletic origin of polyploid wheats. Chromosome Res 1994; 2:59-64.

21. Mukai Y, Nakahara Y, Yamamoto M. Simultaneous discrimination of the three genomes in hexaploid wheat by multicolor fluorescence in situ hybridization using total genomic and highly repeated DNA probes. Genome 1993; 36:489-494.

22. Chen Q, Armstrong K. Genomic in situ hybridization in *Avena sativa*. Genome 1994; 37:607-612.

23. Jellen EN, Gill BS, Cox TS. Genomic in situ hybridization differentiates between A/D- and C-genome chromatin and detects intergenomic translocations in polyploid oat species (genus *Avena*). Genome 1994; 37:613-618.

24. Hizume H. Allodiploid nature of *Allium wakigi* Araki revealed by genomic in situ

hybridization and localization of 5S and 18S rDNAs. Jpn J Genet 1994; 69:407-415.

25. Badaeva ED, Jiang J, Gill BS. Detection of intergenomic translocations with centromeric and non-centromeric breakpoints in *Triticum araraticum*: mechanism of origin and adaptive significance. Genome 1995; 38:976-981.

26. Naranjo T, Roca A, Goicoecha PG, Giraldz R. Arm homoeology of wheat and rye chromosomes. Genome 1987; 29:873-882.

27. Naranjo T. Chromosome structure of durum wheat. Theor Appl Genet 1990; 79:397-400.

28. King IP, Purdie KA, Orford SE, Reader SM, Miller TE. Detection of homoeologous chiasma formation in *Triticum durum* x *Thinopyrum bessarabicum* hybrids using genomic in situ hybridization. Heredity 1993; 71:369-372.

29. Miller TE, Reader SM, Purdie KA, King IP. Determination of the frequency of wheat-rye chromosome pairing in wheat x rye hybrids with and without chromosome 5B. Theor Appl Genet 1994; 89:255-258.

30. Parokonny AS, Kenton A, Gleba YY, Bennett MD. The fate of recombinant chromosomes and genome interaction in *Nicotiana* asymmetric somatic hybrids and their sexual progeny. Theor Appl Genet 1994; 89:488-497.

31. Fernandez-Caivin B, Benavente E, Orellana J. Meiotic pairing in wheat-rye derivatives detected by genomic in situ hybridization and C-banding - a comparative analysis. Chromosoma 1995; 103:554-558.

32. Schmidt R, West J, Love K, Lenehan Z, Lister C, Thompson H, Bouchez D, Dean C. Physical map and organization of *Arabidopsis thaliana* chromosome 4. Science 1995; 270:480-483.

33. Katayose Y, Toyonaga R, Umehara Y, Yano M, Kurata N. Progress in physical map construction. Rice Genome 1995; 4:4-5.

34. Endo TR. Gametocidal chromosomes and their induction of chromosome mutations in wheat. Jpn J Genet 1990; 65:135-152.

35. Endo TR, Gill BS. The deletion stocks of common wheat. J Hered 1996; (in press)

36. Werner JE, Endo TR, Gill BS. Toward a cytogenetically based physical map of the wheat genome. Proc Natl Acad Sci USA 1992; 89:11307-11311.

37. Gill KS, Gill BS, Endo TR. A chromosome region-specific mapping strategy reveals gene-rich telomeric ends in wheat. Chromosoma 1993; 102:374-381.

38. Arumuganathan K, Earle ED. Nuclear DNA content of some important plant species. Plant Mol Biol Rep 1991; 9:208-218.

39. Barton DW. Pachytene morphology of the tomato chromosome complement. Am J Bot 1950; 37:639-643.

40. Ganal MW, Young ND, Tanksley SD. Pulsed field gel electrophoresis and physical mapping of large DNA fragments in the *Tm-2a* region of chromosome 9 in tomato. Mol Gen Genet 1989; 215:395-400.

41. Trask BJ. Mapping of human chromosome Xq28 by two-color fluorescence in situ hybridization of DNA sequences to interphase cell nuclei. Am J Hum Genet 1991; 48:1-15.

42. Brandriff BF, Gordon LA, Tynan KT, Olsen AS, Mohrenweiser HW, Fertitta A, Carrano AV, Trask BJ. Order and genomic distances among members of the carcinoembryonic antigen (CEA) gene family determined by fluorescence in situ hybridization. Genomics 1992; 12:773-779.

43. Senger G, Ragoussis J, Trowsdale J, Sheer D. Fine mapping of the human MHC class II region within chromosome band 6p21 and evaluation of probe ordering using interphase fluorescence in situ hybridization. Cytogenet Cell Genet 1993; 64:49-53.

44. Jiang J, Hulbert SH, Gill BS, Ward DC. Interphase fluorescence in situ hybridization mapping: a physical mapping strategy for plant species with large complex genomes. Theor Appl Genet (submitted).

45. Civardi L, Xia Y, Edwards KJ, Schnable PS, Nikolau BJ. The relationship between genetic and physical distances in the cloned *a1-sh2* interval of the *Zea mays* L. genome. Proc Natl Acad Sci USA 1994; 91:8268-72.

46. Hulbert SH, Bennetzen JL. Recombination at the *Rp1* locus of maize. Mol Gen Genet 1991; 226:377-382.

47. Hong KS, Richter TE, Bennetzen JL, Hulbert SH. Complex duplications in maize lines. Mol Gen Genet 1993; 239:115-121.

48. Haaf T, Ward DC. High resolution ordering of YAC contigs using extended chromatin and chromosomes. Hum Mol Genet 1994; 3:629-633.

49. Houseal TW, Dackowski WR, Landes GM, Klinger KW. High resolution mapping of overlapping cosmids by fluorescence in situ hybridization. Cytometry 1994; 15:193-98.

50. Brandriff BF, Gordan LA, Trask BJ. A new system for high-resolution DNA sequence mapping in interphase pronuclei. Genomics 1991; 10:75-82.

51. Heng HHQ, Squire J, Tsui L-C. High-resolution mapping of mammalian genes by in situ hybridization to free chromatin. Proc Natl Acad Sci USA 1992; 89:9509-9513.

52. Wiegant J, Kalle W, Mullenders L, Brookes S, Hoovers JMN, Dauwerse JG, van Ommen GJB, Raap AR. High resolution in situ hybridization using DNA halo preparations Hum Mol Genet 1992; 1:587-591.

==========CHAPTER 12==========

CLONING OF PLANT GENES BASED ON GENETIC MAP LOCATION

Knut Meyer, Gregor Benning and Erwin Grill

12.1. INTRODUCTION

Recent breakthroughs in plant biology have been accomplished by the combination of genetic analysis of mutations and subsequent cloning of the affected genes.[1-18] The characterization of chromosomal alterations and their phenotypic consequences, the so-called mutational analysis, provides the key to unravelling complex phenomena in the plant's life cycle, such as its reaction to environmental and hormonal signals. Mutational analysis represents a powerful tool since it does not require any a priori knowledge of the molecular mechanisms underlying the biological process to be dissected. Physiological responses which have been recalcitrant to biochemical approaches can thus be elucidated at the molecular level, by cloning the underlying genes based solely on an altered phenotype. In general, two techniques have been successfully employed to clone plant genes based on a mutant phenotype: tagging by insertion mutagenesis (see chapter 13); and positional or "map-based" cloning, isolation of a gene based on its chromosomal location.

12.1.1. RELATIVE STRENGTHS OF INSERTION MUTAGENESIS AND POSITIONAL CLONING FOR ISOLATING PLANT GENES

Insertion mutagenesis in plants has been discussed in a previous chapter (chapter 13) and elsewhere.[19] This chapter will focus on positional cloning. Both techniques have their respective strengths and shortcomings.

Briefly, insertion mutagenesis generates mutants by insertion of a distinct DNA fragment into a gene, disrupting the normal function of the gene. The affected gene can be isolated by exploiting the inserted

---

DNA fragment as a molecular tag. In plants, this procedure has been used successfully, with either transposable elements or the T-DNA of *Agrobacterium tumefaciens* serving as the mutagen. Once cosegregation of the mutant phenotype with the tag is established, gene isolation is much simpler and faster than positional cloning.

However, insertion mutagenesis schemes have several limitations. First, insertion mutagenesis suffers from a low efficiency compared to chemical mutagenesis. Further, this only a narrow spectrum of insertion mutants are readily detected, specifically those which generate recessive loss of function alleles. Thus, the tagging of essential genes may be difficult due to lethality in a homozygous mutant state, and tagging of duplicated genes is difficult due to minimal phenotypic consequences. The development of entrapment techniques alleviate these limitations. An exon trap system based on transposon mutagenesis has recently been successfully employed in *Arabidopsis thaliana*.[20] This approach allows the identification of tagged genes with tissue- and developmental-specific expression patterns as well as the convenient identification of genes in plants heterozygous for the mutations that provide a codominant phenotype.

At present, positional cloning is applicable to a wider spectrum of plant taxa than insertion mutagenesis. Up until now, only a few plant species are amenable to insertion mutagenesis since this technique relies on an efficient transpositional system or on an efficient protocol to generate transgenic plants, e.g., via *Agrobacterium tumefaciens*-mediated gene transfer. Detailed genetic maps are available for most major crops, and megabase DNA libraries for many of these. Further, increasing integration of large, complex genomes, with smaller more facile genomes by comparative mapping (see chapter 3), is extending use of detailed genetic maps from model taxa to positional cloning of genes in more difficult systems.

12.1.2. THE CLONING OF A GENE BY ITS CHROMOSOMAL POSITION

Positional cloning is related to the shotgun complementation approach developed to clone genes in prokaryotes and lower eukaryotes such as yeasts.[21] In both techniques the gene is identified by extragenetic transfer ("transformation") and complementation of the mutant phenotype. The difference between the positional cloning and the shotgun approach lies in the delineation of the target gene to a small segment of the genome. To clone genes from large genomes such as those of higher plants and animals, delineation of the gene's location is necessary due to relatively low transformation and regeneration efficiencies. Nevertheless, attempts were made to establish the "shotgun complementation" approach in plants. Using total genomic DNA, the size of plant genomes requires the generation of thousands of transformed calli or plants to ensure genome coverage in the complementation experiment.[22,23] This method was found to be straightforward, provided a stringent selection scheme for gene complementation (transfer) is available and can be applied to the transgenic plant material, preferably at a callus stage.

In most instances, however, shotgun complementation is not feasible in higher plants, and a positional cloning strategy is the method of choice, using a genetic map to delineate a small DNA segment which can be more easily evaluated by transformation. In contrast to insertion mutagenesis, positional cloning can use phenotypic alteration as a result of any mutagen, including naturally-occurring mutations. Chemical mutagenesis has the advantage of being able to introduce efficiently point mutations, causing subtle modifications in the gene products and their function.[24] These subtle changes are required in screening programs with the objective of isolating leaky alleles of essential genes, second site suppressors or enhancers of known mutations. In addition, chemical mutagens[24] are orders of magnitude more potent than insertion mutagens, making chemical mutagenesis the method of choice to achieve mutational saturation of complete pathways, and to generate multiple alleles varying in the strength of phenotypic effects. Once the gene has been cloned, mutant alleles provide an informative source to identify specific amino acid residues in the gene product essential for protein function.

12.1.3. OBJECTIVES OF THIS CHAPTER

The current status of positional cloning in plants will be summarized. In addition, the individual steps of the approach will be outlined and the experimental tools which are currently available will be described. New methodological developments will be reviewed that hold the potential to improve the cloning technique considerably. Finally, an example of successful positional cloning in plants will be presented in order to illustrate the general scheme. Most successful applications of this technique have so far been performed with *Arabidopsis*, therefore, the article mirrors a bias in favor of this model species.

12.2. THE THREE STEPS OF POSITIONAL CLONING

The procedure of positional cloning can be divided into the steps of: (1) confining the target gene to a chromosomal segment, (2) isolation of the chromosomal region containing the gene of interest, and (3) identification of the gene.

12.2.1. FIRST STEP: CHROMOSOMAL POSITIONING OF THE TARGET GENE

The objective of this step is to identify molecular markers tightly linked to the target gene. Markers are chosen either from a pre-existing high density genetic map of the organism or they are selected by an enrichment technique such as bulked segregant analysis.[25,26] Subsequently, the position of the target gene is determined by genetic linkage analysis using a population of plants segregating for the allele of interest, and for molecular markers.[27,28] This

is generally accomplished by genotyping individual F2 plants,[27] intercrossed plants,[29] recombinant inbred lines,[30] or DNA pools of plants from any of these populations. This analysis not only locates the mutated gene in the genome but also identifies segregants which contain recombinational breakpoints closely-flanking the target locus. These recombination breakpoints are crucial in the subsequent cloning effort, both in order to direct the assembly of a contig spanning the gene, and to pinpoint the gene to a minimal region with a cloned DNA fragment.

12.2.1.1. Molecular marker technology

To date, most high-density genetic maps consist of classical, visible markers (mutant phenotypes) and molecular markers revealing a restriction fragment length polymorphism (RFLP).[31] RFLP markers are reliable tools for genotyping but the analysis requires the comparably labor-intensive isolation of DNA suitable for DNA blot analysis. More sensitive methods have been developed that rely on PCR-amplified molecular markers and several of these markers have been integrated into existing genetic maps.[31] These methods comprise techniques such as randomly amplified polymorphic DNA (RAPD),[32] cleaved amplified polymorphic DNA sequences (CAPS),[33] simple sequence length polymorphism (SSLP),[34,35] amplified fragment length polymorphism (AFLP),[36] representational difference analysis (RDA)[37] and surely, more to come in the future. The RAPD technique uses short primers, 10 nucleotides in length, to detect DNA fragments that are specific for one of the parental lines of the segregating population. The successful amplification of such a marker usually does not permit the differentiation between the heterozygous and homozygous state (e.g., RAPD markers are "dominant"). Consequently, part of the genotypic information is not retrieved. The conversion of RAPD markers into preferable codominant markers, by either subcloning them or isolating larger DNA fragments from genomic libraries probed with the marker, can be difficult due to the fact that many RAPD amplification products contain repetitive DNA sequences.[30,38] CAPS are not subject to this problem since specific primers are used to amplify a sequence that can be genotyped by RFLP assay. The establishment of these codominant markers, however, requires DNA-sequencing of the polymorphic site, or identification of an RFLP within the length of the genomic DNA corresponding to the probe.

A second class of codominant PCR-based markers is represented by SSLP. These are amplified sequences of highly polymorphic microsatellite DNA which seem to be randomly dispersed in the genomes of plant species.[34,35] Similar to CAPS, the design of specific primers requires DNA sequence information.

The prerequisite of DNA sequence information is not required to generate AFLP markers.[36] This elegant method exploits PCR amplification of small DNA fragments derived from restriction fragments. Technically, genomic DNA of the parental lines is cleaved with restriction enzymes and ligated to adaptors. PCR amplification is selectively performed by choosing primers having a few distinct nucleotides at their 3' end, in addition to the sequence corresponding to the ligated adaptor and the adjacent restriction site. The number of amplifiable DNA fragments can easily be adjusted by varying the number of target nucleotides, usually 1-3, of the primers. Separation of the radio labeled AFLP products by polyacrylamide gel electrophoresis allows the examination of hundreds of anonymous DNA fragments. This technology holds promise as a powerful tool to detect polymorphic targets and to perform high volume marker analysis, as recently demonstrated.[39-41]

Another promising, high volume marker technology is based on "representational difference analysis" or RDA.[42,43] RDA represents a sophisticated DNA subtraction technique to identify unique DNA fragments among (genomic) DNA pools.

The basic idea of the procedure is to hybridize restriction fragments from the DNA pool to be analyzed ('tester'), in the presence of excess DNA from the other pool ('driver'). Only the 5' ends of tester DNA fragments contain an overhanging primer sequence that allows amplification by PCR after filling-in of the single-stranded region. DNA fragments present only in the tester DNA sample can self-anneal efficiently during the hybridization whereas other DNA-fragments hybridize preferentially to the excess of homologous driver DNA. Subsequently, those double-stranded tester-tester DNA molecules are filled in and efficiently PCR-amplified. An enrichment of unique fragments in the tester sample is achieved by this means. Several rounds of this procedure allow the identification and cloning of PCR products such as small DNA fragments derived from RFLPs, by subtracting just a subset of the DNA-fragments generated by restriction digest of parental DNA pools. An additional benefit of this procedure is that the cloned difference products represent RFLP markers polymorphic for the enzyme used to generate the tester and driver DNA fragments.

12.2.1.2. Marker targeting techniques

High volume marker technologies like RAPD, AFLP and RDA offer an exciting alternative to the classical approximative approach of gene mapping mentioned for conventional RFLP markers by affording the analysis of combined pools of segregants. The bulked segregant analysis (BSA) employs DNA from segregants homozygous for the gene of interest. This technique is referred to as marker targeting.[44]

An impressive demonstration of the potential of this marker targeting technique is given by the genomically directed RDA (GDRDA), the combination of RDA and BSA.[43] In an example employing a mammalian genome, a DNA pool derived from mouse F2 intercross progenies homozygous for the recessive mutant trait was used as the driver DNA fraction. The parental DNA with the wild-type allele was used as the tracer fraction. The subtraction steps as outlined above resulted in the enrichment of markers linked to the mutationally identified locus since only polymorphic fragments in the vicinity of the wild-type gene are uniquely present in the tracer fraction. Preselection of suitable recombinants for the driver DNA sample can further narrow down the chromosomal segment for which markers are amplified. The successful application of GDRDA with a mammalian genome implies that the technique is appropriate to handle the genome complexities of most, if not all plant species. Similarly, the combination of the AFLP technique and BSA resulted in the identification of markers physically linked as close as 20 kb and flanking the target gene in a mapping experiment in tomato.[40] These markers were identified, among a total of 42,000 AFLP loci analyzed. Thus, in contrast to the approaches based on existing genetic maps with limited numbers of markers, the novel targeted mapping methods can provide a tremendous number of closely-linked molecular markers, facilitating the next step of positional cloning—the isolation of chromosomal fragments containing the target gene.

12.2.2. SECOND STEP: PHYSICAL LOCALIZATION OF THE TARGET GENE

Marker targeting techniques as well as classical mapping approaches using high density maps allow the identification of molecular markers linked to the target locus in the range of 0.1-1 cM genetic distance. This genetic distance may still correspond to a physical distance of several hundred kb, reflecting the relationship between physical and genetic complexity of the genome for a given plant species (see chapter 3).[45] A genetic distance of 1 cM equals an average physical distance close to 1500 kb in maize, while in *Arabidopsis* a value of 200 kb is estimated. The relationship between genetic and physical distance, however, varies not only among different genomes of plant species but also within a

genome, and numbers ranging from 70 to 500 kb per cM were reported for a single *Arabidopsis* chromosome.[46]

Ideally, flanking molecular markers identified in step 1 of the positional cloning procedure allow the identification of genomic fragments that already contain the gene of interest. This is referred to as chromosome landing,[44] while the successive isolation of partly overlapping chromosomal fragments is called chromosome walking.[47,48] Genomic libraries of plants with cloned chromosomal fragments of up to several hundred kb in size provide the means to achieve the goal.[49]

12.2.2.1. Systems for cloning large genomic DNA fragments

The advent of megabase DNA cloning methods with yeast artificial chromosomes (YACs) revolutionized the positional cloning technique.[50] This system allows the stable maintenance of large eukaryotic sequences as linear molecules in yeast, and several YAC libraries of plant genomes have been established with insert sizes up to several hundred kb.[51-58] In most of the plant map-based cloning experiments published to date, YAC libraries were used. YAC libraries of the first generation contain a relatively high proportion (up to 30%) of chimeric clones generated either by false ligation or by in vivo recombination between cotransformed DNA fragments.[59-65] The *rad52* mutant of yeast deficient in recombination suppresses the occurrence of chimeric YACs but unfortunately the replication stability of chromosomes is also negatively affected.[62,63] Second-generation plant YAC libraries exhibit a lower level of chimeric clones due to efficiently pre-sizing the DNA fragments, and reducing the DNA concentration in the yeast transformation during library construction.[51,58] As an alternative to the YAC system, bacterial cloning vectors have been developed that tolerate the stable replication in *E. coli* of DNA molecules of up to 100 kb if based on the P1 phage system[66] or 300 kb if based on the F-plasmid.[67] These prokaryotic cloning systems have been employed to generate large insert genomic libraries for some plant species such as *Arabidopsis*,[68,69] rice,[70] and sorghum.[71] Clones of the F-plasmid based BAC libraries are maintained as a single copy in the host cell, and chimeric clones appear to be rare.[71] The cloning capacity of P1 libraries is restricted to approximately 100 kb due to limits set by the in vitro packaging mechanism of the P1 phage.[66] The linear genomic DNA molecules that are packaged into the phage particles are circularized by virtue of cre/lox-mediated recombination in the host cell and are maintained as single copy plasmids. A possible bias in the cloning of DNA fragments by these systems has not been systematically investigated, however, in one example, clones of an *Arabidopsis thaliana* P1 library were successfully assembled to contiguous chromosomal fragments (contig) that represented more than 600 kb of a chromosomal region.[69] Thus, these systems could promise not to suffer from the shortcomings encountered with cosmids[72] such as cloning bias, stability problems and comparatively small fragment sizes of the cloned DNA.

A major advantage of the bacterial cloning system compared to the YAC system is the possibility to purify and isolate the cloned DNA with standard plasmid isolation protocols. The isolation of YAC DNA generally requires time-consuming preparative runs using pulsed field gel electrophoresis to recover the DNA from the agarose gel. The isolated DNA can either serve to generate subclones for target gene identification in step 3 of the positional cloning procedure, or to identify additional markers for mapping as well as probes for chromosome walking.

12.2.2.2. Integration of genetic and physical maps

The landmarks for gene isolation are the two molecular markers which reveal the minimal number of recombinational events proximal and distal to the target gene. The identification of genomic clones or contigs containing such a landmark, and the mutationally identified target locus forms the second step of the positional cloning procedure. The establishment of a contig is achieved by repeated isolation of DNA fragments using end probes of the cloned genomic DNA and rescreening the YAC, BAC, or P1 libraries. The isolation of such end probes from YAC (or other megabase) inserts is simple, and either based on inverse PCR-amplification[54] or on plasmid rescue from total yeast DNA.[73,74] The currently available plant YAC libraries were constructed with standard vectors that allow the recovery of only one insert-end by plasmid rescue. New vectors overcome this limitation by providing two different selectable markers and origins of replication suitable for rescue in *E. coli* on both vector arms.[75]

The existence of chimeric YACs necessitates verification of the integrity of the cloned genomic DNA. Genetic mapping of the isolated end probes relative to the identified recombination events is a recommended method, provided a RFLP can be detected with this probe (in our hands, about a 50% chance in *Arabidopsis*). The chances to detect a RFLP in the two parental lines are considerably improved (80-90% chance in *Arabidopsis*) by screening genomic cosmid or lambda libraries for larger chromosomal fragments corresponding to the end probe. Another analysis for integrity involves cross-hybridization of end probes, to independent YAC clones. Both procedures are helpful to direct the chromosome walk and to align the genomic fragments. In our hands, alignment of the chromosomal fragments is simply achieved by DNA blot analysis of pulse field-separated YAC chromosomes. Isolated YAC DNA is radiolabeled and used to determine the degree of overlap with other YAC chromosomes by quantitative analysis of the cross-hybridization signals which are normalized by a vector-specific probe. The size of the YAC chromosomes and the normalized signal (in relation to the signal obtained of the YAC from which the probe was derived) elegantly affords the physical alignment of the chromosomal fragments.

In a chromosome walk, genetic mapping of new RFLPs or other molecular markers derived from the established genomic contig is essential to assess the position of the markers relative to the starting point of the walk and to the target gene. Cloning of the gene has been accomplished when a contig covers two molecular markers that are localized by recombinational breakpoints proximal and distal from the target, respectively.

The genomic region flanked by these markers is subsequently referred to as the target region. In some map-based cloning experiments the target region was further focused in order to minimize the amount of work required to identify the gene by selecting more recombinants among the segregants and thereby increasing the mapping resolution.[11,12,76]

To identify the gene of interest on the target region, subfragments need to be analyzed either functionally, by gene transfer and complementation experiments, or indirectly, by the candidate gene approach.

12.2.3. THIRD STEP: GENE IDENTIFICATION

12.2.3.1. Functional gene identification

Unequivocal proof that a target gene has been isolated comes from the combination of two pieces of information: (1) Identification of a mutant transcript which absolutely co-segregates with the phenotype, and (2) The ability of the wild-type transcript to confer the dominant phenotype to transgenic plants homozygous for the mutant allele. Indeed, in most map-based cloning experiments of plant genes these stringent criteria have been met. This analysis requires the generation of transgenic plant material. Plants recalcitrant to DNA transformation and regeneration necessitate resorting to alternatives such as the candidate gene approach, of identifying several independently-derived mutant alleles in the same transcript. Several gene transfer technologies are available for plant transformation.[77,78] Indirect gene transfer, employing binary vectors and *Agrobacterium* is recom-

mended for complementation analysis due to the efficient transfer and integration of DNA fragments in the size range of 10-30 kb.[79]

Efficient subcloning of the target region is achieved by using binary cosmid vectors that provide a reliable selection for the size of the insert DNA. Binary cosmid vectors such as pCLD04513 and pOCA1822 are derived from the broad host range vector RK29080 and contain its complete origin of replication which supports stable replication of large fragments of genomic DNA in *E. coli* as well as in *Agrobacterium tumefaciens*. In addition, both vectors provide the plant selectable marker, a neomycin-phosphotransferase gene, next to the left border of the T-DNA. The unidirectional transfer and integration of the T-DNA strand terminates at the left border and, thus, the presence of the resistance marker in the transgenic plant material implies a complete transfer of the subcloned DNA.

We recently introduced the binary cosmid vector pBIC20 (Fig. 12.1). This vector is also based on the pRK290 replicon with two plant-expressed markers flanking the cloning site. A resistance marker at the right T-DNA border provides a means to assess successful plant transformation. Activity of the glucuronidase gene at the left T-DNA border provides a visual indicator of complete T-DNA transfer and integration of the genomic DNA into the plant genome. The T-DNA topology of pBIC20 turned out to be very useful due to the observed large variations in the efficacy of right and left T-DNA border integration into the *Arabidopsis* genome. The presence of both markers among independently transformed plant material varied from 5 to 95% between individual cosmids with different subcloned genomic fragments. This variation was not correlated with the size of the T-DNA, but seemed to be caused by sequences within the genomic insert. Premature interruption of the T-DNA during the process of generation, transfer or integration could be responsible for the observed phenomenon. pBIC20

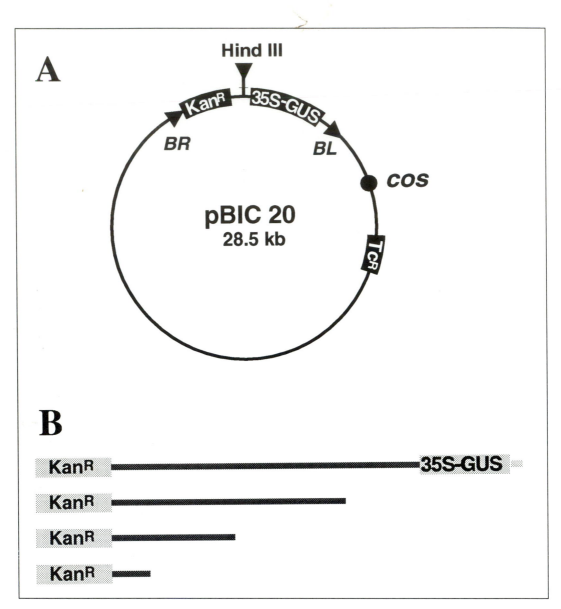

Fig. 12.1. Schematic structure of the binary cosmid vector pBIC20 and its T-DNA.
(A) The topology of the pBIC20 vector. Genomic fragments can be cloned into the unique HindIII site by in vitro packaging of fragments of 8 to 25 kb in size into phage heads using the cos-sequences (cos). The 28.5 kb vector provides tetracycline resistance in E. coli and A. tumefaciens. BR and BL indicate the right and left border sequences of the T-DNA, respectively. The T-DNA confers the expression of two markers in the plant tissue the neomycinphosphotransferase (Kan^R) and the (-glucuronidase under the control of the cauliflower mosaic virus 35S promoter (35S-GUS).
(B) Presentation of T-DNA structures integrated into the plant genome. The genomic DNA cloned between the two plant-expressed markers is indicated by the black line. The complete transfer and integration of the T-DNA into the plant genomes is schematically presented in the upper row, while below, several truncated versions are shown. The truncated T-DNA molecules providing only the resistance marker but not the glucuronidase marker can help to pinpoint the target gene within the cloned DNA in complementation experiments.

permits the recovery and selection of transgenic plants with truncated T-DNA insertions—an important feature in cases where the genomic insert prevents complete T-DNA integration, as well as in cases where truncated T-DNA integration events help to localize the target gene on the T-DNA.

The observation of incomplete transfer and integration of T-DNA needs to be further studied in view of the exciting demonstration that Agrobacterium is able to mediate the mobilization and integration of T-DNA into the plant genome of molecules of up to 200 kb in length, albeit with low efficiency and sometimes associated with structural rearrangements.[81] The modification of P1, BAC, or YAC vectors into a binary plant transformation vehicle should not pose a problem and, in fact, the development of such a binary BAC vector has been reported at a recent meeting (Hamilton CM. Evaluation of a BIBAC library vector designed for transfer of high molecular weight DNA into plants. Poster Abstract #271 presented at Plant Genome IV, San Diego, 1996). If binary vectors turn out to be reliable tools for large DNA complementation experiments, conventional subcloning of the target region could be obsolete in the future.

Currently, however, subcloning of the target region and subsequent plant transformation is required. In most cases, the subcloned DNA fragments of the target region are assembled into a highly redundant (overlapping) contig to ensure the presence of a physically intact target gene with all sequences required for its expression in planta. The method of choice to generate this contig is the direct subcloning of the yeast artificial chromosome(s) or other large DNA fragments comprising the target region into binary cosmid vectors. The contig is assembled by cross-hybridization analyses of available probes that have been isolated in the course of physical mapping and by restriction mapping of the identified cosmid clones. Alternative subcloning strategies consist of screening a genomic library of the plant or of

the YAC-containing yeast with either the purified YAC DNA or probes localized on the YAC. It is quite tedious to establish highly redundant subclone contigs with this kind of approach.

In principle, plant transformation could be performed as a shotgun experiment, particularly if a straightforward selection scheme can be applied for functional identification of the target gene.

Once complementation experiments allow the mutationally identified locus to be tracked down to a few kb of genomic sequence, the gene transcript can be identified by screening of cDNA libraries. The identification of transcripts could pose a problem in case of extremely low abundance of the mRNA. In such a case,[15] sequencing of the genomic region in combination with exon prediction analysis[82] as well as PCR-based methods for transcript detection have been successfully applied.

12.2.3.2. Candidate gene approach

Low abundance of transcript levels could also represent a serious challenge to the candidate gene approach. This strategy involves the identification of all transcripts encoded by the target region. A simple and frequently successful[2,11] approach is to screen cDNA libraries with gel-purified YAC DNA. More sensitive methods have been established in order to ensure the identification of all transcribed regions. Such methods are based on direct selection by immobilization of the genomic DNA and capturing of transcripts or cDNAs via specific hybridization.[83] PCR-mediated amplification of the bound DNA affords the high sensitivity of the methods which have been incorporated into mammalian map-based cloning procedures.[84] Such a technique has also been applied to identify 15 transcribed sequences on a 130 kb YAC with *Arabidopsis* DNA.[85] Subsequently, the identified transcripts are used in RNA blot experiments to examine whether the mutation affects abundance or size of the mRNAs. If not, the search for the mutated gene employs high-throughput sequencing of all candidate gene transcripts from wild type and the mutants or, conceivably, mismatch repair detection methods.[86] The identification of mutations in genes of several independently isolated mutant alleles represents indirect evidence that the gene of interest has been identified. Several independent mutant alleles are frequently not available, especially if the gene of interest confers a trait such as pathogen resistance and has been introgressed into crop plants.

12.3. POSITIONAL CLONING OF THE *ABI1* LOCUS OF *ARABIDOPSIS*

In the previous chapter we attempted to outline the principles and general methods currently available to perform positional cloning. Now, we would like to illustrate such an endeavor by providing the details and results of our map-based cloning of the ABI1 gene which was subsequently shown to encode a protein phosphatase.[9,12]

The *abi1* mutant was isolated by an elegant screening program involving M2 seed material of ethylmethanesulfonic acid-treated *Arabidopsis* seeds.[87] The mutant locus is characterized by conferring a dominant pleiotropic insensitivity towards the plant hormone abscisic acid, and the *ABI1* locus has been integrated into the genetic map of this organism.[88]

Based on the map position of the *ABI1* locus, three RFLP markers[89,90] were selected to fine-map the target gene. Establishment of the genetic fine structure map involved scoring the genotype of more than 300 F2 plants segregating for both ABI1 and the molecular markers and is depicted in Figure 12.2A. The F2 mapping population was obtained by harvesting the seeds of the F1 intercross plants generated between the homozygous mutant of the ecotype Landsberg with wild-type plants of ecotype Niederzenz. DNA samples of the F2 intercross progeny or of pooled F3 segregants from single F2 plants were genotyped with the three RFLP markers polymorphic in the parental lines. The genotype of the *ABI1* locus was scored among the mapping population by testing for ABA-insensitive germination and seedling growth of F3 seeds. The analysis revealed a location of *ABI1* genetically closely linked to marker B, in the interval of the markers B and C. This localization was based on the frequency of recombination events detected among the four markers *ABI1* and RFLP A, B, and C in the mapping population (Fig. 12.2A). The molecular marker B was used to screen genomic YAC libraries of *Arabidopsis*, including a YAC library generated from the *abi1* mutant. The identified YACs were physically aligned by cross-hybridization experiments performed with the YACs and their end probes as shown (Fig. 12.2B). Genetic mapping of the YAC end probe signified as RFLP E confined the location of the *ABI1* gene to a 150 kb target region delimited by the marker B and E. The YACs representing the target region and thus containing the dominant allele were subcloned into the binary cosmid vector pBIC20 and assembled into the contig shown in Figure 12.2C. For instance, a 130 kb YAC was isolated by preparative pulsed field electrophoresis and subcloned into 5000 independent cosmids in a single experiment. The library obtained represents approximately 400 equivalents of the YAC. This level of representation supported the rapid establishment of a highly redundant contig and helped to pinpoint the *ABI1* gene in a single round of transformation experiments. A part of the aligned subcloned fragments that were mobilized via *Agrobacterium* into wild type plant tissue is presented in Figure 12.2D. The alignment of the fragments was based on cross-hybridization of the cosmids with the markers D, G, and H from the target region (Fig. 12.2C) and, particularly, by restriction mapping of the cosmid DNA with the enzyme HindIII that had been used to clone the fragments. Restriction analysis of partly overlapping pBIC20 clones is shown in Figure 12.3.

The identification of the *ABI1* gene involved recovery of transgenic seeds from regenerated plants and testing for the presence of the mutant gene by scoring for

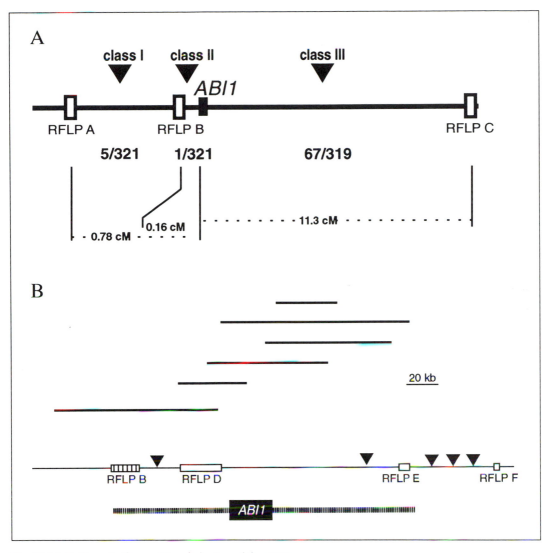

Fig. 12.2A-D. Steps in the positional cloning of the ABI1 gene.

(A) Genetic map showing the position of the ABI1 locus (filled rectangle) relative to the three RFLP markers A, B, and C (open rectangles). Three classes of recombinants are distinguishable: class I revealing a chromosomal breakpoint between ABI1 and RFLP A, class II between ABI1 and marker B, and class III between ABI1 and C. In class I and II five and one recombination event(s), respectively, were identified among 321 individuals of the F2 intercross progeny scored while in class III 67 were detected among 319 F2 plants tested. The observed frequencies of recombination between the markers and ABI1 reflect the genetic distances given below in cM map units.[28]

(B) Alignment of yeast artificial chromosomes comprising the target region. Genomic fragments of the mutant plant cloned in YACs are shown as thick black lines. The YACs were identified by probing the library with the proximal marker B the starting point of the chromosome walk and, subsequently, marker D as well as an additional marker not shown, both derived from YAC end probes. The end probe E revealed a RFLP that mapped distal to the target gene. Thus, marker B and E delimit the target region containing the mutant abi1 gene indicated by a hatched line. The approximate position of the chromosomal breakpoints is signified by filled triangles and is given relative to the markers on the schematic chromosome (thin line). Reprinted with permission from Meyer K, Leube MP, Grill E. A protein phophatase 2C involved in ABA signal transduction in Arabidopsis thaliana. Science 264: 1452-1455.

(C and D continued on next page)

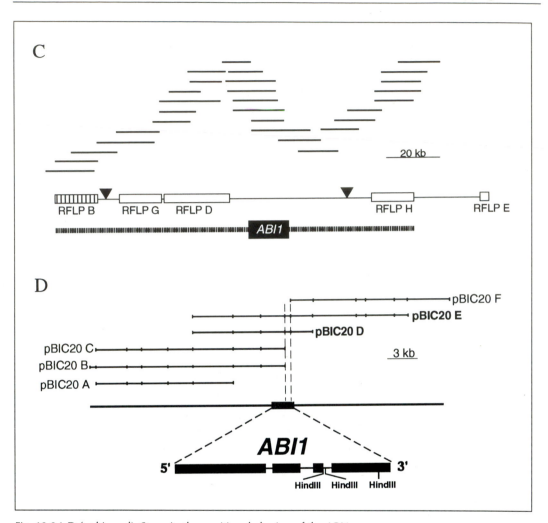

Fig. 12.2A-D (coltinued). Steps in the positional cloning of the ABI1 gene.
(C) The contig of the subcloned target region. The target region introduced already in B was recovered as subcloned fragments of the YACs in the binary cosmid vector pBIC20. Cloned DNA fragments (short lines) were assembled to the contig by cross-hybridization analysis and restriction mapping. The markers B, D, H, and I served as molecular probes.
(D) Identification of the ABI1 gene. A fraction of the contig shown in C consisting of DNA fragments derived from the mutant was introduced by Agrobacterium-mediated transformation into wild type Arabidopsis. Analysis of transgenic seed material from regenerated plants documented that the genomic fragments of pBIC20 D and pBIC20 E conferred the dominant trait of the mutant gene while the other fragments failed to mediate ABA-insensitivity. The comparison of the physically aligned fragments (ticks indicate restriction sites for HindIII) revealed the presence of a fragment in pBIC20 D and pBIC20 E that is crucial for the gene function and is marked by vertical hatched lines. Subsequently, the cDNA of the transcript was identified as well as the position of the exons of the ABI1 gene (shown as thick black lines). Reprinted with permission from Meyer K, Leube MP, Grill E. A protein phophatase 2C involved in ABA signal transduction in Arabidopsis thaliana. Science 264: 1452-1455.

ABA-insensitive germination. Two fragments of the contig, pBIC20D and pBIC20E, conferred ABA-insensitivity while other fragments lacking a small DNA fragment failed. Thus, the *ABI1* gene was unequivocally identified and DNA analysis of mutant and wild-type sequences documented that a single point mutation in the *ABI1* gene results in the dominant mutant phenotype.

12.4. OUTLOOK

The availability of abundant molecular markers and genomic libraries of large DNA fragments form the basis of positional cloning. While in a few plant species, such as *Arabidopsis*, tomato, rice, maize and some nitrogen-fixing plants these basic tools are already available or in the progress of being established, many other plant species with interesting and valuable traits will not attract the manpower and resources needed to provide these tools. However, methods such as marker targeting techniques, techniques for construction of suitable genomic libraries and high-throughput DNA sequencing seem to offer the means required to apply positional cloning to all those species provided they are amenable to classical genetics.

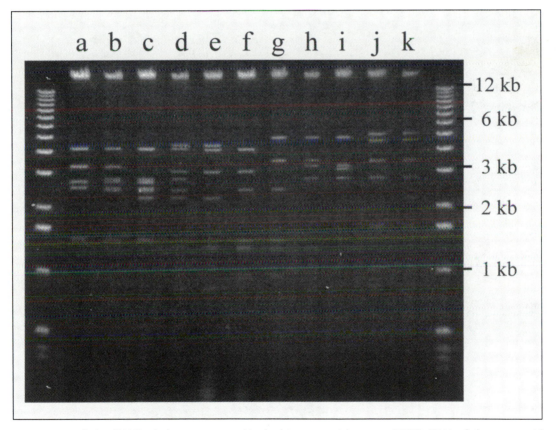

Fig. 12.3. Analysis of YAC subclones generated in the binary cosmid vector pBIC20. DNA of eleven cosmid clones (a-k) containing fragments of the ABI1 target region from a 130 kb YAC was digested with HindIII restriction enzyme and analyzed by agarose gel electrophoresis. The restriction digest liberates the HindIII-fragments of the cloned DNA and the 28.5 kb vector. The presence of identical HindIII restriction fragments in physically overlapping clones (verified by cross-hybridization) allows the rapid assembly of the contig. The binary cosmids analyzed in this experiment constitute a contig encompassing 85 kb. The fragment sizes of standard DNA molecules are given.

ACKNOWLEDGMENT

This contribution was supported by the Swiss National Science Foundation.

REFERENCES

1. Ahmad MGM and Cashmore AR. HY4 gene of Arabidopsis thaliana encodes a protein with characteristics of a blue-light photoreceptor. Nature 1993; 715-717.

2. Arondel V, Lemieux B, Hwang I, Gibson S, Goodman HM, and Somerville CR. Map-based cloning of a gene controlling omega-3 fatty acid desaturation in Arabidopsis. Science 1992; 258:1353-1355.

3. Bent AF, Kunkel BN, Dahlbeck D, Brown KL, Schmidt R, Giraudat J, Leung J and Staskawicz BJ. RPS2 of Arabidopsis thaliana: A leucine-rich repeat class of plant disease resistance genes. Science 1994; 265:1856-1860.

4. Chang C, Kwok SF, Bleecker AB, and Meyerowitz EM. Arabidopsis ethylene response gene ETR1: Similarity of product to two-component regulators. Science 1993; 262:539-544.

5. Giraudat J, Hauge BM, Valon C, Smalle J, Parcy F and Goodman HM. Isolation of the Arabidopsis ABI3 gene by positional cloning. Plant Cell 1992; 4:1251-1261.

6. Grant MR, Godiard L, Straube E, Ashfield T, Lewald J, Sattler A, Innes RW and Dangl JL. Structure of the Arabidopsis RPM1 gene enabling dual specificity disease resistance. Science 1995; 269:843-846.

7. Jones DA, Thomas CM, Hammond-Kosack KE, Balint-Kurti PJ and Jones JDG. Isolation of the tomato Cf-9 gene for resistance to Cladosporium fulvum by transposon tagging. Science 1994; 266:789-793.

8. Kieber JJ, Rothenberg M, Roman G, Feldmann KA and Ecker, JR. CTR1, a negative regulator of the ethylene response pathway in Arabidopsis, encodes a member of the raf family of protein kinases. Cell 1993; 72:427-441.

9. Leung J, Bouvier-Durand M, Morris PC, Guerrier D, Chefdor F, Giraudat J. Arabidopsis ABA response gene ABI1: features of a calcium-modulated protein phosphatase. Science 1994; 264:1448-1452

10. Leyser HMO, Lincoln CA, Timpte C, Lammer D, Turner J and Estelle M. Arabidopsis auxin-resistance gene AXR1 encodes a protein related to ubiquitin-activating enzyme E1. Nature 1993; 364: 161-164.

11. Martin GB, Brommonschenkel SH, Chunwongse J, Frary A, Ganal MW, Spivey R, Wu T, Earle ED and Tanksley SD. Map-based cloning of a protein kinase gene conferring disease resistance in tomato. Science 1993; 262:1432-1436

12. Meyer K, Leube MP and Grill E. A protein phosphatase 2C involved in ABA signal transduction in Arabidopsis thaliana. Science 1994; 264:1452-1455

13. Mindrinos M, Katagiri F, Yu GL and Ausubel FM. The A. thaliana disease resistance gene RPS2 encodes a protein containing a nucleotide-binding site and leucine-rich repeats. Cell 1994; 78:1089-1099.

14. Pepper A, Delaney T, Washburn T, Poole D and Chory J. DET1, a negative regulator of light mediated development and gene expression in Arabidopsis, encodes a novel nuclear-localized protein. Cell 1994; 78:109-116.

15. Putterill J, Robson F, Lee K, Simon R and Coupland G. The CONSTANS gene of Arabidopsis promotes flowering and encodes a protein showing similarities to zinc finger transcription factors. Cell 1995; 80:847-857.

16. Song W-Y, Wang G-L, Chen H-S, Kim H-S, Pi L-Y, Holsten T, Gardner J, Wang B, Zhai W-X, Zhu L-H, Fauquet C, Ronald P. A receptor kinase like protein encoded by the rice disease resistance gene Xa21. Science 1995; 270:1804-1807.

17. Vollbrecht E, Veit B, Sinha N and Hake S. The developmental gene Knotted-1 is a member of a maize homeobox gene family. Nature 1991; 350:241-243.

18. Weigel D, Alvarez J, Smyth DR, Yanofsky MF, Meyerowitz E. M. Leafy controls floral meristem identity in Arabidopsis. Cell 1992; 69:843-859.

19. Walbot, V. Strategies for mutagenesis and gene cloning using tranposon tagging and T-DNA inserional mutagenesis. Annu Rev Plant Physiol Plant Mol Biol 1992; 43:49-82.

20. Nussaume L, Harrison K, Klimyuk V,

Martienssen R, Sundaresan V, Jones JDG. Analysis of splice donor and acceptor site function in a transposable gene trap derived from the maize element Activator. Mol Gen Genet 1995; 249:91-101.

21. Beach D, Piper M and Nurse P. Construction of a Schizosaccharomyces pombe gene bank in a yeast shuttle vector and its use to isolate genes by complementation. Mol Gen Genet 1982; 187:326-329.

22. Olszewski NE, Martin FB and Ausubel FM. Specialized binary vector for plant transformation: expression of the Arabidopsis thaliana AHAS gene in Nicotiana tabacum. Nucleic Acids Res 1988; 16:10765-10782.

23. Klee HJ, Hayford MB and Rogers RG. Gene rescue in plants: A model system for ìshotgunî cloning by retransformation. Mol Gen Genet 1987; 210:282-287.

24. Redei GP and Koncz, C. Classical mutagenesis. In: Koncz C, Chua NM, and Schell J. eds. Methods in Arabidopsis Research. Singapore: World Scientific Publishing, 1992:16-82.

25. Michelmore RW, Paran I and Kesseli RV. Identification of markers linked to disease-resistance genes by bulked segregant analysis: a rapid method to detect markers in specific genomic regions by using segregating populations. Proc Natl Acad Sci USA 1991; 88:9828-9832.

26. Williams JGK, Reiter RS, Young RM and Scolnik PA. Genetic mapping of mutations using phenotypic pools and mapped RAPD markers. Nucleic Acids Res. 1993; 21: 2697-2702.

27. Koornneef M and Stam P. Genetic analysis. In: Koncz C, Chua NH, and Schell J. eds. Methods in Arabidopsis Research. Singapore: World Scientific Publishing, 1992:83-99.

28. Stam P. Construction of integrated genetic linkage maps by means of a new computer package: JoinMap Plant Journal 1993; 3:739-744.

29. Liu SC, Kowalski SP, Lan, TH, Feldmann K and Paterson A. Genome wide high resolution mapping by recurrent intermating using Arabidopsis thaliana as a model. Genetics 1996; 142:247-258.

30. Reiter RS, Williams JGK, Feldmann KA, Rafalski JA, Tingey SV and Scolnik PA. Global and local genome mapping in Arabidopsis thaliana by using recombinant inbred lines and random amplified polymorphic DNAs. Proc Natl Acad Sci USA 1992; 89:1477-1481.

31. Paterson AH and Wing RA. Genome mapping in plants. Current opinion in biotechnology. 1993; 4:142-147.

32. Williams JGK, Kubelik AR, Livak KJ, Rafalski JA and Tingey SV. DNA polymorphisms amplified by arbitrary primers are useful as genetic markers. Nucleic Acids Res 1990; 18:6531-6535.

33. Konieczny A and Ausubel FM. A procedure for mapping Arabidopsis mutations using codominant ecotype specific PCR-based markers. Plant J 1993; 4:403-410.

34. Akkaya MS, Bhagwat AA and Cregan PB. Length polymorphisms of simple sequence DNA in soybean. Genetics 1992; 132: 1131-39.

35. Bell CJ and Ecker JR. Assignment of 30 microsatellite loci to the linkage map of Arabidopsis. Genomics 1994; 19:137-44.

36. Vos P, Hogers R, Bleeker M, Reijans M, van de Lee T, Hornes M, Frijters A, Pot J, Peleman J, Kuiper M and Zabeau M. AFLP: a new technique for DNA fingerprinting. Nucleic Acids Res 1995; 23:4407-4414.

37. Lisitsyn N. Representational difference analysis: finding the differences between genomes. TIG 1995; 11:303-307.

38. Martin GB, Williams JGK and Tanksley SD. Rapid identification of markers linked to a Pseudomonas resistance gene in tomato by using random primers and near-isogenic lines. Proc Natl Acad Sci USA 1991; 88:2336-2340.

39. Meksem K, Leister D, Peleman J, Zabeau M, Salamini F and Gebhardt C. A high-resolution map of the vicinity of the R1 locus on chromosome V of potato based on RFLP and AFLP markers. Mol Gen Genet 1995; 249:74-81.

40. Thomas CM, Vos P, Zabeau M, Jones DA, Norcott KA, Chadwick BP and Jones JDG. Identification of amplified restriction fragment polymorphism (AFLP) markers tightly linked to the tomato Cf-9 gene for resistance to Cladosporium fulvum. Plant J 1995; 8:785-794.

41. Becker J, Vos P, Kuiper M, Salamini F and Heun M. Combined mapping of AFLP and RFLP markers in barley. Mol Gen Genet 1995; 249:65-73.

42. Lisitsyn N, Lisitsyn NM and Wigler M. Cloning differences between two complex genomes. Science 1993. 259:946-951.

43. Lisitsyn N, Segre JA, Kusumi K, Lisitsyn NM, Nadeau JH, Frankel WN, Wigler MH, and Lander ES. Direct isolation of polymorphic markers linked to a trait by genetically directed representational difference analysis. Nature Genetics 1994; 6:57-63.

44. Tanksley SD, Ganal MW, and Martin GB. Chromosome landing a paradigm for map-based cloning in plants with large genomes. TIG 1995; 11:63-68.

45. Dean C and Schmidt R. Plant genomes: A current molecular description. Ann Rev Plant. Physiol Plant Mol Biol 1995; 46:395-418.

46. Schmidt R, West J, Love K, Lenehan Z, Lister C, Thompson H, Bouchez D, and Dean C. Physical map and organization of Arabidopsis thaliana chromosome 4. Science 1995; 270:480-483.

47. Meyerowitz E. Arabidopsis thaliana. Ann Rev Gen 1987; 61:93-111.

48. Grill E, Somerville CR. Development of a system for efficient chromosome walking in Arabidopsis; In: Jenkins GI, Schuch W, eds. Molecular Biology of Plant Development. Cambridge: The Company of Biologists, 1991:57-63.

49. Schlessinger D. Yeast artificial chromosomes: tools for mapping and analysis of complex genomes. TIG 1990; 6 (8): 251-258.

50. Burke DT, Carle GT, Olson MV. Cloning of large segment of exogenous DNA into yeast by means of artifical chromosome vectors. Science 1987; 236:806-812.

51. Creusot F, Fouilloux E, Dron M, Lafleuriel J, Picard G, Billault A, Le Paslier D, Cohen D, Chaboute M, Durr A, Fleck J, Gigot C, Camilleri C, Bellini C, Caboche M, and Bouchez D. The CIC library: a large insert YAC library for genome mapping in Arabidopsis thaliana. Plant J. 1995; 8:763-770.

52. Umehara Y, Inagaki A, Tanoue H, Yasukochi T, Nagamura Y, Saji S, Otsuki Y, Fujimura T, Kurata N, and Minobe Y. Construction and characterization of a rice YAC library for physical mapping. Plant Breeding, 1996; in press

53. Ecker, JR. PFGE and YAC analysis of the Arabidopsis genome. Methods 1990; 1:186-194.

54. Grill E, Somerville CR. Construction and characterization of a yeast artificial chromosome library of Arabidopsis which is suitable for chromosome walking. Mol Gen Genet 1991; 226:484-490

55. Martin GB, Ganal MW, and Tanksley SD. Construction of a yeast artifcial chromosome library of tomato and identification of cloned segments of genomic DNA linked to two disease resistance loci. Mol Gen Genet 1992; 233:25-32.

56. Ward ER and Jen GC. Isolation of single-copy-sequence clones from a yeast artificial chromosome library of randomly-sheared Arabidopsis thaliana DNA. Plant Mol Biol 1990; 14:561-568.

57. Edwards KJ, Thompson H, Edwards D, de Saizieu A, Sparks C, Thompson JA, Greenland AJ, Eyers M, and Schuch W. Construction and characterization of a yeast artificial chromosome library containing three haploid maize genome equivalents. Plant Mol Biol 1992; 19:299-308.

58. Bonnema G, Hontelez J, Verkerk R, Zhang YQ, van Dealen, RV, van Kammen, A, and Zabel, P. An improved method of partially digesting plant megabase DNA suitable for YAC cloning: application to the construction of a 5.5 genome equivalent YAC library of tomato. Plant J 1996; 9:125-133.

59. Schmidt R, Putterill J, West J, Cnops G, Robson F, Coupland G and Dean C. Analysis of clones carrying repeated DNA sequences in two YAC libraries of Arabidopsis thaliana DNA. Plant J 1994; 5:735-744.

60. Bradshaw SM, Bollekens JA, and Ruddle FH. New vector for recombination-based cloning of large DNA fragments from yeast artificial chromosomes. Nucleic Acids Res 1995; 23:4850-4856.

61. Larionov V, Kouprina N, Graves J, Chen X-N, Korenberg JR, and Resnik M. Specifc

cloning of human DNA as yeast artificial chromosomes by transformation-associated recombination. Proc Natl Acad Sci USA 1996; 93:491-496.

62. Dunford R, Vilageliu L, and Moore G. Stabilization of a yeast artificial chromosome containing plant DNA using a recombination-deficient host. Plant Molecular Biology 1993; 21:1187-1189.

63. Kouprina N, Eldarov M, Moyzis R, Resnik M, and Larionov V. A model system to assess the intergrity of mammalian YACs during transformation and propagation in yeast. Genomics 1994; 21:7-17.

64. Green ED, Riethman HC, Dutchik JE, and Olson MV. Detection and characterization of chimeric yeast artificial-chromosome clones. Genomics 1991; 11:658-669.

65. Wada M, Abe K, Okumura K, Taguchi H, Kohno K, Imamoto F, Schlessinger D, and Kuwano M. Chimeric YACs were generated at unreduced rates in conditions that suppress coligation. Nucleic Acids Res 1994; 22:1651-1654.

66. Sternberg NL. Cloning high molecular weight DNA fragments by the bacteriophage P1 system. TIG 1992; 8:11-16.

67. Shizuya H, Birren B, Kim U, Mancino V, Slepak T, Tachiiri Y, and Simon M. Cloning and stable maintenance of 300-kilobase-pair fragments of human DNA in Escherichia coli using an F-factor-based vector. Proc Natl Acad Sci USA 1992; 89:8794-8797.

68. Choi S, Creelman RA, Mullet JE, and Wing R. Construction and characterization of a bacterial artificial chromosome library of Arabidopsis thaliana. Weeds World 1995; 2(I).

69. Liu Y, Mitsukawa N, Vazquez -Tello A, and Whittier RF. Generation of a high quality P1 library of Arabidopsis suitable for chromosome walking. Plant J 1995; 7:351-358.

70. Wang G, Holsten TE, Song W, Wang H, and Ronald PC. Construction of a rice bacterial artificial chromosome library and identification of clones linked to the Xa-21 disease resistance locus. Plant J 1995; 7:525-533.

71. Woo S, Jiang J, Gill BS, Paterson AH, and Wing R. Construction and characterization of a bacterial artificial chromosome library of Sorghum bicolor. Nucleic Acids Res 1994; 22(22):4922-4931.

72. Coulson A, Kozono Y, Lutterbach B, Shownkeen R, Sulston J, and Waterston R. YACs and the C. elegans genome. BioEssays 1991; 13:413-417.

73. Gibson SI, and Somerville CR. Chromosome walking in Arabidopsis thaliana using yeast artificial chromosomes; In: Methods in Arabidopsis Research. Koncz C, Chua, NH and Schell, J eds. Singapore: World Scientific Publishing 1992:119-143.

74. Schmidt R and Dean C. Physical mapping of the Arabidopsis thaliana genome. In: Davies KE and Shirley MT, eds. Genome Analysis, Vol.4: Strategies for Physical Mapping. Cold Spring Harbor Laboratory Press 1992:71-98.

75. Traver CN, Klapholz S, Hyman RW, and Davis, RW. Rapid screening of a human genomic library in yeast artificial chromosomes for single copy sequences. Proc Natl Acad Sci USA 1989; 86:5898-5902.

76. Putterill J, Robson F, Lee K, and Coupland G. Chromosome walking with YAC clones in Arabidopsis: isolation of 1700 kb of continous DNA on chromosome 5, including a 300 kb region containing the flowering time gene CO. Mol Gen Genet 1993; 239:145-157.

77. Potrykus I. Gene transfer to plants: Assessment of published approaches and results. Ann Rev Plant Physiol Plant Mol Biol 1991; 41:205-225.

78. Walden R and Wingender R. Gene-transfer and plant-regeneration techniques. TiBtech 1995; 13:324-331.

79. Citovsky V, and Zambrisky P. Transport of nucleic acids through membrane channels: snaking through small holes. Ann Rev Microbiol 1993; 47:176-197.

80. Schmidhauser T and Helinski D. Regions of broad host range plasmid RK2 involved in replication and stable maintenance in nine species of gram-negative bacteria. J Bacteriol 1985; 164:446-455.

81. Miranda A, Janssen G, Hodges L, Peralta E, and Ream W. Agrobacterium tumefaciens transfers extremely long T-DNAs by

a unidirectional mechanism. J Bacteriol 1992; 174:2288-2297.

82. Xu Y, Mural R, Shah M, and Uberbacher E. Recognizing exons in genomic sequence using GRAIL II. In: Setlow JK, ed. Genetic Engineering. New York: Plenum Press, 1994; 16:241-255.

83. Gardiner K and Mural RJ. Getting the message: identifying transcribed sequences. TIG 1995; 11:77-79.

84. Lovett M. Fishing for complements: finding genes by direct selection. TIG 1994; 10:352-357.

85. Hayasida N, Sumi Y, Wada T, Handa H, and Shinozaki K. Construction of a cDNA library for a specific region of a chromosome using a novel cDNA selection method utilizing latex particles. Gene 1995; 165:155-161.

86. Faham M and Cox DR. A novel in vivo method to detect DNA sequence variation. Genome research 1995; 5:474-482.

87. Koornneef M, Reuling G, and Karssen CM. The isolation and characterization of abscisic acid-insensitive mutants of Arabidopsis thaliana. Physiol. Plant. 1984; 61:377-383.

88. Koornneef M and Hanhart CJ Arabidopsis Inf Serv 1984; 21:5.

89. Chang C, Bowman JL, DeJohn AW, Lander ES, and Meyerowitz EM. Restriction fragment length polymorphism linkage map for Arabidopsis thaliana. Proc Natl Acad Sci USA 1988; 85:6856-6860.

90. Nam H-G, Giraudat J, den Boer B, Moonan F, Loos WDB, Hauge BM, and Goodman HM. Restriction fragment length polymorphism linkage map of Arabidopsis thaliana. Plant Cell 1989; 1:699-705.

CHAPTER 13

CLONING GENES BY INSERTION MUTAGENESIS

Matthew A. Jenks and Kenneth A. Feldmann

13.1. INSERTION MUTAGENESIS: INTRODUCTION

Mutants have played an important historic role in agriculture. For example, many important horticultural plants such as apples and roses arose as spontaneous mutations (or sports) that have been propagated asexually. To increase the amount of variation, chemical and physical mutagens have been used to create a wide variety of unique plant mutants.[1-3] Many of these mutants have attracted the attention of plant molecular biologists as a means to identify and clone desirable genes.

Perhaps the most popular mutagens in use by plant molecular biologists for gene cloning have been the insertion mutagens, transfer-DNAs (T-DNAs) and transposable elements (transposons). The advantage of these mutagens is that they both disrupt the gene, and also serve as vehicles for recovery of the plant genic DNA. This process, known as gene-tagging, provides a means to examine the biochemical and developmental consequences of mutations in the gene of interest, as well as a relatively facile means for isolating the affected gene. This approach has several advantages over other gene cloning strategies which require information about the protein or location of the gene in the genome.

In this chapter, we will describe the two types of insertion mutants, discuss protocols for cloning genes from identified mutants and give examples of cloned genes from each system.

13.2. INSERTION MUTAGENESIS USING T-DNAs

The transfer DNA, or T-DNA, of *Agrobacterium tumefaciens* is a defined segment of the tumor inducing plasmid, delimited by 25 base pair border repeats, which, upon infection, is transferred to a susceptible plant cell where it can integrate into the plant genome (Fig. 13.1).[4] Upon integration into the plant genome, genes contained within the T-DNA will usually be expressed. Depending on where integration occurs (e.g.,

Genome Mapping in Plants, edited by Andrew H. Paterson. © 1996 R.G. Landes Company.

into a gene or into non-coding DNA), a mutation may be induced, resulting in the cosegregation of the introduced gene(s) in the T-DNA and the mutation.

T-DNA insertion mutagenesis has been most successfully employed in *Arabidopsis* where more than 20,000 transformants have been reported and more than two dozen genes have been cloned. Thousands of transformants have also been generated in *Nicotiana*, *Petunia* and *Lycopersicon*, but the intent was to study the expression of foreign genes, rather than to do insertion mutagenesis. Still, in light of the findings by Koncz et al (1989) that 30% of the T-DNAs insert into regions which are potentially transcriptionally active in *Nicotiana*,[5] it is interesting that no T-DNA tagged genes have been cloned from this genus. As *A. thaliana* is the only species where T-DNA tagging has resulted in a large number of cloned genes, it will be discussed in detail below.

Several methods have been developed for introducing T-DNA into *Arabidopsis*. These include various tissue culture and whole plant techniques.[5-9] However, most of the tissue culture-based transformation protocols developed for *Arabidopsis* were not directed toward insertion mutagenesis.[8,9] As a result, these populations contain many somaclonal variants, in addition to insertion mutants, and consequently many mu-

tants are not tagged with inserted DNA. In spite of this shortcoming, thousands of transformants have been generated with various tissue culture-based procedures, and several genes have been isolated from these populations (e.g., *CH42*[10] and *AP2*[11]). Clearly, many more genes will be isolated from tissue culture-derived populations.[12] As the vast majority of T-DNA tagged genes have been isolated from populations of transformants generated with whole plant transformation protocols, these procedures will be described in more detail.

13.2.1. WHOLE PLANT TRANSFORMATION

Feldmann and Marks[13] were the first to develop a protocol for plant transformation in *Arabidopsis* that avoided the tissue culture step (for reviews see refs. 3, 14 and 15). Briefly, wild-type seeds (T_1) were incubated with *Agrobacterium* (containing a marker for kanamycin resistance in the T-DNA) and the infected plants grown to maturity (Fig. 13.2). The resulting progeny (T_2) were screened for Kan[R] seedlings. These Kan[R] seedlings were grown to maturity and the progeny (T_3) were collected from single plants and numbered chronologically. More than 14,000 independent transformants have been harvested in this manner.[3,13,16] The average number of functional T-DNA inserts, as assayed on kana-

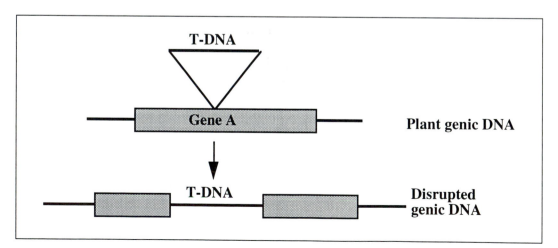

Fig. 13.1. Gene disruption by T-DNA insertion.

mycin-containing medium is 1.5. Of the lines tested, 57% segregated for one insert (3 Kan^R:1 Kan^S), 25% for two linked (3:1 < x < 15:1) or unlinked inserts (15:1), 10% for 3 or more linked or unlinked inserts (x ≥ 63:1), and 9% segregated < 3:1 (Kan^R:Kan^S; exceptional lines). These exceptional lines, in subsequent generations, were found to be segregating for 1-3 independent inserts (unpublished results). Thus in the population of 14,000 transformants there are an estimated 21,000 T-DNA inserts (discussed in section on saturation mutagenesis, below). Southern analyses have shown that these inserts are concatamers of T-DNAs in direct and inverted repeats. Numerous types of rearrangements have also been observed in the individual T-DNAs.[3,13,14]

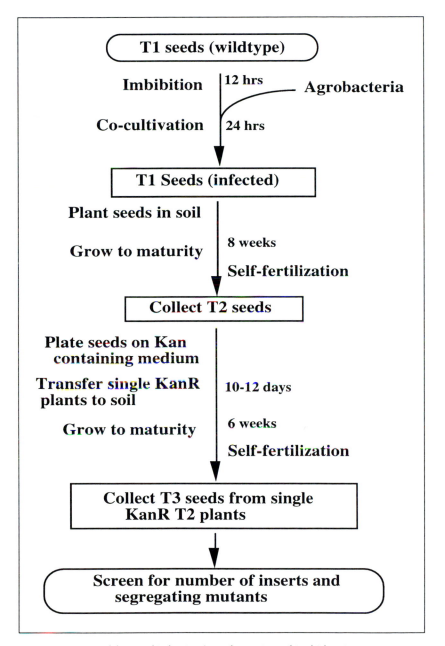

Fig. 13.2. Protocol for seed infection/transformation of Arabidopsis.

These 14,000 segregating families have been screened for visible alterations in phenotype under two environmental regimes, on soil in the greenhouse and in vertically-oriented agar plates.[3] The agar screen was used primarily to identify mutants in root and early seedling development, and pigmentation defects. The greenhouse screen was employed to find mutants in seedling, mature plant, and embryo development. The mutants were placed into one of eight classes: seedling-lethal, size variant, embryo-defective, reduced fertility, pigment, dramatic, physiological and other (Table 13.1). Interestingly, Koncz et al[12] screened 1,340 tissue culture-derived transformants under similar conditions such that a comparison of the frequency of different types of mutants can be made (see section on mutation spectrum).

Another whole plant transformation technique designed both for the introduction of new constructs as well as insertion mutagenesis is the in planta transformation protocol of Chang et al.[17] In this system, young inflorescences are excised near the base of soil-grown plants and the wounded surfaces of the plant are inoculated with *Agrobacterium.* After the secondary inflorescences arise they are again excised near the base and *Agrobacterium* is reapplied. The seeds are collected at maturity from the newly-arising inflorescences and tested on the appropriate selective medium.

With this technique, transformants have been generated utilizing both C58 and LBA4404 strains of *Agrobacterium.*[18] This protocol is being used successfully by a number of labs. R. Whittier has isolated 400 independent lines utilizing this technique and has identified several visible alterations in phenotype (R. Whittier, personal communication). However, this technique is probably not going to be sufficiently useful for insertion mutagenesis. There are at least two reasons for this prediction. First, because new meristems are organized at the base of the plant after infection with *Agrobacterium*, which must ultimately give rise to the new inflorescences, there is the possibility of the incorporation of a transformed cell in a newly forming meristematic region, resulting in the generation of large clones of transformed tissue in a single plant. This will mean the seeds from each treated plant will have to be collected separately to prevent contamination of the putative insertion population with siblings. Secondly, this technique is very labor intensive.

A whole plant procedure which will be very powerful for saturating the *Arabidopsis* genome is one that was published by Bechtold et al.[19] For this protocol, plants (ecotype Wassilewskija) are submerged in an *Agrobacterium* solution and put under vacuum. The plants are allowed to recover

Table 13.1. Frequency of various classes of visible mutant phenotypes observed in transformed populations from seed infection/transformation and tissue culture procedures[12,16]

Phenotype	Seed transformation(%)	Tissue Culture(%)
Seedling-lethal	1.1	0.59
Size Variant	4.4	5.52
Embryo-defective	3.6	5.52
Reduced-fertility	1.1	0.36
Pigment Mutants	3.86	4.47
Dramatic Mutants	4.22	4.54
Physiological Mutants	0.72	0.81
Other	0.72	3.20
Total	19%	25%

in the greenhouse and after several weeks progeny are collected. The transformed progeny are selected and grown to maturity. The authors estimate that they have generated > 50,000 transformants utilizing this procedure and are in the process of characterizing the first 12,000 (M. Caboche, personal communication). These transformants are very similar to those generated by seed infection/transformation, i.e., there are approximately 1.5 inserts/transformant, the inserts are concatamers of T-DNAs and there is a significant percentage of untagged mutants (M. Caboche, personal communication). This technique may prove not only to have solved the problem of saturation mutagenesis but should also be useful for routine transformation.

13.2.2. MUTATION SPECTRUM OF T-DNA GENERATED POPULATIONS

In Table 13.1, the data from screening the last 5,000 lines generated by seed/infection are reported.[16] Seedling-lethals are mutants in which the seeds germinate but fail to develop beyond the cotyledon stage. Size variants are small plants most likely due to reduced root systems or metabolic deficiencies. Embryo-defectives are mutants that had abnormal seed shape in green siliques. Reduced-fertility mutants are those that make dramatically fewer seeds per silique and as a result continue to flower later than wildtypes. Pigment mutants are those that were yellow-green to albino. Dramatic mutants are those with morphological alterations that are likely due to perturbations in developmental pathways. The dwarfs, leaf, root, and flower mutants were placed in this class. Physiological mutants included those altered in epicuticular waxes, turgor, senescence, etc.

The data in Table 13.1 reflect the most accurate frequencies for the seed-derived transformants, as the screen was conducted in Arizona under winter greenhouse conditions in which light, temperature, and humidity were well-controlled. These phenotypic frequencies are compared to a population of 1,340 tissue culture-derived

transformants generated by C. Koncz and his colleagues.[12] A comparison of the mutational spectra and frequency in these two populations indicates that for many of the phenotypes for which we could do a direct comparison, the frequencies are surprisingly similar, e.g., size variants, embryo-defectives, and pigment mutants. The slight variability in some of these classes could be due to the different environmental conditions used by each group to screen the transformants or the specific criteria used to classify a mutant. On the other hand, these influences may actually be minimizing the true differences in the two populations. One class which is probably less subject to environmental conditions is reduced fertility. The percentage of reduced fertiles found among the seed transformants (1.1%) was not different among the subpopulations screened.[3,16] Not included in this population were the dwarf or dwarf-type mutants and the dramatic flower mutants. In the tissue culture derived transformants, only 0.36% of the lines were identified as being reduced fertiles. With the possibility of somaclonal variation, this percentage would be expected to be much higher. It is thus interesting to speculate that the position of T-DNA insertion is dependent on the transformation procedure with a much higher percentage of inserts going into transcriptional units in whole plant-generated transformants.

In summary, the two different transformation regimes result in populations with similar mutants and mutational spectra. However, preliminary data indicate that a larger percentage of the mutants in the seed-derived population are tagged (~40%) in comparison to the tissue culture-derived populations (~10%).[3,10,12,20]

13.2.3. GENE ISOLATION FROM T-DNA TAGGED MUTANTS

There are an increasing number of T-DNA insertion mutants being characterized; this is primarily due to the availability of 6,400 transformants (available from the Arabidopsis Stock Centers). A number of mutants and cloned genes, in various pathways, have been characterized (a partial list is given in Table 13.2). The isolated genes are involved in a variety of biochemical and developmental pathways. For the majority of these cloned genes the sequence gives some clues about the function of the gene in the plant. This is in contrast to randomly sequenced ESTs (expressed sequenced tags) reported by Newman et al[36] (also see chapter 14).

Several techniques have been used to isolate the disrupted plant flanking DNA including: (1) the generation of genomic libraries from the mutants and screening with sequences homologous to the right or left border region,[29] (2) plasmid rescue, utilizing the selectable markers and origin of replication in the T-DNA to isolate T-DNA-plant junctions in *E. coli*,[21] (3) inverse polymerase chain reaction (IPCR), utilizing primers made from the left or right border sequences[23,37] and (4) thermal asymmetric (TAIL)-PCR, using nested border specific primers and arbitrary degenerate primers.[38] For this latter protocol, Liu et al[38] used RB (right border) primers with

three nested degenerate primers and were able to recover plant flanking DNA from 183 of 190 transformants generated with an in planta transformation procedure.[17] This would suggest that the inserts are less complex that those generated by seed infection/transformation. From the Feldmann population, a number of mutants characterized thus far lack RB plant flanking DNA.[21,39]

Proof that the correct gene has been tagged and isolated has been accomplished by introducing the wild-type sequence into the mutants to show that the mutant is reverted to wild type, or sequencing a number of mutant alleles to correlate a change in the mRNA sequence with the severity of the mutant phenotype.[22,24,40]

Along with these tagged mutants many mutants have been found that do not cosegregate with the kanamycin-resistance marker and are not tagged. These include a large number of embryo-defective mutants,[41,42] a flower mutant,[43] epicuticular wax mutants[44] and several dwarfs (Feldmann KA, Wu Y, unpublished). In fact,

Table 13.2. Genes cloned from T-DNA tagged mutants

Gene Symbol	Function	Developmental/ Biochemical Pathway	Reference
AG	transcription factor—MADS	flower	21
CH42	chloroplast protein	photosynthesis	10
CHL1	nitrate transporter	nitrate uptake	22
COP1	transcriptional regulator	light-regulated	23
CTR1	serine-threonine kinase	ethylene	24
DWF1	novel	cell elongation	unpubl. results
EMB30	similarity to Yeast Sec7	embryo	25
FAD2	fatty acid desaturase	f.a. desaturation	26
FAD3	fatty acid desaturase	f.a. desaturation	27
FUS6	novel	embryo	28
GL1	transcription factor—*myb*	trichome	29
GL2	trans. factor—homeodomain	trichome	30
HY3	phytochrome B	light-regulated	31
HY4	microbial DNA photolyase	light-regulated	32
LD	bipartate nuclear local. sig.	flower timing	33
SAB	novel	root	34
TSL1	serine-threonine kinase	flower	35

for embryo-defectives and epicuticular waxes, only ~40% of the mutants found in this population appear to be due to disruption by a functional T-DNA; 6 of 13 (46%) for epicuticular wax,[44] and 41 of 115 (36%) for embryo-defective mutants.[45]

13.2.4. TOWARD SATURATION MUTAGENESIS UTILIZING T-DNAs

For T-DNA insertion mutagenesis to be useful for isolating all genes in the *Arabidopsis* genome, it will be necessary to saturate the genome with T-DNAs. Given that (1) the *Arabidopsis* nuclear genome is approximately 120,000 kb in length,[46] (2) there are 1.5 inserts per transformant (shown above), (3) insertion is random, and (4) the average length of an *Arabidopsis* gene is 4 kb, a population of 65-70,000 (90-100,000 inserts) transformants will need to be generated to have a 95% probability of having an insert in an average gene.

The first two assumptions seem to be well supported.[3,13,46] It will take considerable time and data to definitively prove that T-DNA insertion is random but several pieces of data give strong evidence that this is, in fact, the case. At the genome level, mapping of > 70 T-DNAs or T-DNA tagged mutants has shown that these are randomly distributed over the five chromosomes of *Arabidopsis* (references in Table 13.2).[5,38] Within the gene the insertions also appear to be random. For more than 20 T-DNA-tagged mutants which have been characterized, inserts were found in introns and exons as well as in the 5' and 3' ends of the gene (references in Table 13.2). Finally, when the frequency of mutations at various loci are examined, hotspots have not been found for the T-DNA.

The final assumption, that the average gene is 4 kb in length, is supported by the cloning of more than 20 T-DNA tagged genes. For these genes the average number of introns was 6-7 and the average total exon plus intron length was 4.0 kb. This length does not include the 5' and 3' regions of the gene which, if disrupted, can also result in an altered phenotype. Thus, the present data appear to support these four assumptions.

Still, while T-DNA insertion mutagenesis has been a powerful method in *Arabidopsis* for gene isolation, it is not likely to be a practical method for saturation mutagenesis for all genes in the genome simply because of the large number of lines that need to be maintained and distributed. Toward this end an insertion mutagen which may be of greater utility is the transposable element.

13.3. TRANSPOSABLE ELEMENTS

Transposable elements were first observed, in maize, by McClintock.[47] They are now powerful tools for genetic and molecular analyses of several plant species including maize (*Zea mays*),[48] snapdragon (*Antirrhinum majus*),[49] and *Arabidopsis thaliana*.[50] Maize and snapdragon have active endogenous transposable elements that excise and reinsert into new, usually proximal, chromosomal locations.[49,51] It remains unclear what function these "jumping genes" play in plant growth and survival. Recently, *Arabidopsis* and *Lycopersicon* have been engineered with active heterologous transposons from maize.[52,53]

13.3.1. ENDOGENOUS TRANSPOSABLE ELEMENTS

Two types of endogenous transposable elements have been identified in plants: autonomous and non-autonomous. Transposable elements that are autonomous code for a transposase that cleaves the element from its insertion site in the chromosome (Fig. 13.3). Transposable elements that are non-autonomous lack transposase function but are competent to transpose if transposase is supplied.

13.3.1.1. Transposable element systems in *Zea Mays*

In maize, the three major transposon systems are the *Activator* (*Ac*)/*Dissociation* (*Ds*), the *Mutator* (*Mu*) system and the *Suppressor-mutator* (*Spm*)/*dSPM* system. The autonomous *Ac* element is 4.6 kb in length and has an 11 bp inverted repeat required for transposition while the non-autonomous *Ds* element has the 11 bp inverted repeats but is shorter than *Ac*. *Ds* elements are only capable of transposition when activated by the transposase activity of *Ac*. The *Ds1* element was the first plant transposon to be sequenced.[54]

The *Mutator* system of transposable elements was first described in maize as a system inducing high mutation frequencies.[55] The *Mu1* transposon was sequenced from an *Adh1* allele and shown to be 1.4 kb in length.[56] All six members of the *Mu* transposon family share a similar terminal inverted repeat of ~200 bp but differ in regions between the terminal sequences.[57] Transposition of non-autonomous *Mu* elements are under the control of autonomous *Mu* elements.[58] Lines harboring autonomous *Mu* elements are referred to as *Active Mutator* lines.

The *Spm* system was first described by McClintock.[59] *Spm* was cloned from a disrupted *A* gene and is approximately 8.3 kb

with 13 bp terminal inverted repeats.[60] Recent work suggests that mobility of *Spm* is determined by *Spm*'s TrpA protein and is regulated by methylation in the TrpA promoter region.[61] Non autonomous *Spm* elements are called *defective Spm* (*dSpm*) and, like the *Ds* elements, are slightly shorter than the autonomous *Spm* elements.

13.3.1.2. Transposable element system in *Antirrhinum*

Snapdragon (*Antirrhinum majus*) has been shown to possess at least four transposable elements designated *Tam* (*transposon Antirrhinum majus*).[49] *Tam* elements have been most intensely studied in genes whose products are involved in anthocyanin biosynthesis. These genes provide a good model for studying transposon behavior because mutations result in easily identified phenotypes through both germinal and somatic transposition events. The *Tam1*, *Tam2* and *Tam3* elements are approximately 15, 5 and 3.6 kb in length, respectively. The *Tam* elements have 12 or 13 bp inverted repeats.[62,63] High sequence simi-

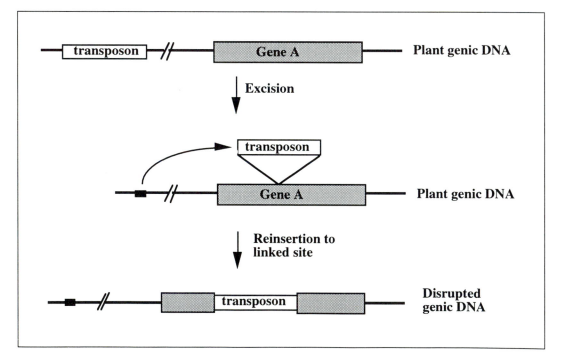

Fig. 13.3. Gene disruption by transposon insertion.

larities between *Tam*, *Ac* and *Spm* elements suggest that they excise and transpose by similar mechanisms.

13.3.1.3. Genes cloned via transposon insertion mutagenesis

The first gene cloned using an endogenous transposon insertion in higher plants was the *BRONZE1* gene of maize.[64] Many other genes have been cloned from maize utilizing transposon-tagged mutants including *C2*, encoding chalcone synthase,[65] *OPAQUE2*,[48] *Y1*, involved in carotenoid biosynthesis,[66] alleles of *HCF106*,[67] *KNOTTED1*,[68] *VIVIPAROUS1*,[69] *PALLIDA*,[70] and *SUGARY1*,[71] among others.

In *Antirrhinum*, most of the genes that have been isolated are involved in flower development or pigmentation. For example, the *FIMBRIATA*, *FLORICAULA*, and *SQUAMOSA* genes involved in floral development, were all isolated from transposon-tagged alleles.[72-74]

While it has been tedious in the past to demonstrate which one of the multiple transposons caused a mutation, new molecular techniques are making it a less laborious process. Thus, the number of cloned genes is increasing rapidly. Another approach to simplify gene cloning has been to transfer transposon systems of one species, to other species which are more amenable to genetic and molecular studies.

13.3.2. HETEROLOGOUS TRANSPOSABLE ELEMENTS

While many plant species harbor transposon-like sequences, active transposon systems have not been identified. Therefore, the movement of transposon systems into model plant species could provide a facile system for mutant identification and gene cloning. The maize *Ac* element was first transformed into *Nicotiana tabacum* by Baker and coworkers.[75] They showed that *Ac* catalyzed its own transposition in tobacco cells.

Although tobacco and other plant species can provide excellent systems for gene tagging and biological studies, *Arabidopsis* has attributes that make it a preferred model system for plant genetic research. The small size of *Arabidopsis* (25 to 40 cm height) means that less growth space is required. The short life cycle (6 to 8 weeks), self pollination, and large seed production (5,000-10,000 seeds per plant) improves the efficiency of genetic analysis. The small haploid genome (120,000 kb)[46] and minimal quantity of interspersed repetitive DNA[76] increases the likelihood for a transposon to tag functional genes, and also reduces the work required for gene isolation. Finally, *Arabidopsis* is a flowering plant with typical cellular physiology and development such that genes obtained can generally be assumed to have similar functions in most other higher plants.

Arabidopsis has been shown to possess an endogenous but non-autonomous transposon called *Tag1*.[77] However, this element is apparently unable to produce its own transposase and therefore cannot be used for gene tagging unless the transposase activity is provided by another active transposon in the same genome. The maize *Ac* element was first introduced into *Arabidopsis* in 1987 by Van Sluys and coworkers.[78] They showed that *Ac*, inserted via the Agrobacterium T-DNA, would actively transpose in *Arabidopsis*. Transposition was verified by inserting the transposon sequence into the 5' untranslated leader of the *nptII* marker gene and then screening for restoration of marker gene activity by *Ac* excision.[78,79] As in maize,[51] the *Ac* element usually transposes to genetically-linked sites in *Arabidopsis*.[80]

Recently, a number of genes have been cloned from *Arabidopsis* using the maize *Ac* as a tag. These genes include *DEFORMED ROOTS AND LEAVES*,[81] *FATTY ACID ELONGATION1*,[82] *LATERAL ROOT PRIMORDIUM1* (*LRP1*)[83] and *PROLIFERA* (*PRL*).[84] *LRP1* and *PRL* were cloned using a novel transposon-borne enhancer trap method.

Heterologous transposons have also introduced into plant species other than *Arabidopsis*, including petunia (*Petunia hybrida*),[85] *Lotus japonicus*[86] and tomato (*Lycopersicon esculentum*)[87] and genes have been cloned from petunia and tomato.[85,87] The *PH6* gene involved in anthocyanin biosynthesis was cloned using an *Ac* tagged mutant in petunia.[85] In tomato, Jones et al identified a novel P450 involved in cell elongation.[87]

13.3.3. APPROACHES TO CLONING GENES FROM TRANSPOSON-TAGGED MUTANTS

One approach to creating insertion mutants in heterologous systems is the introduction of *Ds* and the *Ac* transposase into two separate lines. After crossing these lines an F_1 hybrid is produced in which the *Ac* transposase activates the *Ds* element in trans.[88,89] Since introduction of these heterologous elements requires transformation, the use of dicots rather than monocots has the greatest potential at present. Studies with tomato[90] and *Arabidopsis*[50,81] suggest that the two component *Ac/Ds* system will provide a powerful tool for gene tagging in many plant species.

Another transposon mutagenesis approach for endogenous or heterologous systems involves self-pollination of large numbers of transposon-carrying plants and screening the progeny for homozygous recessive mutations.[49] Mutants identified can be correlated with molecular markers to show that they are due to an insertion event. During the maintenance of the transposon-induced mutants, consideration should be given to conditions that activate and inactivate the transposition of specific elements. For instance, the *Mu* and *Spm* elements produce a higher frequency of heritable transposition events when passed through the male germ-line than the female germ-line.

Once mutants are identified, genetic analysis (if necessary) should be performed to determine the type of element insertion and whether an active or inactive element caused the mutation. Test crosses with active *Spm*, *Ac*, and *Mu* lines when working with maize mutants and tests with active *Tam* lines when working with snapdragon mutants can be performed to determine characteristics of the element. Since maize will generally have a high copy number of the respective transposon, backcrossing to non-transposon carrying lines can reduce the element's copy number. Cone et al[91] got around the need for backcrossing by taking advantage of the undermethylation of sequences in and flanking active transposable elements. In their procedure, DNA is digested using restriction enzymes sensitive to methylation which leaves the bulk of DNA undigested. Segregation of transposon probes with the mutable phenotype can then be used to identify the correct restriction fragment.

As noted earlier, many of the transposons have been sequenced, including various *Ac*, *Mu* and *Spm* transposons, and are available for use as molecular tags for gene cloning.[60,92,93] PCR or Southern analyses can be used to test whether a phenotype cosegregates with the transposon. Once isolated, plant DNA, putatively flanking the transposon, can be used to probe DNA from revertants to determine whether these plants lack the element.

The most commonly used method for cloning transposon tagged genes involves the use of well characterized deletion mutants[64,94] or an allelic series.[66] A transposon-specific probe is used to screen a genomic library of DNA from one line to isolate a collection of transposon containing clones. These clones are then hybridized to RNA from wild-type plants and plants having a deletion or other mutation at the locus of interest. The clones represented in wild type but not mutant are likely to be the gene of interest.

13.4. SUMMARY

Insertion mutagenesis is a well established, efficient approach for the identification and cloning of important plant genes. Transposon tagging and T-DNA insertion have already led to the cloning of numerous genes. A potentially more powerful approach that should see more and more application in the future is the use of heterologous transposon tagging systems in *Arabidopsis* and other plant species. However, these mutagens will lead primarily to loss-of-function mutations and thus will not be useful in identifying all genes. Other technologies described in this book will prove valuable for the isolation of many of these genes.

REFERENCES

1. Koornneef M, Dellaert L, van der Veen J. EMS- and radiation-induced mutation frequencies at individual loci in *Arabidopsis thaliana* (L.) Heynh. Mut Res 1982; 93:109-23.

2. Chandlee JM. The utility of transposable elements as tools for the isolation of plant genes. Physiol Plant 1990; 79:105-115.

3. Feldmann KA. T-DNA insertional mutagenesis in *Arabidopsis*: mutational spectrum. Plant J 1991; 1:71-82.

4. Koncz C, Schell J, Redei GP. T-DNA transformation and insertion mutagenesis. In: Koncz C, Chua N, Schell J, eds. Methods in *Arabidopsis* Research. New Jersey:World Scientific, 1992:224-73.

5. Koncz C, Martini N, Mayerhofer R et al. High-frequency T-DNA mediated gene tagging in plants. Proc Natl Acad Sci USA 1989; 86:8467-71.

6. Feldmann KA, Malmberg RG, Dean C. Mutagenesis in *Arabidopsis*. In: Meyerowitz EM, Somerville CR, eds. *Arabidopsis*. New York:Cold Spring Harbor Laboratory Press, 1994:137-72.

7. Morris PC, Altmann T. Tissue culture and transformation. In: Meyerowitz EM, Somerville C, eds. *Arabidopsis*. Cold Spring Harbor: Cold Spring Harbor Laboratory Press, 1994:173-222.

8. Lloyd AM, Barnason AR, Rogers SG et al. Transformation of *Arabidopsis thaliana* with *Agrobacterium tumefaciens*. Science 1986; 234:464-466.

9. Valvekens K, Van Montagu M, Van Lusebettens M. *Agrobacterium tumefaciens*-mediated transformation of *Arabidopsis thaliana* root explants by using kanamycin selection. Proc Natl Acad Sci USA 1988; 85:5536-40.

10. Koncz C, Mayerhofer R, Kalman Z et al. Isolation of a gene encoding a novel chloroplast protein by T-DNA tagging in *Arabidopsis thaliana*. EMBO J 1990;

5:1137-46.

11. Jofuku KD, den Boer BGW, Van Montagu M et al. Control of *Arabidopsis* flower and seed development by the homeotic gene *APETALA2*. Plant Cell 1994; 5:1211-1225.

12. Koncz C, Nemeth K, Redei G et al. T-DNA insertional mutagenesis in *Arabidopsis*. Plant Mol Biol 1992; 20:963-76.

13. Feldmann KA, and Marks M. *Agrobacterium*-mediated transformation of germinating seeds of *Arabidopsis thaliana*: a non-tissue culture approach. Mol Gen Genet 1987; 208:1-9.

14. Feldmann KA. T-DNA insertion mutagenesis in *Arabidopsis*: Seed infection/transformation. In: Koncz C, Chua N, Schell J, eds. Methods in *Arabidopsis* Tesearch. London: World Scientific, 1992:274-89.

15. Feldmann KA, Meyerowitz EM. Tagging floral structure genes. In: Mol J, Harding J, Sing F, ed. Genetics and Breeding of Ornamental Species. Dordrecht, The Netherlands: Kluwer Academic Publishers, 1991:271-83.

16. Forsthoefel N, Wu Y, Schulz B et al. T-DNA insertion mutagenesis in *Arabidopsis*: Prospects and perspectives. Aust J Plant Physiol 1992; 19:353-66.

17. Chang S, Park S, Kim B et al. Stable genetic transformation of *Arabidopsis thaliana* by *Agrobacterium* inoculation in planta. Plant J 1994; 5:551-558.

18. Coomber SA, Feldmann KA. Gene tagging in transgenic plants. In: Kung SD, Wu R, eds. Transgenic Plants: Engineering and Utilization. Vol 1. New York:Academic Press, 1993:225-240.

19. Bechtold N, Ellis J, Pelletier G. *In planta Agrobacterium* mediated gene transfer by infiltration of adult *Arabidopsis thaliana* plants. CR Acad Sci 1993; 316:1194-99.

20. Van Lijsebettens M, Vanderhaeghen R, Van Montagu M. Insertion mutagenesis in *Arabidopsis thaliana*. Plant Sci 1991; 80:27-37.

21. Yanofsky M, Ma H, Bowman J et al. The protein encoded by the *Arabidopsis* homeotic gene *agamous* resembles transcription factors. Nature 1990; 346:35-39.

22. Tsay Y-F, Schroeder J, Feldmann K et al The herbicide sensitivity gene *CHL1* of *Arabidopsis* encodes a nitrate-inducible ni-

trate transporter. Cell 1993; 72:705-13.

23. Deng X-W, Matsui M, Wei N et al. *COP1*, an *Arabidopsis* regulatory gene, encodes a protein with both a zinc-binding motif and a G$_\beta$ homologous domain. Cell 1992; 71:791-801.

24. Kieber J, Rothenberg M, Roman G et al. *CTR1*, a negative regulator of the ethylene response pathway in *Arabidopsis*, encodes a member of the Raf family of protein kinases. Cell 1993; 72:427-41.

25. Shevell DE, Leu W-M, Gillmor CS et al. *EMB30* is essential for normal cell division, cell expansion, and cell adhesion in Arabidopsis and encodes a protein that has similarity to Sec7. Cell 1994; 77:1051-62.

26. Yadav N, Wierzbicki A, Agatier M et al. Cloning of higher plant *omega-3* fatty acid desaturases. Plant Physiol 1993; 103: 467-76.

27. Okuley J, Lightner J, Feldmann K et al. The Arabidopsis *FAD2* gene encodes the enzyme that is essential for polyunsaturated lipid synthesis. Plant Cell 1994; 6:147-158.

28. Castle L, Meinke DW. A *FUSCA* gene of Arabidopsis encodes a novel protein essential for plant development. Plant Cell 1994; 6:25-41.

29. Oppenheimer D, Herman P, Sivakumaran S et al. A *myb* gene required for leaf trichome differentiation in *Arabidopsis* is expressed in stipules. Cell 1991; 67:483-93.

30. Rerie WG, Feldmann KA, Marks MD. The *Glabra2* gene encodes a homeodomain protein required for normal trichome development in *Arabidopsis thaliana*. Genes and Devel 1994; 8:1388-1399.

31. Reed JW, Nagpal P, Poole DS et al. Mutations in the gene for the red/far-red light receptor phytochrome B alter cell elongation and physiological responses throughout Arabidopsis development. Plant Cell 1993; 5:147-157.

32. Ahmad M, Cashmore AR. *HY4* gene of *A. thaliana* encodes a protein with characteristics of a blue-light photoreceptor. Nature 1993; 366:162-164

33. Lee I. Aukerman MJ, Gore SL et al. Isolation of *Luminidependens*: A gene involved in the control of flowering time in Arabidopsis. Plant Cell 1994; 6:75-83.

34. Aeschbacher RA, Hauser M-T, Feldmann KA et al. The *SABRE* gene is required for normal cell expansion in Arabidopsis. Genes and Devel 1995; 9:330-340.

35. Roe J, Rivin C, Sessions R et al. The *tousled* gene in *A. thaliana* encodes a protein kinase homolog that is required for leaf and flower development. Cell 1993; 75: 939-50.

36. Newman T, de Bruijn FJ, Green P et al. Genes galore: a summary of methods for accessing results from large-scale partial sequencing of anonymous *Arabidopsis* cDNA clones.Plant Phys 1994; 106:1241-1255.

37. Gasch A, Aoyama T, Foster R et al. Gene isolation with the polymerase chain reaction. In: Koncz C, Chua N, Schell J, eds. Methods in *Arabidopsis* Research. New Jersey:World Scientific, 1992: 242-356.

38. Liu Y-G, Mitsukawa N, Oosumi T et al. Efficient isolation and mapping of Arabidopsis thaliana T-DNA insert junctions by thermal asymmetric interlaced PCR. Plant J 1995; 8:457-463.

39. McKinney EC, Ali N, Traut A et al. Sequence based identification of T-DNA insertion mutations in *Arabidopsis*: actin mutants *act2-1* and *act4-1*. Plant J 1995; 7: 613-622.

40. Herman P, Marks MD. Trichome development in *Arabidopsis thaliana*. II. Isolation and complementation of the GLABROUS1 gene. Plant Cell 1989; 1:1051-1055.

41. Errampalli D, Patton D, Castle L et al. Embryonic lethals and T-DNA insertional mutagenesis in *Arabidopsis*. Plant Cell 1991; 3:149-57.

42. Meinke D. Embryo-lethal mutants and the study of plant embryo development. Oxford Survey of Plant Molecular Biology 1992; 3:122-65.

43. Weigel D, Alvarez J, Smyth DR et al. *LEAFY* controls floral meristem identity in *Arabidopsis*. Cell 1992; 69:843-859.

44. McNevin J, Woodward W, Hannoufa A et al. Isolation and characterization of *eceriferum* (*cer*) mutants induced by T-DNA insertions in *Arabidopsis thaliana*. Genome 1993; 36:610-18.

45. Castle L, Errampalli D, Atherton TL et al. Genetic and molecular characterization of embryonic mutants identified following seed

transformation of Arabidopsis. Mol Gen Genet 1993; 241:504-541.

46. Meyerowitz EM. Structure and organization of the *Arabidopsis thaliana* nuclear genome. In: Meyerowitz EM, Somerville CR, eds. Arabidopsis. New York:Cold Spring Harbor Laboratory Press, 1994:21-36.

47. McClintock B. Mutable loci in maize. Carnegie Inst Wash 1948; 47:155-169.

48. Schmidt RJ, Burr FA, Burr B. Transposon tagging and molecular analysis of the maize regulator locus *opaque2*. Science 1987; 238:960-963.

49. Coen ES, Robbins TP, Almeida J et al. Consequences and mechanisms of transposition in *Antirrhinum majus*. In: Berg DE, Howe MM, eds. Mobile DNA. Washington DC:Amer Soc Micro, 1989:413-436.

50. Aarts M, Corzann P, Stiekema W et al. A two-element *Enhancer-Inhibitor* transposon system in *Arabidopsis thaliana*. Mol Gen Genet 1995; 247:555-64.

51. Greenblat IM. A chromosome replication pattern deduced from pericarp phenotypes resulting from movements of the transposable element, Modulator, in maize. Genetics 1984; 108:471-485.

52. Yoder J, Palys J, Alpert K et al. *Ac* transposition in transgenic tomato plants. Mol Gen Genet 1988; 213:291-96.

53. Bancroft I, Bhatt A, Sjodin C et al. Development of an efficient two-element transposon tagging system in *Arabidopsis thaliana*. Mol Gen Genet 1992; 233:449-61.

54. Sutton WD, Gerlach WL, Schwartz, D et al. Molecular analysis of *Ds* controlling element mutations at the *Adh1* locus of maize. Science 1984; 223:1265-1268.

55. Robertson DS. Characterization of a *mutator* system in maize. Mutat Res 1978; 51:21-28.

56. Barker RF, Thompson DV, Talbot DR et al. Nucleotide sequence of the maize transposable element *Mu1*. Nucl Acids Res 1984; 12:5955-5967.

57. Lisch D, Chomet P, Freeling M. Genetic characterization of the *Mutator* system in maize: behavior and regulation of *Mu* transposons in a minimal line. Genetics 1995; 139:1777-96.

58. Chomet PS. Transposon tagging with *Mutator*. In: Freeling M, Walbot V, eds.

The Maize Handbook. New York:Springer-Verlag, 1994:243-249.

59. McClintock B. The *suppressor-mutator* system of control of gene action in maize. Carnegie Inst Wash 1958; 57:415-452.

60. Banks J, Kingsbury J, Raboy V et al. The *Ac* and *Spm* controlling element families in maize. Cold Spring Harbor Symp Quant Biol 1985; 50:307-311.

61. Fedoroff N, Schlappi M, Raina R. Epigenetic regulation of the maize *Spm* transposon. BioEssays 1995; 17:291-97.

62. Sommer H, Carpenter R, Harrison BJ et al. The transposable element *Tam3* of *Antirrhinum majus* generates a novel type of sequence alteration upon excision. Mol Gen Genet 1985; 199:225-231.

63. Luo D, Coen ES, Doyle S et al. Pigmentation mutants produced by transposon mutagenesis in *Antirrhinum majus*. Plant J 1991; 1:59-69.

64. Federoff N, Furtek D, Nelson Jr O. Cloning of the *Bronze* locus in maize by a simple and generalizable procedure using the transposable controlling element *Ac*. Proc Natl Acad Sci USA 1984; 81:3825-3829.

65. Wienand U, Weydemann V, Niesbach-Klosgen, U et al. Molecular cloning of the *c2* locus of *Zea mays*, the gene coding for chalcone synthase. Mol Gen Genet 1986; 203:202-207.

66. Buckner B, Kelson TL, and Robertson DS. Cloning of the *y1* locus of maize, a gene involved in the biosynthesis of carotenoids. Plant Cell 1990; 2:867-876.

67. Das L, Martienssen R. Site-selected transposon mutagenesis at the *hcf106* locus in maize. Plant Cell 1995; 7:287-94.

68. Hake S, Volbrecht E, Freeling M. Cloning *Knotted*, the dominant morphological mutant in maize using *Ds2* as a transposon tag. EMBO J 1989; 8:15-22.

69. McCarty DR, Carson CB, Stinard PS et al. Molecular cloning of *viviparous1*: an abscisic acid sensitive mutant of maize. Plant Cell 1989; 1:523-532.

70. Martin C, Carpenter R, Sommer H et al. Molecular analysis of instability of flower pigmentation of *Antirrhinum majus*, following isolation of the *pallida* locus by transposon tagging. EMBO J 1985; 4:

1625-30.

71. James M, Robertson D, Myers A. Characterization of the maize gene *sugary1*, a determinant of starch composition in kernels. Plant Cell 1995; 7:417-29.

72. Simon R, Carpenter R, Doyle S et al. *Fimbriata* controls flower development by mediating between meristem and organ identity genes. Cell 1994; 78:99-107.

73. Coen ES, Romero JM, Doyle S et al. *floricaula*: a homeotic gene required for flower development in *Antirrhinum majus*. Cell 1990; 63:1311-1322.

74. Huijser P, Klein J, Lonnig W-E et al. *Bracteomania*, an inflorescence anomaly, is caused by the loss of function of the MADS-box gene *squamosa* in *Antirrhinum majus*. EMBO J 1992; 11:1239-1249.

75. Baker B, Schell J, Lorz H et al. Transposition of the maize controlling element "*Activator*" in tobacco. Proc Natl Acad Sci USA 1986; 83:4844-4848.

76. Leutwiler L, Hough-Evans B, Meyerowitz E. The DNA of *Arabidopsis thaliana*. Mol Gen Genet 1984; 219:225-234.

77. Tsay Y-F, Frank MJ, Page T, Dean C, Crawford NM. Identification of a mobile endogenous transoposon in Arabidopsis thaliana. 1993; 260:342-344.

78. Van Sluys M, Tempe J, Fedoroff N. Studies on the introduction and mobility of the maize *Activator* element in *Arabidopsis thaliana* and *Daucus carota*. EMBO J 1987; 6:3881-89.

79. Schmidt R, Willmitzer L. The maize autonomous element *Activator* (*Ac*) shows a minimal germinal excision frequency of 0.2%-0.5% in transgenic *Arabidopsis thaliana* plants. Mol Gen Genet 1989; 220:17-24.

80. Keller J, Lim E, Dooner H. Preferential transposition of *Ac* to linked sites in *Arabidopsis*. Theor Appl Genet 1993; 86:585-88.

81. Bancroft I, Jones J, Dean C. Heterologous transposon tagging of the *DRL1* locus in Arabidopsis. Plant Cell 1993; 5:631-38.

82. James D, Lim E, Keller J et al. Directed tagging of the *Arabidopsis FATTY ACID ELONGATION1* (*FAE*) gene with the maize transposon *Activator*. Plant Cell 1995;

7:309-19.

83. Smith D, Fedoroff N. *LRP1*, a gene expressed in lateral and adventitious root primordia of Arabidopsis. Plant Cell 1995; 7:735-45.

84. Springer P, McCombie R, Sundaresan V et al. Gene trap tagging of *PROLIFERA*, an essential *MCM2-3-5*-like gene in *Arabidopsis*. Science 1995; 268:877-80.

85. Chuck G, Robbins T, Nijjar C et al. Tagging and cloning of a petunia flower color gene with the maize transposable element *Activator*. Plant Cell 1993; 5:371-378.

86. Thykjaer T, Stiller J, Handberg K et al. 1995. The maize transposable element *Ac* is mobile in the legume *Lotus japonicus*. Plant Mol Biol 1995; 27:981-93.

87. Bishop G, Harrison K, Thomas C et al. Transposon tagging and characterization of the tomato dwarf. J Cell Biochem 1995; 21A:462.

88. Long D, Martin M, Sundberg E et al. The maize transposable element system *Ac/Ds* as a mutagen in *Arabidopsis*: identification of an albino mutation induced by *Ds* insertion. Proc Natl Acad Sci USA 1993; 90:10370-74.

89. Osborne BI, Wirtz U, Baker B. A system for insertional mutagenesis and chromosomal rearrangement using the *Ds* transposon and *Cre-lox*. Plant J 1995; 7:687-701.

90. Carroll BJ, Klimyuk VI, Thomas CM et al. Germinal transpositions of the maize element *Dissociation* from T-DNA loci in tomato. Genetics 1995; 139:407-420.

91. Cone KC, Schmidt RJ, Burr F et al. Advantages and limitations of using *Spm* as a transposon tag. In: Nelson OE, ed. Plant Transposable Elements. New York:Plenum Publishing Corp, 1988:149-160.

92. Federoff N, Wessler S, Shure M. Isolation of the transposable maize controlling elements *Ac* and *Ds*. Cell 1983; 35:243-251.

93. Pereira A, Schwarz-Sommer Z, Gierl A et al. Genetic and molecular analysis of the *Enhancer* (*En*) transposable element system of *Zea mays*. EMBO J 1985; 4:17-23.

94. McLaughlin M, Walbot V. Cloning of a mutable *bz2* allele of maize by transposon tagging and differential hybridization. Genetics 1987; 117:771-776.

=CHAPTER 14=

ANALYSIS OF GENE EXPRESSION AND GENE ISOLATION BY HIGH-THROUGHPUT SEQUENCING OF PLANT cDNAS

Andrew N. Nunberg, Zhongsen Li and Terry L. Thomas

14.1. INTRODUCTION

Understanding the complexity and pattern of gene expression programs is fundamental to elucidating the regulatory mechanisms that direct development and growth in diverse multicellular organisms, including plants. Analysis of gene expression patterns by gene isolation and RNA/DNA hybridization and associated technologies is an ongoing, productive process. However, the advent of rapid and efficient DNA sequencing strategies, incorporating PCR, automated sequencing methods and associated informatics, has allowed an entirely different approach in the analysis of the diversity, and in some cases the prevalence, of mRNAs expressed in specific developmental stages, organs, tissues or cell types.

Sequencing of anonymous cDNAs from cDNA libraries is a powerful tool for the identification of novel expressed genes and for the comparison of mRNA populations. This approach, frequently referred to as EST (expressed sequence tag) analysis, has been extensively applied in humans[1] and in *Caenorhabditis elegans*.[2] More recently, significant progress has been made in identifying ESTs in plants.[3,4] Two groups (one at Michigan State/DOE and the other in France) have identified numerous *Arabidopsis* ESTs.[4,5] Table 14.1 summarizes results of several EST studies of varying magnitude. It is noteworthy that there are approximately the same number of ESTs for *Arabidopsis* as there are for the well studied nematode *C. elegans*, and rice does not lag too far behind.

Genome Mapping in Plants, edited by Andrew H. Paterson. © 1996 R.G. Landes Company.

Thus, a substantial number of ESTs have been generated in at least two plant systems. The precise nature of the sequence information obtained by EST analysis and the ever increasing number of gene sequences of known function accessible in various databases make it possible and productive to identify specific genes by sequence similarity. In addition, novel ESTs can be used to study gene expression programs during plant growth and development and as probes in large scale mapping projects. Below we will briefly review "global" sequencing strategies and give some examples where these strategies have been applied, and in the end, we will speculate on future directions of this rapidly changing field. Throughout this review we will emphasize the accessibility of information in the DNA sequencing community through the World Wide Web (WWW).

14.2. GLOBAL cDNA SEQUENCING STRATEGIES

In its most basic form, EST analysis is straightforward and almost instantaneously yields useful results. cDNAs representing the target mRNA population are partially sequenced, usually in a single pass, to generate an expressed sequence tag or EST; this information can then be used to screen extant sequence data bases to determine if the sequence has a known congener in the database. In addition, the frequency of recovering the same or similar sequences gives an estimate of the prevalence of the sequence in the mRNA population. Furthermore, by sequencing cDNAs from libraries representing mRNA populations of different developmental states, it is possible to estimate the sequence diversity and sequence overlaps of these different states. Although straightforward, some details of the basic strategy warrant discussion.

14.2.1. LIBRARY CONSTRUCTION

Construction of a representative cDNA library is an important step in any EST analysis. The source of the mRNA for the cDNA library in an EST analysis is critical and will vary depending on the goal of the study. If the goal is to estimate the diversity of mRNAs expressed in a given plant, the mRNA should represent most plant tissues and organs. On the other hand, if the objective is to define the diversity of mRNAs represented in a tissue, organ or developmental stage, then the library should be prepared from the most highly defined source feasible. The efforts of Karrer et al[6] on the isolation of mRNA from individual plant cells are particularly noteworthy. It is better to invest the time to harvest sufficient quantities of scarce tissue for a library rather than using material which will contain a significant fraction of extraneous messages.

Table 14.1. Summary of some EST studies in plants and animals

Organism	Number of ESTs[a]	Reference[b]
Homo sapiens (human)	317,246	Adams et al[12], Adams et al[13]
Caenorhabditis elegans (nematode)	23,950	Waterston et al[2], McCombie et al[14]
Arabidopsis thaliana (thale cress)	22,624	Hoeffte et al[4], Newman et al[5]
Oryza sativa (rice)	11,301	Uchimiya et al[3], Sasaki et al[15]
Zea mays (corn)	1,183	Keith et al[16]
Brassica napus (oilseed rape)	1,021	Park et al[17]
Brassica campestris (field mustard)	965	Lim, Hwang, and Cho, unpublished results
Ricinus communis (castor bean)	750	Van-de-Loo et al[9]
Borago officinalis (borage)	450	Nunberg et al, unpublished results

[a] Number of ESTs is according to the January 4, 1996 release of dbEST with the exception of the borage entry
[b] Cited studies are representative and do not refer to all the ESTs deposited

There are a number of considerations when choosing an appropriate cloning vector. Usually it is of interest to generate a library that may also be used for applications other than EST analysis, as well as a system that affords easy cloning into bacterial vectors. Two common vectors are λZAP™ (Stratagene) and λZipLox™ (Gibco BRL). Both are efficient bacteriophage λ insertion vectors with multiple cloning sites compatible with a number of cDNA cloning strategies; these vectors allow directional cloning and subsequent excision of the cDNA insert into a high copy number plasmid containing the cloned cDNAs. λZAP™ employs in vivo excisions via single stranded M13 bacteriophage while λZipLox™ uses Cre-Lox in vivo excision; both can be used to perform mass excisions of the library. Directional cloning of the cDNA insert allows the investigator to choose which end of the cDNA to sequence, by taking advantage of unique primer sites flanking the multiple cloning site. In two studies described below, it was determined that ESTs generated from the 5' end were of greater value, since sequences generated from the 3' end of the cloned cDNA contained untranslated regions thus limiting the amount of coding sequence that could be obtained in a single pass. Consequently, in some studies[2] random primed cDNAs were cloned and used to generate EST data; in principle, this should enhance the probability of hitting a protein coding sequence in a single pass. However, cloning of random primed cDNAs precludes directional cloning; also a random-primed library will be deficient in full length clones. Furthermore, ESTs obtained from random primed libraries will overestimate the complexity of mRNA populations, since ESTs from different regions of the same anonymous mRNA would be scored as unique ESTs. However, this bias is correctable using a known, full length cDNA as an internal control.

If large quantities of RNA are available, it is possible to create a plasmid library directly. This is particularly feasible since electroporation transformation efficiencies are so high. Plasmid libraries may or may not be directional, and are easily arranged in an ordered array. Constructing plasmid libraries directly avoids any sequence bias, including internal deletion and *trans* recombination, that may occur during the excision process.

14.2.2. ANALYSIS OF ESTs

Once a library is in hand, sequences from randomly selected, or preselected, cDNAs can be determined. ESTs are generated by one-pass sequencing of cDNAs. Although manual sequencing is feasible, this approach is most efficient when automated sequencing strategies are applied. Any sequencing machine that generates sequence data in an automated mode, i.e., software-driven base calling written directly to disk, with limited downstream editing required, will suffice; however, the most widely used systems are those provided by Perkin Elmer®/Applied Biosystems®, the current state-of-the-art machine being the ABI 377. Effective utilization of high-throughput DNA template preparation strategies, e.g., those provided by Qiagen® and ACGT®, and robotics for reaction assembly and gel loading is critical to the success of any large scale EST project. A single-pass read of a cDNA sequence of 400-500 nucleotides with accuracy approaching 99% is optimal. Depending on the overall configuration of the sequencing workstation(s), particularly the extent of peripheral automation, and staffing, an average, minimum throughput rate of two hundred to three hundred cDNAs per week per automated workstation is a reasonable expectation.

Each EST generated must be checked for quality, edited to remove vector sequences, and analyzed for similarity to other sequences. To facilitate and automate these necessary tasks, investigators at the University of Minnesota are developing software to process raw EST sequences. This sequence analysis system uses a graphical interface in which raw EST sequence data is checked for quality and trimmed to give the highest quality sequence, edited for vector sequences, compared against a database, and the results assembled into a hyper text markup language (HTML) formatted file. The system is also capable of identifying regions, via search algorithms like BLASTP in any reading frame, that may give rise to low complexity sequences and mask them from analysis. These low complexity regions are identified in the HTML output. An overview and example of this system can be found on the World Wide Web (WWW) at the Arabidopsis Data Base (AAtDB) home page (http://genome-www.stanford.edu/Arabidopsis/). Table 14.2 summarizes major databases and infor-

matics nodes and provides WWW site information.

Once edited, ESTs can be compared to known sequences contained in GenBank or to unknown or anonymous expressed sequences in dbEST. Assigning an identity to an EST requires in depth analysis of database comparisons. The BLAST algorithm gives a score and a Poisson statistic that reflects the probability that the match could have been assigned randomly (Altschul et al[7]). In some cases, a high BLAST score may result from a large stretch of sequence with a relatively low identity or from a short segment with high identity. It is important to consider these possibilities; Newman et al[5] gives an in depth description of the BLAST applica-

Table 14.2. Useful sites on the World Wide Web (WWW)

Page Name and URL	Comments[a]
NCBI home page http://www.ncbi.nlm.nih.gov	Genbank and other database searches including dbEST.
Baylor College of Medicine Human Genome Center http://kiwi.imgen.bcm.tmu.edu:8088	YAC database searches. BCM search launcher - database searches using BLAST and BEAUTY, also allows pairwise and multiple sequence alignment as well as secondary protein structure prediction. Batch client for BCM search launcher - application for Unix or Mac based systems for submitting multiple sequence searches. Output is in HTML format.
Arabidopsis Data Base (AtDB) http://genome-www.stanford.edu	Searchable database including *Arabidopsis* ESTs. Links to *Arabidopsis* EST analysis database and the *Arabidopsis* Genome Center.
Agricultural Genome Information Server (AGIS) http://probe.nalusda.gov:8000/index.html	Plant genome databases including *Arabidopsis*, rice, cotton, maize, and soybean. Includes links to other plant genome web pages.
The Institute for Genomic Research (TIGR) http://www.tigr.org	For-profit company with human EST database. Access requires a written agreement.
The Sanger Centre http://www.sanger.uk	Human, *C. elegans*, and yeast databases. Links to other databases (EMBL, EMBnet).

[a] All home pages listed contain links to other related topics

tions. In addition, BEAUTY at The Baylor College of Medicine Human Genome Center (see Table 14.2) provides a robust set of applications including BLAST that facilitate detailed analysis. Of course, detailed analyses can only be carried out on a select number of ESTs or complete cDNAs once a larger EST project has been completed or is ongoing.

Abundant sequences pose a problem for large scale EST projects. Redundancy is important in terms of determining the abundance of different messages, but it can impede sequencing efforts. If the library in use is ordered, then abundant sequences may be removed from the sequencing population by hybridization methods. Once a sequence, or its close relative has been sequenced five or more times, it should be removed from the sequencing population. Based on the frequency of occurrence of a given sequence (number of hits/total number of ESTs determined to that point), the prevalence of the message can be calculated with > 95% confidence. The usual statistical arguments apply.

14.3. DIRECTED EST SCREENS TO IDENTIFY SPECIFIC GENES

It is often necessary to isolate genes from a particular organism based on marginal similarities to known genes or predicted similarities. Frequently, the similarities are too distant for hybridization or PCR-based methods to be useful. In this case, a directed EST screen may prove useful. EST data is obtained and searched for characteristic sequences to identify a specific EST as a candidate gene. Maximal information about the characteristic sequence motifs and their approximate locations in the target protein sequence is critical to the success of this exercise. Two examples follow; both involved searches for genes encoding enzymes involved in lipid metabolism.

14.3.1. SEARCH FOR A BORAGE Δ^6-DESATURASE

The enzyme Δ^6-desaturase converts the fatty acid linoleic acid (18:2 $\Delta^{9,12}$) to γ-li-

nolenic acid (18:3 $\Delta^{6,9,12}$) or GLA. GLA is an important fatty acid in the human diet. No agronomically significant oilseed plant accumulates this fatty acid in their seed oils; however, borage, a temperate zone ornamental, does. Because Δ^6-desaturase has significant biotechnology implications, a major objective of our laboratory has been the isolation and expression of Δ^6-desaturase genes in oilseed crops. Our first objective, was to clone a gene encoding Δ^6-desaturase. We initially cloned a Δ^6-desaturase gene from the unicellular cyanobacterium *Synechocystis*.[8] Using the sequence of the cyanobacterial gene and a relatively small number of plant fatty acid desaturase genes, three conserved, histidine-rich motifs were identified; these motifs, referred to as metal or lipid boxes, were used to "screen" ESTs generated from a directionally cloned borage seed cDNA library constructed in λZAP™. Identity was verified by the accumulation of GLA in transgenic plants following expression of a Δ^6-desaturase transgene (Nunberg et al, in preparation). Expression of the full-length putative desaturase cDNA was driven by the cauliflower mosaic virus (CaMV) 35S promoter; GLA represented greater than 20% of C18 fatty acids in leaves of transgenic tobacco and approximately 3% of C18 fatty acids in seeds. In this study, approximately 500 ESTs (Table 13.1) were generated by sequencing from the 5' end of the directionally cloned inserts. It is noteworthy that highly abundant seed protein mRNAs were "subtracted" from the sequencing population prior to the initiation of the EST analysis; these abundant mRNAs were initially identified in a pilot random sequencing experiment.

14.3.2. SEARCH FOR A CASTOR Δ^{12}-HYDROXYLASE

Somerville and colleagues[9] conducted a similar exercise to isolate the castor (*Ricinus communis*) gene encoding the oleate 12-hydroxylase responsible for hydroxylating oleic acid (18:9) at the 12 position to produce ricinoleic acid. The oleate hydroxylase shares several superficial similarities with the microsomal ω-3 (Δ^{12}) desaturase.[9] In contrast to the previous study, moderately abundant and abundant seed-specific sequences were selected prior to sequencing by hybridization with first-strand cDNA probes prepared from seeds and leaves. Approximately, 400 bp of sequence was obtained from the 5' end of 468 anonymous cDNAs; additional castor ESTs were subsequently added to the data base. Two cDNAs were obtained that shared significant sequence similarity with a microsomal ω-3 linoleate desaturase (Δ^{12}-desaturase) from *Arabidopsis*. Subsequent comparisons using a full-length clone revealed approximately 67% sequence similarity to the *Arabidopsis* gene. Expression of the full-length putative hydroxylase gene driven by the CaMV 35S promoter revealed limited (approximately 0.1%) but significant expression of ricinoleic acid in transgenic tobacco seeds but not in leaves of corresponding plants.

Both directed EST studies were successful because several assumptions were basically correct: (1) Adequate sequence information from related species allowed reliable predictions on the conservation of specific motifs and the ability to recognize these and overall sequence similarities even with the relatively limited sequence data available. (2) As predicted, the mRNAs of the targeted genes were moderately abundant (≈ 0.1-0.5%) and were either enriched in seeds or seed-specific. Based on the two studies described here, the directed EST approach can be successful when the above assumptions are met. Thus, this should be the method of choice for the isolation of new genes based on sequence motif conservation or the isolation of poorly conserved known genes from new species. In these situations, directed EST screens should be intrinsically more rapid and reliable than hybridization- or PCR-based screens.

14.4. EST SCREENS TO IDENTIFY NOVEL *ARABIDOPSIS* EMBRYO GENES

We have initiated an EST analysis focusing on early embryos in *Arabidopsis*.

We limited the sequencing population using a method we developed (Li and Thomas, in preparation) called "virtual subtraction." In this procedure, a random primed-PCR (RP-PCR) probe representing a subtractive RNA population is hybridized with a cDNA library representing the target tissue plated at low density (800-1000 recombinants per 15 cm plate). RP-PCR simultaneously amplifies and labels trace amounts of double-stranded cDNA using random hexamer primers and PCR. This method can reliably generate more than one microgram of cDNA probe from less than 20 nanograms of cDNA template. Because of the high specific activity and probe mass, RP-PCR probes detect sequences in the mid- to low-prevalence range as well as abundant sequences typically identified with less robust probes. In our initial experiments, cDNAs from an immature seed cDNA library were hybridized with a leaf + root RP-PCR cDNA probe. Seed cDNAs that failed to hybridize were selected; these represent tissue- or cell-specific cDNAs or cDNAs that represent mRNAs that are rare and are shared with the subtractive population. Approximately one hundred cDNAs were selected following virtual subtraction; these were partially sequenced and also were characterized by RNA gel blot analysis. Table 14.3 is representative of the results obtained with this approach. Patterns of expression were initially determined by RNA gel blots and subsequently, in some case, refined using in situ hybridization. The latter results demonstrated, for example, that AtS35 is expressed primarily in the early embryo through heart stage. It is clear that screens, such as virtual subtraction, of cDNA libraries prepared from already well defined organs, tissues and cell types can significantly limit the sequencing population and the resulting ESTs can be a rich source of new genes.

14.5. FUTURE DIRECTIONS

Although a significant number of ESTs have been generated for *Arabidopsis* and rice, there is substantial need for additional efforts in maize, several of the *Brassicas* and perhaps some other novel crops and congeneric wild species. Continued EST studies could be exceptionally useful in addressing issues such as systematics and plant evolution, particularly if coupled to more global studies on chromosomal organization and evolution. Continued efforts in *Arabidopsis* and rice are also justified, but with more defined targets. For example, additional EST studies in *Arabidopsis*, like the one described above for early embryos, focusing on more defined organs, tissues or cell-types will be very productive in identifying subsets of expressed sequences, some that intersect and others that do not.

Some additional technical innovations will make the overall EST process more facile. Certain advances can be anticipated, and these are described below, but there will be unanticipated advances as well:

(1) cDNA Library Construction. Continued improvements in cDNA library construction, e.g., Karrer et al,[6] in either lambda vectors capable of efficient mass excision or in plasmid vectors is anticipated. Ordered libraries of > 10^4 cDNA recombinants can be easily stored in a manageable number of 384 well microtiter dishes. With off-the-shelf robotics, high density replica filters can be prepared rapidly and with great precision. Furthermore, unique state-of-the-art robotics will become commonplace allowing all manipulations to become more facile. For example, the density that recombinant cDNAs can be applied to novel substrates will be greatly increased.[10]

(2) Restricting the Sequencing Population. Replica filters, or other substrates, can be screened with various labeled probes to "virtually subtract" subsets of sequences from the sequencing population (see above). Using differentially detectable substrates for labeling cDNA populations, these "virtual subtractions" can be multiplexed and automated. If the library is ordered, all that is necessary to perform this subtraction are a set of plate coordinates and appropriate software, and any cDNA can be "subtracted" from the sequencing pool. This approach could be quite powerful in establishing ESTs that are specific for certain developmental stages, organs, tissues or cell types. Using the same approach, redundant sequences can also be "subtracted" from the ordered library during the course of the EST analysis.

Table 14.3. Arabidopsis *embryo cDNAs identified in a limited EST search following virtual subtraction*

Name	Sequence Similarity	RNA Expression[a]					
		F	L	R	S1	Se	Si
AtS20	Anonymous (EST ATTS0219)	–	–	–	±	+	+
AtS29	Anonymous	–	–	–	±	+	–
AtS35	His$_3$Cys zinc finger protein	–	–	–	±	+	–
AtS120	*Arabidopsis ABI3* Maize *VP1*	+	+	+	+	+	+
AtS144	DNA-binding protein	+	–	+	+	+	–
AtS177	Yeast Ser/Thr protein kinase	+	+	+	+	+	+
AtS212	CAAT-box-binding protein	+	+	–	+	+	+
AtS214	Receptor-like protein kinase	+	+	–	+	+	+
AtS372	Protein phosphatase 2C (*ABI1*)	+	+	+	+	+	+
AtS478	Glutamic acid-rich protein	–	–	–	–	+	–

[a] F, flower; L, leaf; R, root; Se, seed; S1, one-day silique; Si, three-day silique

(3) Sequencing. Just as microcomputers have improved in speed and other capabilities on nearly an annual basis over the past decade; overall sequencing technologies, both instrumentation and accompanying techniques, have also been improving at a rapid rate. The third generation of capillary electrophoresis based automated sequencers or other microchannel devices will soon come on line; this will greatly reduce the run time and sample size required to obtain accurate and reproducible sequence. Downstream analytical software will enhance the length and accuracy of each read. New, more radical sequencing approaches are also being explored, and some such as sequence-on-a-chip are beginning to be considered as serious prototypes for future automated machines.

(4) Informatics. Continued enhancement of existing sequence analysis software and development of new applications will greatly enhance the information that can be gleaned from limited sequence data. This will allow sequence relationships to be established between highly divergent coding sequences. What will be particularly exciting will be the capability to scan predicted protein sequences for characteristic three dimensional protein structures determined by emerging predictive models. Much of this will be available in remote locations by virtue of the World Wide Web.

(5) Further Analysis. Although not strictly part of an EST project, it is important to consider how EST information can be used. First novel ESTs, especially those derived from highly specific libraries or subtracted sequencing populations, will provide a rich source of new genes to study the patterns of gene expression during plant growth and development. These studies will in turn lead to a greater understanding of the cell and molecular biology and biochemistry of plants as well as regulatory mechanisms that direct development and growth in plants. In addition, novel genes and promoters identified in EST studies will have important biotechnology applications. One advance needed to enhance the initial analysis of newly identified genes is a reliable whole mount or high-throughput thick section in situ hybridization method.

ESTs can easily be incorporated into ongoing mapping projects. High throughput mapping will be facilitated when ordered yeast and bacterial artificial chromosome (YAC/BAC) libraries are available for most major plant species. Assigning functions to genes identified through EST analysis is problematic. As the density of genetic maps for a few plants increase, it is possible that mapping will identify mutations in target genes; however, this will remain a low yield exercise for some time to come. Consequently, antisense or co-suppression approaches will be the method of choice for the short term. Hopefully, targeted knockouts, analogous to those available in mouse and in yeast, are in the future at least for *Arabidopsis*. Lacking that, recent progress on facile screening of T-DNA and transposon tagged lines in *Arabidopsis* is encouraging.[11]

EST analysis is rapidly becoming the method of choice to characterize gene expression patterns and to isolate defined or novel genes in plants. The technologies are improving rapidly, and this should continue for some time. In addition, instrumentation, including robotics capabilities, required for large scale EST projects are becoming more widespread. Furthermore, the World Wide Web makes data analysis and information gathering and sharing commonplace; this almost instantaneous connectivity greatly expands participation in cutting edge research and will contribute to the rapidly expanding sequence information event horizon.

ACKNOWLEDGMENTS

Research in the author's lab was supported by grants from the USDA/NRI Competitive Grants Program No. 94-

37034-1228 and from the BioAvenir Program sponsored by Rhône-Poulenc, the Ministry in Charge of Research and the Ministry in Charge of Industry (France).

REFERENCES

1. Adams MA, Dubnick M, Kerlavage AR et al. Sequence identification of 2,375 human brain genes. Nature 1992; 355:632-34.

2. Waterston R, Martin C, Craxton M et al. A survey of expressed genes in *Caenorhabditis elegans*. Nature Genet 1992; 1:114-23.

3. Uchimiya H, Kidou S-i, Shimazaki T et al. Random sequencing of cDNA libraries reveals a variety of expressed genes in cultured cells of rice (*Oryza sativa* L.). Plant J 1992; 2:1005-1009.

4. Höfte H, Desprez T, Amselem J et al. An inventory of 1152 expressed sequence tags obtained by partial sequencing of cDNAs from *Arabidopsis thaliana*. Plant J 1993; 4:1051-61.

5. Newman T, de Bruijn FJ, Green P et al. Genes galore: a summary of methods for accessing results from large-scale partial sequencing of anonymous *Arabidopsis* cDNA clones. Plant Phys 1994; 106:1241-25.

6. Karrer EE, Lincoln JE, Hogenhout S et al. In situ isolation of mRNA from individual plant cells: creation of cell-specific cDNA libraries. Proc Natl Acad Sci USA 1995; 92:3814-18.

7. Altschul SF, Gish W, Miller W et al. Basic local alignment search tool. J Mol Biol 1990; 215:403-10.

8. Reddy AS, Nuccio ML, Gross LM et al. Isolation of a Ø6-desaturase gene from the cyanobacterium *Synechocystis* sp. strain PCC 6803 by gain-of-function expression in *Anabaena* sp. strain PCC 7120. Plant Mol Bio 1993; 27:293-300.

9. Van-de-Loo FJ, Bruon P, Turner S et al. An oleate 12-hydroxylase from *Ricinus communis* L. is a fatty acyl desaturase homolog. Proc Natl Acad Sci USA 1995; 92:6743-47.

10. Schena M, Shalon D, Davis RW et al. Quantitative monitoring of gene expression patterns with a complementary DNA microarray. Science 1995; 270:467-70.

11. McKinney EA, Ali N, Traut A et al. Sequence-based identification of T-DNA insertion mutations in *Arabidopsis*: actin mutants act2-1 and act4-1. Plant J 1995; 8:613-22.

12. Adams MD, Kerlavage AR, Fields C et al. 3,400 new expressed sequence tags identify diversity of transcripts in human brian. Nat Genet 1993; 4:256-67.

13. Adams MD, Soares AR, Kerlavage AR et al. Rapid cDNA sequencing (expressed sequence tags) from a directionally cloned human infant brain cDNA library. Nat Genet 1993; 4:373-80.

14. McCombie RW, Adams MD, Kelley JM et al. *Caenorhabditis elegans* expressed sequence tags identify gene families and potential disease gene homologues. Nat Genet 1992; 1:124-31.

15. Sasaki T, Song J, Koga-Ban Y et al. Toward cataloguing all rice genes: large-scale sequencing of randomly chosen rice cDNAs from a callus cDNA library. Plant J 1994; 6:615-24.

16. Keith CS, Hoang DO, Barret BM et al. Partial sequence analysis of 130 randomly selected maize cDNA clones. Plant Phys 1993; 101:329-332.

17. Park YS, Kwak JM, Kim YS et al. Generation of expressed sequence tags of random root cDNA clones of *Brassica napus* by single-run partial sequencing. Plant Phys 1993; 103:359-70.

CHAPTER 15

THE COMPOSITAE: SYSTEMATICALLY FASCINATING BUT SPECIFICALLY NEGLECTED

Richard V. Kesseli and Richard W. Michelmore

15.1. INTRODUCTION

The Compositae (Asteraceae) is one of the largest and most diverse families of flowering plants. It is easily-recognized by the compound inflorescence that has the appearance of a single "composite" flower from which it derives its name. The Compositae is divided into three subfamilies with 12 to 18 tribes, 1,100 to 2,000 genera and 20,000 or more species.[1-3] The family probably originated 30 to 100 million years ago in South America[4,5] and underwent rapid and extensive diversification producing a cosmopolitan array of taxa. Members of the Compositae are present in nearly every ecological niche. They are predominantly herbaceous although woody species also exist. The size and adaptive success of the family have stimulated considerable research into its systematics and evolution. The Compositae contains several economically important species including food (lettuce), oil (sunflower), medicinal (chamomile) and many horticultural (chrysanthemum, dahlia, zinnia, marigold) crops (Table 15.1). In addition, the family is the source of several important chemicals such as insecticides (pyrethrum) and rubber (guayule) as well as detrimental weeds (dandelion, ragweed, thistle).[6] The proceedings of two international conferences have been devoted to the evolution as well as the utilization and economic impact of the family.[7,8]

The taxonomy of the Compositae has relied on extensive classical and, recently, molecular data. A wealth of classical studies describing variation in the capitulum (composite flower), corolla forms, growth habit, chemical composition, pollen structure, habitat, distribution and chromosome number in the Compositae provide a broad base of information for systematic comparisons.[6] Based on the presence of several specialized

Genome Mapping in Plants, edited by Andrew H. Paterson. © 1996 R.G. Landes Company.

floral characteristics, the Compositae is considered a monophyletic and distinct family. Recent molecular studies also support monophyly (for example, see ref. 9). Several schemes have been proposed for taxonomic divisions within the family. At the higher levels of classification, recent studies relying heavily on molecular characteristics of the chloroplast genome have recognized three sub-families, two major (Asteroideae and Cichorioideae) taxa that possess a large 22 kb inversion, and one small primitive taxon (Barnadesioidae) that retains the primitive chloroplast genome arrangement of most other flowering plants.[10,11] This molecular work was instrumental in clarifying the deeper branches of the family's phylogeny. Classical studies have provided little consensus on the distribution of taxa within the sub-families, the number of tribes, or the relationships among the tribes.[2,6,12-14] Molecular systematics based on chloroplast (cpDNA) genomes and, more recently, nuclear ribosomal DNA will continue to unravel the relationships among and within many tribes.[3,9,15-17] These two categories of mark-

ers by themselves may not be sufficient, however, to resolve all major systematic controversies in the family. Additional markers that are informative at intermediate taxonomic levels are required before tribal and sub-tribal relationships are clear. At the species level, a variety of molecular markers (RFLPs, RAPDs and SSRs) have been used for phylogenetic, ecological and evolutionary inferences.[18-20]

Despite the size of the family and the extensive investigations of systematics and evolutionary phenomena, no species in the Compositae has been studied extensively using classical genetics. The Compositae lags behind several other families in molecular dissection of individual species. With two exceptions, there has been little research involving molecular markers focused on genome organization and manipulation of species in this family, largely because the Compositae possesses few major crop species. Although there are more than 40 crop, medicinal and horticultural species spread throughout most tribes in the family, the majority have relatively low economic value and have restricted use. The

Table 15.1. Economically important crop and weed species in the Compositae

Species	Common Name	Importance	Tribe
Lactuca sativa	Lettuce	Food	Lactuceae
Cichorium endivia	Endive	Food	Lactuceae
Cichorium intybus	Chicory	Food	Lactuceae
Cynara scolymus	Artichoke	Food	Cardueae
Helianthus annuus	Sunflower	Food, oil	Heliantheae
Helianthus tuberosus	Jerusalem artichoke	Food	Heliantheae
Carthamus tinctorius	Safflower	Oil	Cardueae
Chrysanthemum spp.	Mums	Horticulture	Anthemideae
C. cinerariifolium	Pyrethrum	Insecticide	Anthemideae
Gerbera spp.	Gerbera	Horticulture	Mutisieae
Tagetes spp.	Marigolds	Horticulture	Tageteae
Taraxacum officinale	Dandelion	Weed	Lactuceae
Hieracium spp.	Hawkweeds	Weed	Lactuceae
Senecio spp.	Ragwort, groundsel	Weed	Senecioneae
Ambrosia spp.	Ragweed	Weed	Heliantheae
Cirsium spp.	Thistle	Weed	Cynareae
Sonchus spp.	Thistle	Weed	Lactuceae

Additional information is provided by Heywood et al.[7]

two exceptions are *Lactuca sativa* (lettuce) and *Helianthus annuus* (sunflower). These are the most economically important species and have been the subjects of separate molecular investigations of genetic structure and diversity as discussed below. Detailed or partial genetic maps have now been constructed in four genera: *Lactuca*, *Helianthus*, *Cichorium* (chicory and endive), and *Microseris*. These studies have laid the foundations for more detailed investigations of the family as a whole.

15.2. ORIGIN AND GENETIC DIVERSITY IN DOMESTICATED LETTUCE AND ITS WILD RELATIVES

Lactuca sativa (lettuce; n = 9) belongs to the Lactuceae, often considered the most distinctive tribe.[21,22] Within the tribe, different groupings have been proposed based mostly upon morphological characters.[23,24] The basal chromosome number for the family and the tribe is probably nine[23] or possibly four and five.[25] The center of diversity for the genus is the Mediterranean region and engravings on ancient Egyptian tombs suggest that lettuce was domesticated more than 4500 BC, possibly as an oil seed crop.[26] Analysis of 67 diverse cultivars and related species with RFLP markers suggested that domesticated lettuce evolved directly from wild *L. serriola*.[20]

Genetic variation in *Lactuca* spp. has been analyzed using several types of markers. Early isozyme surveys detected extremely low levels of genetic variation within cultivated accessions. A screen of 70 putative loci identified only 12 which showed differences among cultivars and at most five loci varied between any pair of accessions.[27] This narrow genetic base was confirmed by RFLP data; probes from leaf cDNA and genomic libraries detected a low level of polymorphism (< 20%) between a diverse set of accessions representing the major plant types (crisphead, butterhead, latin, and a land-race that may have been partially domesticated as a seed crop).[28] cDNA probes detected higher levels of polymorphism than single copy ge-

nomic clones. Two diverse cultivars (Calmar, a crisphead, and Kordaat, a butterhead) that also differed for multiple resistance genes, were chosen as parents for the basic mapping population (see below). In further analyses of this pair of cultivars, approximately 10% of the loci screened for RFLPs, RAPDs and AFLPs are polymorphic.[29,30] While domesticated accessions possess little diversity, the progenitor wild species, *L. serriola*, which is fully sexually compatible with *L. sativa*, exhibits great diversity. Both RFLP and AFLP data showed that the small collection (eight accessions) of *L. serriola* studied possessed 50% more total diversity (HT = 0.45) than all 47 diverse accessions of *L. sativa*.[20,30] The markers used in this study were biased as they were polymorphic in our mapping population. Therefore, comparisons to diversity statistics from other studies should be made with caution. The *L. serriola* germplasm is a critical resource for agronomically important genes, such as pathogen resistance, and represents a large genetic reservoir for the relatively depauperate cultivated *L. sativa*.

15.3. ORIGIN AND GENETIC DIVERSITY OF DOMESTICATED SUNFLOWER

The genus *Helianthus* is distantly related to lettuce and belongs to the Heliantheae tribe in the subfamily Asteroideae. Sunflower (*H. annuus*; n = 17) seeds and Jerusalem artichoke (*H. tuberosus*; n = 17, 34, etc.) are two major food sources in the genus. The centers of diversity for both species are North America.[31] Sunflower is one of the few major food crops that is native to temperate North America and was likely domesticated in western United States as early as 1000 BC.[31]

Two recent studies examined the diversity of sunflower germplasm with RFLPs. Gentzbettel et al[32] examined ten maintainer and seven cytoplasmic male-sterility restorer (CMS) lines; 62% cDNA probes derived from seedling, leaf, ovary, and anther tissues detected polymorphism; polymorphism levels detected by single-copy genomic clones could not be determined from their data but they also appear high. Excluding high-copy and complex patterns in the data of Berry et al,[33] an initial screen of six inbred accessions found 56% of the genomic and 64% of the cDNA probes were polymorphic. In subsequent analysis of the diverse parents chosen for the mapping population, 55% of the probes detected polymorphism.[34] These values are approximately three times higher than that found in lettuce.

Gentzbettel et al[32] estimated Neiís genetic diversity for the sample of 17 accessions (HT = 0.21). This value is comparable to that found in natural outcrossing populations.[35] These values must be cautiously interpreted, however, since studies vary in their inclusion of monomorphic loci. Both studies estimated genetic distances and examined the relationships of the surveyed accessions. Gentzbettel et al[32] included monomorphic and polymorphic loci and Berry et al[33] included only polymorphic loci. While the distance values cannot be directly compared, the dendrograms constructed from the data of both studies showed maintainer and restorer lines generally forming distinct branches; a likely consequence of breeders retaining separate breeding programs for the two groups of accessions.[32,33]

15.4. GENETIC MAPS OF *LACTUCA* SPP.

Prior to the advent of molecular markers, there was no genetic map of lettuce (n = 9). A basic mapping population for *L. sativa* was generated in 1980, comprised of 68 F2 individuals from a cross between a crisphead (Calmar) and a butterhead (Kordaat) cultivar. An expanded population of over 400 F2 individuals has been used for fine-scale genetic mapping of specific characters. These 'eternal' mapping populations have been maintained by producing large F3 families (generally more than 1,000 seeds are produced per individual). Samples of these families can be pooled to reconstruct the F2 genotype for continued molecular analyses and used to score the segregation of morphological, physiological or disease resistance traits. The first genetic map comprised 41 RFLPs, four isozyme, three morphological markers and five disease resistance genes.[36] From the proportion of linked to unlinked markers, we estimated the size of the genetic map to be 2,000 cM after 160 markers had been placed on the map.[37] At that time, we estimated that less than doubling the number of markers would coalesce the map into the nine chromosomal linkage groups. Some gaps were closed using RAPD markers and bulked segregant analysis[38] but even tripling the number of markers still has not saturated the map, possibly due to a non-random distribution of polymorphic markers, hot spots of recombination or regions of monomorphism between the parents.[29] The latest published map contains 319 randomly-identified markers (152 RFLPs, 130 RAPDs, 7 isozymes, 19 disease resistance genes, 11 morphological markers).[29] These are distributed over eight major linkage groups of more than 14 loci; the major linkage groups span 1160 cM and the estimated genome size remains at approximately 1950 cM. However, this may be an overestimate and must be treated with caution until the map condenses into nine chromosomal groups. More recently, over 120 RAPD and 200 AFLP random markers have been added as well as numerous RAPDs and AFLPs that had been targeted to regions containing resistance genes (H. Witsenboer et al unpublished). RFLP and RAPD markers had similar distributions throughout the genome and identified similar levels of polymorphism. While AFLPs identified many loci per reaction, they did not detect a higher level of polymorphism per locus than RFLPs or RAPDs.

Maps are currently being generated from several other populations. To facilitate consolidation of the map, a second detailed map is being generated from an interspecific cross between *L. sativa* cv. Vanguard 75 x *L. saligna* (UC82US1). This is as wide a cross as can be readily made and the progeny are highly polymorphic.

The 158 individuals of this F2 population cannot, however, be maintained as eternal F3 families due to high levels of sterility and segregation distortion. The current interspecific map possesses 491 markers (H. Witsenboer et al, unpublished). These markers include: 70 RFLPs most of which were also mapped in the intraspecific cross, 274 AFLPs, 133 RAPDs, 4 SCARs linked to resistance genes, and 10 isozyme or morphological markers. The map consists of eight large linkage groups (110-186 cM) and a number of smaller groups. The linkage groups contain nearly 1200 cM and the estimated map size is 1490, approximately 400 cM less than the intraspecific map. This map compression is probably due to increased heterozygosity and therefore reduced recombination in the hybrid population. Again, the map failed to coalesce into the expected nine chromosomal groups; the frequency of polymorphism between the parents of this interspecific cross is very high (> 50%) indicating that regions of monomorphism between the parents of the intraspecific cross were not the sole reason for gaps in the map. Several other intra- and interspecific crosses have been analyzed to map genes of interest that did not segregate in the basic mapping populations. Local maps have been generated to characterize the clusters of disease resistance genes.[39-41]

15.5. GENETIC MAPS OF *HELIANTHUS* SPP.

The construction of genetic maps in sunflower (n = 17) has accelerated recently. Four groups have published mapping studies in the past two years. Berry et al[34] examined a population of 289 individuals from the cross HA89 x ZENB8. These are diverse maintainer lines of Russian and Argentinean origin (respectively) that had been previously assayed for polymorphisms.[33] This population can be self-pollinated and maintained as F3 families or recombinant inbred lines. The 213 probes detected 235 loci which mapped to 17 linkage groups of 10 to 19 markers. These markers cover 1380 cM and the cross dis-

played little segregation distortion (10%).

Gentzbettel et al[42] generated a series of three F2 and two backcross populations with 80 to 150 individuals in each. One of the accessions (PAC2) used to generate the segregating populations has *H. petiolaris* in its pedigree. The other three accessions with origins in the U.S., France and Russia were derived from diverse breeding programs involving wild and cultivated *H. annuus*. Two of the F2 crosses (CX x RHA266 and PAC2 x RHA266) segregated for 123 and 146 markers and allowed the construction of genetic maps with 90 and 113 linked markers. The two maps contained five and eight linkage groups of 7 to 12 markers, respectively. These maps were based on Haldane's mapping function that gives higher values than Kosambi's function that was used in the other studies, and covered 582 and 763 cM, respectively. The authors did not include loci with distorted segregation ratios (8% and 7%, respectively); this may have reduced the number of linkage groups and total map distance. These two maps have recently been extended with 241 and 282 AFLP markers, most of which were scored as codominant markers.[43]

Jan et al[44] have constructed a map from a F2 population between restorer and maintainer lines, RHA 271 x HA 234, that had been shown to be highly polymorphic. The reciprocal cross is also being analyzed. Large F3 families have been created and RILs are being generated to produce an eternal mapping population. The 229 cDNA probes detected 270 loci and 16 linkage groups of seven or more markers were identified. This map covered 1129 cM. Combined, these three mapping studies have identified nearly 600 polymorphic probes for sunflower. More than 90% of these in the first two studies and all of the probes of the last study are from cDNA libraries. These will be useful markers for studies of synteny within the Compositae.

In addition to these RFLP based maps, Rieseberg et al[45,46] have used up to 400 RAPDs in an investigation of hybridization and introgression in domesticated sunflower and its wild relatives. The published map of a backcross population of 56 individuals of *H. anomalus* with 162 markers, showed 15 linkage groups of seven or more markers.[45] The map covered 2,338 cM which is twice the size of that reported for *H. annuus* using larger samples of codominant RFLPs. This discrepancy could be because *H. anomalus* has a larger genome, because of scoring difficulties in RAPD data, or possibly spurious linkages when using LOD threshold for linkage of 3.0 with dominant loci. The authors acknowledge this last problem; therefore, some linkages may be tenuous and the genetic size inflated. Additional maps have now been constructed for related species; *H. annuus*, *H. petiolaris* and the putative hybrid species *H. anomalus*. These studies have documented the chromosomal rearrangements associated with speciation events.[20] The data suggest that bursts of chromosomal rearrangements occur after hybridization, that these rearrangements may not be random, and that these chromosomal changes may function as reproductive isolating mechanisms.

15.6. GENETIC MAPS OF OTHER COMPOSITAE SPECIES

A genetic map of chicory (*Cichorium intybus*; n = 9) was recently constructed from 46 segregating F1 individuals of a cross between outbreeding chicory and inbreeding endive (*Cichorium endivia*). This pseudo-testcross population segregated 1:1 for 274 AFLPs, 16 RAPDs and 74 SAMPL markers of the chicory parent.[47] Ten major linkage groups were identified and markers flanked 1330 cM. The map size was estimated to be 1400 cM which is smaller than the intraspecific maps of lettuce and sunflower, but is comparable to the interspecific map in *Lactuca*.

Bachmann and colleagues have extensively studied the systematics, evolution and quantitative traits in the predominantly North American genus *Microseris*. *Microseris* is a relative of lettuce and chicory; each belonging to a separate subtribe in the tribe Lactuceae (see ref. 48 for

a discussion of the classification of this tribe). Recently, Hombergen and Bachmann[49] examined 289 RAPDs and mapped quantitative trait loci affecting trichome formation in a F2 population of an interspecific cross involving *M. douglasii* (B14) and *M. bigelovii* (C94).

15.7. GENOME ORGANIZATION AND STRUCTURE

Chromosome counts have been made for more than 7,900 species in the family.[50] Chromosome numbers range from n = 2 for *Haplopappus gracilis* (in the Lactuceae tribe) to greater than 100. The most common chromosome number is n = 9 and 78% of the species have haploid counts between 4 and 18.50 There has been controversy over whether the base chromosome number for the family is n = 9 or n = 4 or 5.[23,25] The latter value might suggest that lettuce and chicory are ancient tetraploids and that sunflower is an ancient octaploid. Conversely, lettuce and chicory could be diploids and sunflower could be a tetraploid or, through chromosome breakage, a diploid. Recent phylogenetic studies in the Lactuceae suggest that n = 9 is the basal number.[48] The molecular mapping data also suggest that lettuce and its relatives (n = 9) are diploids. Nearly 90% of the cDNA probes hybridized to single loci.[29] Of the cDNA probes that hybridize to multiple loci, several identify multigene families whose members are scattered throughout the genome. Their duplications appear to be restricted to a few specific genes and do not seem to have resulted from a genome-wide event. In contrast to lettuce, hybridization data for sunflower suggest a higher ploidy level. Extrapolating from Berry et al[33] 36% of the polymorphic cDNA probes produced complex hybridization patterns. In addition, the class of 'simple' probes often detected two to four bands and therefore could also be detecting multiple loci. The hybridization data from Gentzbettel et al[32] were more definitive. There was always more than one band detected per probe, with an average of three. The accessions were likely homozygous inbred lines in both of these studies; therefore sunflower seems to be a tetraploid.

Several studies have examined the genome sizes of species in the family. Genomes of Composite species tend to be large. Estimates for over 30 species range tenfold, from 1.1 picograms (pg)/1C (*Senecio pterophorus*) to greater than 10 pg.[51-54] No species with small genomes have been reported; even though some, for example, *Senecio vulgaris*, have life cycles of only a few weeks. *Happlopappus gracilis*, with only two chromosomes, still has a substantial genome of 1.8 pg. Lettuce and sunflower are close to the modal value in the range for the family with 2.6 and 2.3 to 3.3 pg, respectively.

Genomes in the Compositae studied to date all appear to have extensive levels of high-copy and methylated DNA. In lettuce, genomic libraries constructed with methylation insensitive endonucleases show less than 10% low copy sequences.[28,29] Libraries constructed after digestion with a methylation sensitive endonuclease, PstI, increased the frequency of low copy sequences five-fold to 45-50%. In sunflower, 37% of the clones were low copy in PstI derived genomic libraries (data extrapolated from Berry et al[33]). Extrapolation of the data reported by Gentzbettel et al,[32] indicates that the proportion of all low-copy genomic clones was approximately 50% after digestion with PstI and 20% after digestion with HindIII. These values are similar to those found in lettuce and give a preliminary indication of the repetitive nature of the genomes in the family.

There is little information on the relationship between genetic and physical distance. Assuming a genetic map of approximately 1950 cM for lettuce,[29] the average length of 1 cM is approximately 1 Mb. A comparable figure is not available for sunflower, as accurate estimates of total genetic size has not been reported. However, great variation is likely between different regions of the genome and these estimates are of little utility for predicting local physical distances from segregation data. Markers do not obviously cluster on the map of lettuce, in a manner that could be indicative of reduced recombination around the centromere. Our inability to consolidate the lettuce map may be indicative of hot spots of recombination (see above). Cytological studies have yet to be done to relate the genetic maps to the cytological maps.

Clustering of resistance genes in plant genomes is becoming increasingly well documented. Lettuce has been a model for studies on the genetic and physical distribution of resistance genes and was one of the first species in which clustering of genes for resistance to diverse pathogens was demonstrated. At least 26 of the 28 known resistance specificities to seven diseases are clustered in five linkage groups in lettuce (Table 15.2).[39-41,55-57] On the basis of this clustered distribution and by

Table 15.2. Clusters of disease resistance genes in lettuce

I	Dm1, Dm2, Dm3, Dm6, Dm14, Dm15, Dm16, Dm 31, R18, Ra, R30
II	Dm5, Dm8, Dm10, R17, Tu, Mo2, plr
III	Dm4, Dm7, Dm11, R24
IV	Dm13, R26
V	R23, R25
VI	mo1

Dm#, R# : resistance to *Bremia lactucae* (Resistance is designated as an R-factor until a single Dm gene is demonstrated). Ra : resistance to root aphid. Tu : resistance to turnip mosaic virus. plr : resistance to *Plasmopara lactucae-radicis*. mo# : resistance to lettuce mosaic virus. Ant1 : resistance to anthracnose (loosely linked).

inference from other cell-cell recognition systems, resistance genes were hypothesized to be functionally and evolutionary related.[58] The tight linkage of sequences related to *Tpi* to all three of the major clusters of resistance genes in lettuce provided evidence for the evolution of these clusters by duplication; the presence of introns in the duplicated *Tpi*-related sequences precludes amplification via a RNA intermediate.[59,41] The largest cluster of resistance genes is a highly complex array of duplicated sequences. It contains at least ten resistance genes (Table 15.2) and spans 20 cM covering several megabases (P. Anderson et al, unpublished). There are several levels of sequence duplication: locally around Dm3, around multiple resistance genes in the major cluster, and with other clusters of resistance genes.[60] Cloned resistance gene-like sequences (see below) hybridize to small multigene families and preliminary genetic evidence indicates that these are clustered in the genome. Therefore, resistance gene clusters seem to be composed of arrays of related genes, potentially each with different specificities. These clusters may be in a state of dynamic flux, expanding and contracting over time. Resistance loci should be considered as dynamic, complex haplotypes rather than a series of discrete, alternate alleles of individual genes.

To determine the prevalence of microsatellite sequences in lettuce, many simple sequence oligonucleotides were hybridized to a panel of DNA from animals, plants, and fungi. This demonstrated that the microsatellites are present in varying abundance within plant species and that the relative abundance of each sequence varies from species to species. $(CT)_n$ and $(CA)_n$ were both prevalent in lettuce and at least as abundant as in humans (D. Zungri and R. Michelmore, unpublished). Prevalence of microsatellites in random genomic libraries and in BAC clones of lettuce genomic DNA also demonstrated that microsatellites were frequent. Together, these results indicate that dinucleotide repeats are present every 25 to 40 kb in the lettuce genome.

This is more frequent than has been reported for other plant species.

15.8. MAPPING OF PHENOTYPIC AND QUANTITATIVE TRAIT LOCI

Nearly 80 phenotypic characters have been described in lettuce (reviewed in ref. 61). Of these, 11 morphological characters and 19 disease resistance genes have been linked to molecular markers on either intra- or interspecific maps (Witsenboer, unpublished).[29,41] The morphological genes affect seed color, anthocyanin production, bolting, production of spines, leaf color and shape, among others. Most of the recent specific mapping efforts have focused on using NILs[62] and bulked segregant analysis[38] to map resistance genes and developing reliable markers for them. Three new dominant genes for resistance to downy mildew were mapped using bulked segregant analysis to previously characterized clusters of resistance genes (Keyser et al, unpublished)[40] This was interesting as two of these genes originated from *L. serriola* and one from *L. virosa*. The recessive gene for resistance to the root infecting downy mildew, *plr*, was also mapped by bulked segregant analysis to the second largest cluster of resistance genes,[39] *Tu* and *Mo2*, that each provide resistance to a different potyviruses (turnip mosaic virus and lettuce mosaic virus, LMV), are also within this cluster.[41,57] A recessive LMV resistance gene, *mo1*, was mapped to a unique position in the genome (S. Irwin and R. Michelmore, unpublished). Therefore, although most resistance genes are clustered, not all are (Table 15.1); this would be expected if there are several distinct mechanistic classes of resistance genes (reviewed in ref. 63).

In order to increase the reliability of markers linked to resistance genes, polymorphic RAPD fragments were excised from gels, cloned, and sequenced.[64] Longer primers were designed so that only a single Sequence Characterized Amplified Region (SCAR) locus was amplified. Absolute linkage of each SCAR locus with the progenitor RAPD marker was confirmed by segregation analysis. SCARs have been developed for markers flanking several of the recently mapped genes as well as for previously characterized *Dm* genes (Table 15.3). RAPD fragments can be converted to SCARs with varying ease. For some, it was necessary to try several primers before satisfactory amplification was obtained. In some cases, the same-sized fragment was amplified from both of the mapping parents; this indicated that the original RAPD polymorphism was due to sequence differences at one of the priming sites that were no longer detected by the longer primers. In this situation, the amplified fragment was digested with a range of four base cutters to search for restriction site polymorphisms or small length polymorphisms. Codominant SCARs were developed in this manner at several loci.

Since genetic maps have only recently been constructed in sunflower, few phenotypic characters have been precisely mapped. Gentzbettel et al[42] were able to map a male sterility restorer gene (*Rf1*) in one F2 population and a branching gene (*b1*) in a second cross. The cross of Berry et al[34] segregates for genes affecting oil seed content and these quantitative trait loci are expected to be mapped in the future. There are numerous genes for resistance to downy mildew in sunflower; RFLP and RAPD markers have been identified to the *P/1* locus.[65]

There have been few QTL analyses on Compositae species. In *Microseris* (Lactuceae), flanking RAPDs have been identified for three QTLs that affect the formation of trichomes.[66,49] The low levels of DNA polymorphism within *L. sativa*, particularly within horticultural types, have obstructed QTL studies. The currently available markers will not provide complete genome coverage for populations from intraspecific crosses. The development of microsatellite markers may detect adequate allelic variation and QTL analyses of horticulturally-useful traits may become possible.

15.9. PHYSICAL MAPPING AND MAP-BASED CLONING

Map-based cloning efforts have been directed at cloning genes for resistance to downy mildew from lettuce.[67] The major cluster of disease resistance genes is being characterized in detail at the molecular level, concentrating on *Dm3*. Many markers absolutely cosegregate with this gene in over 400 progeny; there are numerous mutants at this locus and many recombinants in the region. Several markers are missing in all deletion mutants of *Dm3* (Anderson et al 1996; unpublished).[68] A genomic BAC library of over 50,000 clones has been generated with an average insert size of 111 Kb clones (approximately two genome equivalents) and 30 BAC clones identified from the *Dm3* region (A. Frijters et al, unpublished). A genomic contig is being constructed using AFLP fingerprinting of the clones. These clones cover approximately 2 Mb of the *Dm3* region. A comparison of genetic and physical distances indicates that this region is not highly recombinogenic. Two sequences (RLGs: resistance-like genes) have been identified using PCR with degenerate

---

Table 15.3. RAPD-derived markers linked to Dm genes

CLUSTER	RAPDs within 5cM	# sequenced	SCARs dominant	SCARs codom.	# with high copy seqs.	Hybridization probe[a]
Dm 1 + 3	21	17	4	3	13	9
Dm 5 + 8	11	9	7	2	7	4
Dm 4 + 7	4	2	2	0	–	–

[a]Hybridization probes were developed from sub-fragments if the RAPD fragment contained high copy sequences.

oligonucleotide primers homologous to sequences that are conserved between resistance genes cloned from other species (K. Shen et al, unpublished). These RLG sequences are as similar to each other as they are to heterologous resistance genes and contain characteristic conserved domains. Clones carrying candidate resistance gene sequences are being introduced into lettuce by *Agrobacterium*-mediated transformation.

15.10. FUTURE DIRECTIONS

The molecular markers that are now available provide the opportunity for studies on synteny within the Compositae. Preliminary studies have already started to compare the genomic organization of lettuce, sunflower, and chicory. The RFLP markers that form the basis of the lettuce and sunflower maps are predominantly detected by cDNA probes and therefore are good candidates for anchor points. The majority of cDNA probes from lettuce hybridized well to most accessions tested of 12 other Composite species from the Asteriodeae and Cichorioideae subfamilies under moderately stringent conditions; the copy number was not generally variable (Kesseli et al, unpublished). In addition, some SCAR and microsatellite primers developed from lettuce amplify related sequences in different taxa, including *Helianthus annuus*. These data indicate substantial levels of sequence conservation across the family. Therefore, over the next five years we can expect a detailed picture to emerge on the genome rearrangements that have accompanied the expansion of this diverse and successful family.

A major focus will be the evolution of resistance gene clusters in the Compositae. Several resistance genes are close to being cloned from lettuce. If these cross-hybridize to sunflower or strategies employing degenerate oligonucleotide primers are successful, they can be mapped rapidly, as segregating populations already exist that have been scored for disease resistance, particularly downy mildew. It will be interesting to see if resistance genes are orga-

nized in clusters of similar complexity, as in lettuce.

There is a need for more informative markers at several taxonomic levels. Markers that differentiate among the intermediate taxonomic levels are critical to further resolution of tribal relationships. Existing marker technologies are probably sufficient for most interspecific studies. However, several species lack sufficient polymorphism for detailed intraspecific analyses. It remains to be seen whether microsatellites can be developed in a sufficiently economical manner to allow such studies, particularly QTL analyses.

Studies on synteny and the physical organization of the genome would be greatly aided by the discovery and characterization of a species with a small genome (the size of *Arabidopsis* or less). All the species so far characterized have genomes greater than 1.1 pg and therefore none are candidates for development as a model for the Compositae. However, given the number of species and their diversity within this family, particularly as many are small annual herbs, it is difficult to believe that some species do not have small genomes. A survey of Compositae species, particularly those more closely related to sunflower or lettuce to identify a small genome model, is an outstanding need.

Another need is for an electronic database covering the whole Compositae family. There is a wealth of fragmented information for numerous species and detailed molecular information is being generated for lettuce and sunflower. To exploit the potential of synteny between family members, this information needs to be readily accessible. Hopefully, the creation of such a database will be initiated in the near future.

REFERENCES

1. Stebbins GL. Flowering Plants: Evolution Above the Species Level. Second ed. Cambridge MA: Belknap Press Harvard, 1974.
2. Cronquist A. An Integrated System of Classification of Flowering Plants. Vol 1. New

York: Columbia University Press, 1981.

3. Jansen RK, Michaels HJ, Palmer JD. Phylogeny and character evolution in the Asteraceae based on chloroplast DNA restriction site mapping. Syst Bot 1991; 16:98-115.

4. Raven PH, Axelrod DI. Angiosperm biogeography and past continental movements. Annals Missouri Bot Garden 1974; 61:539-673.

5. Turner BL. Fossil history and geography. In: Heywood Vh, Harborne JB, Turner BL eds. The Biology and Chemistry of the Compositae. 1977; 1:21-39.

6. Heywood VH, Harbourne JB, Turner BL. An overture to the Compositae. In: Heywood VH, Harbourne JB, Turner BL Eds. The Biology and Chemistry of the Compositae. Academic Press 1977b:1-20.

7. Heywood VH, Harbourne JB, Turner BL. The Biology and Chemistry of the Compositae. Vols 1 & 2. Academic Press 1977:1189.

8. Hind N. Proc Int Compositae Conf, Kew 1994; 1995; in press.

9. Kim K, Jansen RK, Wallace RS et al. Phylogenetic implications of rbcL sequence variation in the Asteraceae. Annals Missouri Bot Garden 1992; 79:428-455.

10. Palmer JD, Jansen RK, Michaels HJ et al. Chloroplast DNA variation and plant phylogeny. Annals Missouri Bot Garden 1988; 75:1180-1206.

11. Bremer K, Jansen RK. A new subfamily of the Asteraceae. Annals Missouri Bot Garden 1992; 79:414-415.

12. Carlquist S. Tribal interrelationships and phylogeny of the Asteraceae. Aliso 1976; 8:465-492.

13. Jeffrey C. Corolla forms in Compositae—some evolutionary and taxonomic speculations, In: Heywood VH, Harborne JB, Turner BL eds. The Biology and Chemistry of the Compositae. London: Academic Press, 1977; 1:111-118.

14. Bremer K. Tribal interrelationships of the Asteraceae. Cladistics 1987; 3:210-253.

15. Karis PO, Kallersjo M, Bremer K. Phylogenetic analysis of the Cichorioideae (Asteraceae) with emphasis on the Mutisieae. Annals Missouri Bot Garden

1992; 79:416-27.

16. Baldwin BG. Molecular phylogenetics of Calycadenia (Compositae) based on ITS sequences of nuclear ribosomal, chromosomal and morphological evolution reexamined. Am J Bot 1993; 80:222-238.

17. Kim KJ, Jansen RK. NDHF sequence evolution and the major clades in the sunflower family. Proc Natl Acad Sci USA 1995; 92:10379-10383.

18. Riesberg L, Soltis D, Palmer JD. A molecular reexamination of introgression between Helianthus annuus and H. bolanderi (Compositae). Evolution 1988; 42:227-238.

19. Riesberg LH, Vanfosswen C, Desrochers AM. Hybrid speciation accompanied by genomic reorganization in wild sunflowers. Nature 1995a; 375:313-316.

20. Kesseli RV, Ochoa O, Michelmore RW. Variation at RFLP loci in Lactuca spp. and origin of cultivated lettuce. Genome 1991; 34:430-436.

21. Tomb AS. Lactuceae—systematic review. In: Heywood VH, Harborne, Turner BL, eds. The Biology and Chemistry of the Compositae. London: Academic Press, 1977; 2:1066-1079.

22. Cronquist, A. The Compositae revisited. Brittonia 1977; 29:137-153.

23. Stebbins GL. A new classification of the tribe Cichorieae, family Compositae. Madrono 1953; 12:65-81.

24. Jeffrey C. Notes on Compositae: I. The Cichorieae in east tropical Africa. Kew Bull 1966; 18:427-486.

25. Turner BL, Elliston WL, King RM. Chromosome numbers in the Compositae IV: North American species, with phylogenetic interpretations. Am J Bot 1961; 48:216-23.

26. Ryder EJ, Whitaker TW. Lettuce. In: Simmonds NW, ed. Evolution of Crop Plants. 1976:39-41.

27. Kesseli RV, Michelmore RW. Genetic variation and phylogenies detected from isozyme markers in species of Lactuca. J Hered 1986; 77:324-331.

28. Landry BS, Kesseli R, Leung H et al. Comparison of restriction endonucleases and sources of probes for their efficiency in detecting restriction fragment length polymorphisms in lettuce (Lactuca sativa L.). Theor

Appl Genet 1987a; 74:646-653.

29. Kesseli RV, Paran I, Michelmore, RW. Analysis of a detailed genetic linkage map of Lactuca sativa (lettuce) constructed from RFLP and RAPD markers. Genetics 1994; 136:1435-1446.

30. Hill M, Witsenboer H, Zabeau M et al. AFLP fingerprinting as a tool for studying genetic relationships in Lactuca spp. Theor Appl Genet; in press.

31. Hieser Jr CB. Sunflowers. In: Simmonds NW, ed. Evolution of Crop Plants. 1976:36-38.

32. Gentzbittel L, Zhang YX, Vear F et al. RFLP studies of genetic relationships among inbred lines of the cultivated sunflower, Helianthus annuus L—evidence for distinct restorer and maintainer germplasm pools. Theor Appl Genet 1994; 89:419-425.

33. Berry ST, Allen RJ, Barnes SR et al. Molecular maker analysis of Helianthus annuus L. 1. Restriction fragment length polymorphism between inbred lines of cultivated sunflower. Theor Appl Genet 1994; 89:435-41.

34. Berry ST, Leon AJ, Hanfrey CC et al. Molecular marker analysis of Helianthus annuus L. 2. Construction of an RFLP linkage map for cultivated sunflower. Theor Appl Genet 1995; 91:195-199.

35. Hamrick JL. Gene flow and distribution of genetic variation in plant populations. In: Urbanska K, Differentiation Patterns in Higher Plants. Academic Press 1987:53-67.

36. Landry BS, Kesseli RV, Farrara B et al. A genetic map of lettuce (Lactuca sativa L.) with restriction fragment length polymorphism, isozyme, disease resistance and morphological markers. Genetics 1987b; 116:331-337.

37. Kesseli RV, Paran P, Michelmore RW. Genetic linkage map of lettuce (Lactuca sativa, 2n = 18). In: OíBrien SJ, ed. Genetic Maps, Cold Spring Harbor Laboratory Press 1990; 6:100-102.

38. Michelmore RW, Paran I, Kesseli RV. Identification of markers linked to disease resistance genes by bulked segregant analysis: a rapid method to detect markers in specific regions by using segregating populations. Proc Natl Acad Sci 1991; 88:9828-9832.

39. Kesseli RV, Witsenboer H, Stanghellini M et al. Recessive resistance to Plasmopara lactucae-radicis maps by bulked segregant analysis to a cluster of dominant disease resistance genes in lettuce. Molec Pl Microbe Interact 1993; 6:722-728.

40. Maisonneuve B, Bellec Y, Anderson P et al. Rapid mapping of two genes for resistance to downy mildew from Lactuca serriola to existing clusters of resistance genes. Theor Appl Genet 1994; 89:96-104.

41. Witsenboer H, Kesseli RV, Fortin M et al. Sources and genetic structure of a cluster of genes for resistance to three pathogens in lettuce. Theor Appl Genet 1995; 91:178-88.

42. Gentzbittel L, Vear F, Zhang YX et al. Development of a consensus linkage RFLP map of cultivated sunflower (Helianthus annuus L). Theor Appl Genet 1995; 90:1079-1086.

43. Peerbolte R, Peleman J. The Cartisol sunflower RFLP map (146) extended with 291 AFLP markers. Poster Abstr. #234 Plant Genome IV, San Diego 1996.

44. Jan CC, Vick BA, Miller JF et al. Construction of an RFLP linkage map for cultivated sunflower. Poster Abstr. #253 Plant Genome IV, San Diego 1996.

45. Rieseberg LH, Choi HC, Chan R et al. Genomic map of a diploid hybrid species. Heredity 1993; 70:285-293.

46. Rieseberg LH, Linder CR, Seiler GJ. Chromosomal and genic barriers to introgression in Helianthus. Genetics 1995b; 141: 1163-71.

47. DeSimone M, Marocco A, Lucchini, et al. A saturated linkage map of chicory (Cichorium intybus) using a pseudotestcross and AFLP, SAMPL, and RAPD markers. Poster Abstr. #259 Plant Genome IV, San Diego 1996.

48. Whitton J, Wallace RS, Jansen RK. Phylogenetic relationships and patterns of character change in the tribe Lactuceae (Asteraceae) based on chloroplast DNA restriction site variation. Can J Bot 1995; 73:1058-1073.

49. Hombergen EJ, Bachmann K. RAPD mapping of three QTLS determining trichome formation in Microseris hybrid H27 (Asteraceae, Lactuceae). Theor Appl Genet 1995; 90:853-858.

50. Solbrig OT. Chromosomal cytology and evolution in the family Compositae. In: Heywood VH, Harborne JB, Turner BL, eds. The Biology and Chemistry of the Compositae. London: Academic Press, 1977; 1:267-282.

51. Bennett MD. Nuclear DNA content and minimum generation time in herbaceous plants. 1972; Proc R Soc Lond 181: 109-135.

52. Bennett MD, Smith JB, Heslop-Harrison JS. Nuclear DNA contents in angiosperms. 1982; Proc R Soc Lond 216:179-199.

53. Bennett MD, Smith JB. Nuclear DNA amounts in angiosperms. Phil Trans Roy Soc Lond B 1991; 334:309-345.

54. Arumuganathan K, Earle ED. Estimation of nuclear DNA content of plants by flow cytometry. Plant Mol Biol Reptr 1991; 9:208-218.

55. Farrara B, Illott TW, Michelmore RW. Genetic analysis of factors for resistance to downy mildew (Bremia lactucae) in lettuce (Lactuca sativa). Plant Pathol 1987; 36:499-514.

56. Bonnier FJM, Reinink K, Groenwald R. Genetic analysis of Lactuca accessions with new major gene resistance to lettuce downy mildew. Phytopathology 1994; 84:462-468.

57. Robbins MA, Witsenboer H, Michelmore RW et al. Genetic mapping of turnip mosaic virus resistance in Lactuca sativa. Theor Appl Genet 1994; 89:583-589.

58. Michelmore RW, Hulbert SH, Landry BS et al. Towards a molecular understanding of lettuce downy mildew. In: Day PR, Jellis GJ, eds. Genetics and Plant Pathogenesis. Oxford: Blackwell Scientific Publications. 1987:221-231.

59. Paran I, Kesseli RV, Westphal L et al. Recent amplification of triose phosphate isomerase related sequences in lettuce. Genome 1992; 35:627-635.

60. Anderson PA, Okubara PA, Arroyo-Garcia R et al. Molecular analysis of irradiation-induced and spontaneous deletion mutants at a disease resistance locus in Lactuca sativa. Mol Gen Genet 1996; in press.

61. Michelmore RW, Kesseli RV, Ryder EJ. Genetic mapping in lettuce. In: Phillips RL, Vasil IK eds. DNA-Based Markers in Plants. Kluwer Publ. 1993:223-239.

62. Paran I, Kesseli RV, Michelmore RW. Identification of restriction fragment length polymorphism and random amplified polymorphic DNA markers linked to downy mildew resistance genes in lettuce using near-isogenic lines. Genome 1991; 34:1021-27.

63. Michelmore RW. Molecular approaches to manipulation of disease resistance genes. Ann Rev Phytopathol 1995; 15:393-427.

64. Paran I, Michelmore RW. Development of reliable PCR-based markers linked to downy mildew resistance genes in lettuce. Theor Appl Genet 1993; 85:985-993.

65. Mouzeyar S, Roeckeldrevet P, Gentzbittel L et al. RFLP and RAPD mapping of the sunflower P/1 locus for resistance to Plasmopara halstedii race 1. Theor Appl Genet 1995; 91:733-737.

66. Van Houten W, van Raamsdonk L, Bachmann K. Intraspecific evolution of Microseris pygmaea (Asteraceae, Lactuceae) analyzed by cosegregation of phenotypic characters (QTLS) and molecular markers (RAPDS). Pl Syst Evol 1994; 190:49-67.

67. Michelmore RW, Kesseli RV, Francis DM et al. Strategies for cloning plant disease resistance genes. Molecular Plant Pathology—A Practical Approach. Ed. S Gurr, MJ MacPherson, DJ Bowles. IRL Press, Oxford. 1992; 2:233-288.

68. Okubara P, Anderson P, Michelmore RW. Mutants of downy mildew resistance in Lactuca sativa (lettuce). Genetics 1994; 137:867-874.

======CHAPTER 16======

CURRENT STATUS OF GENOME MAPPING IN THE CRUCIFERAE

Wing Y. Cheung and Benoit S. Landry

16.1. INTRODUCTION

The diverse family Cruciferae consists of 360 genera, organized into 13 tribes,[1] but the traditional classification based mainly on morphology may not accurately reflect evolutionary relationships.[2,3] The best studied members of this family are *Arabidopsis thaliana* (n = 5) and the cultivated species in the genus *Brassica*. The former is a model in plant biochemistry, physiology, classical and molecular genetics,[4] and has been chosen as a dicotyledonous model for genome analysis. Genome analysis of *Arabidopsis* has been reviewed recently,[5] and readers are advised to consult it and the references therein. There have been many genetic, breeding and molecular studies into the more economically important Brassicas. In this brief review, we summarize the various approaches to genome analysis of these Brassicas.

Based on interspecific hybridizations and cytogenetic data, relationships were proposed among three cultivated diploid species: *B.rapa* (syn. *campestris*) (2n = 20, genome AA), *B.nigra* (2n = 16, genome BB) and *B.oleracea* (2n = 18, genome CC); and three amphidiploids: *B.napus* (2n = 38, genome AACC), *B.juncea* (2n = 36, genome AABB) and *B.carinata* (2n = 34, genome BBCC), and are summarized in the "U triangle"[6] (Fig. 16.1). Due to the presence of extensive duplications, the diploids *B.nigra*, *B.oleracea* and *B.rapa* are thought to be an evolved ascending aneuploid series from an ancient progenitor with six basic pairs of chromosomes.[7] The three amphidiploid species, *B.napus*, *B.juncea* and *B.carinata* were naturally derived from each pairwise combination of the three diploid ancestors. New synthetic amphidiploids can be generated by hybridizations between modern members of the three diploid species followed by genome doubling, and are normally completely fertile in crosses with established members of the appropriate natural amphidiploid species. Furthermore, amphihaploid and amphidiploid hybrids can

Genome Mapping in Plants, edited by Andrew H. Paterson. © 1996 R.G. Landes Company.

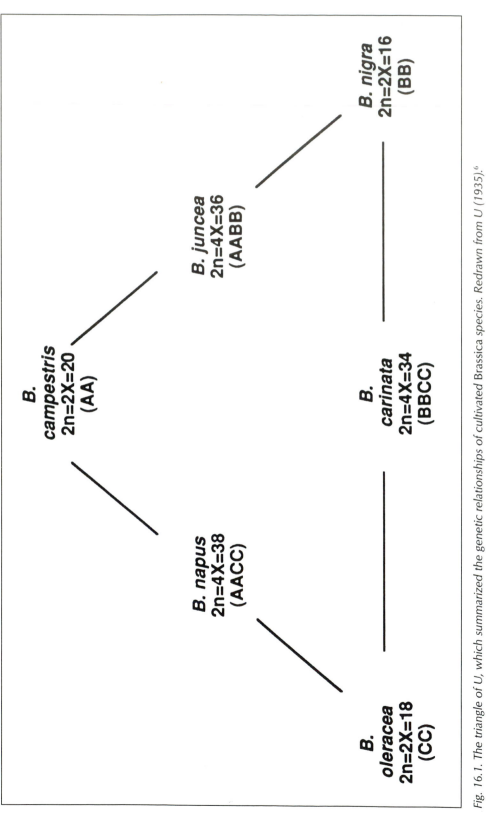

Fig. 16.1. The triangle of U, which summarized the genetic relationships of cultivated Brassica species. Redrawn from U (1935).[6]

be generated from virtually any crosses between other members of the Cruciferae family belonging to the genera *Brassica*, *Sinapis*, *Diplotaxis*, *Raphanus* and *Eruca*.[7]

The genus *Brassica* consists of a wide diversity of morphotypes which are important sources of vegetables, oilseeds and fodder crops throughout the world. *B.oleracea* includes a diversity of vegetables, for example, cabbage, broccoli, cauliflower, Brussel sprouts, kale, collard, Chinese kale and kohlrabi. *B.rapa* comprises economically important oilseed rape, and vegetables such as turnip and Chinese cabbage. The amphidiploid *B.napus* also includes a diversity of oilseed and fodder rapes and the vegetable rutabaga. *B.nigra* (black mustard), *B.juncea* (Brown mustard) and *B.carinata* (Ethiopian mustard) are traditionally grown for condiments. *B.juncea*, which is an oilseed crop in Asia, is emerging also as an oilseed crop in the west.

Cytology, traditionally used to characterize individual chromosomes and genome organization, is difficult in the *Brassicas*, due to their small chromosomes. There is also limited information on the organization of repetitive sequences in the *Brassica* species. Nevertheless, the high degree of natural polymorphisms detected by molecular genetic markers, the gene homology and the conservation of the general genome structure in *Brassica* and related genera have stimulated active research in: (1) construction of genetic maps; (2) genome comparisons through comparative mapping; (3) identification of molecular markers associated with important agronomic traits and (4) intense phylogenetic studies among germplasms, cultivars and breeding lines.

The genome sizes of representatives of the *Brassica* species have been estimated both by Feulgen microdensitometry,[8] and more recently by flow-cytometry.[9] The diploid *Brassicas* have estimated genome sizes between 0.97 pg/2C (468 Mb/1C) and 1.37 pg/2C, (662 Mb/1C) with *B.nigra* having the smallest genome, followed by *B.rapa* and *B.oleracea*. The genome sizes of the amphidiploids are between 2.29 pg/2C (1105 Mb/1C) to 2.56 pg/2C (1235 Mb/1C).

The genomes of diploid *Brassicas* are only 3-4 times larger than that of *Arabidopsis*, while the genomes of the amphidiploids are only 10-20% bigger than the tomato genome. Map-based cloning has been proven possible in both *Arabidopsis* and tomato (see chapter 12). Therefore, with the availability of densely populated molecular genetic maps, the *Brassica* genomes should be amenable to map-based cloning for species-specific alleles that are absent in *Arabidopsis*.

16.2. GENETIC MAPPING USING MOLECULAR MARKERS

Easily scored genetic markers are useful to monitor segregating genes or chromosome segments for the evaluation of their genetic potential. Initially, simply-inherited phenotypic markers were identified for various cultivars. However, their paucity has made it difficult to establish linkages.[10] Isozymes and protein markers have also been explored,[11-14] but their limited numbers and low degree of polymorphisms restricted the possibility of constructing genetic maps covering the whole genome. Polymorphisms detected at the DNA level have the advantages that they are essentially unlimited in number, not tissue-specific, neutral to environmental conditions, and can be assessed at any time during plant development. High levels of polymorphisms are detected by DNA markers in *Brassica* species. The most commonly used DNA markers for *Brassica* genetic mapping are restriction fragment length polymorphism (RFLP) markers. More recently, polymerase chain reaction (PCR)-based markers, such as randomly amplified polymorphic DNA (RAPD) and their derived sequence characterized amplified region (SCAR) markers and sequence tagged sites (STS) have been incorporated into genetic mapping programs.

Genetic linkage maps are useful tools for studying genome structure and evolution, and for identifying introgressions and molecular markers closely associated to important agronomic traits, which can be used for marker-assisted selection in breeding programs. With the high degree of natural DNA polymorphism in *Brassica*, and the almost unlimited source of RFLP probes offered by anonymous cDNA clones, genomic DNA clones and cloned genes of known functions from different Brassicas and *Arabidopsis*, it is not surprising that molecular genetic maps have been constructed for all the cultivated *Brassica* species summarized in the "U triangle" except *B.carinata*. The total number of markers, the total genetic distance covered by each map, the average marker interval and the type(s) of probes or markers used in each case are summarized in Table 16.1.

All the three *B.rapa* maps are based on RFLP markers.[15-17] Genomic probes from *B.rapa*, *B.oleracea* and *B.napus* were used, as were cDNA and the genes of known-functions from *Arabidopsis* and spinach in the most recent map.[17] The homology of the genes and DNA sequences across the related *Brassica* species greatly enhances the efficiency of genetic mapping across the different species with the same sets of probes. Despite the different types of cultivated *B.rapa* used as parents in the crosses, oleiferous[16-17] or vegetable cultivars,[15] the total map distances covering the 10 linkage groups (Lgs) do not vary substantially, even though the numbers of marker loci vary more than two-fold. This suggests that the *B.rapa* maps have an almost complete coverage of the whole genome.

The RFLP map for *B.nigra*[18] comprises 288 marker loci, organized into eight Lgs covering a total distance of 855 cM. The majority of the markers are revealed by genomic DNA (gDNA) probes with a few *Brassica* and *Arabidopsis* cDNA clones. The genome is extensively duplicated with virtually every chromosomal region in three copies. Clustering of loci are observed in seven out of the eight Lgs.

A number of RFLP maps have been published for *B.oleracea*.[19-21] Genomic DNA and/ or cDNA probes were used to generate the marker loci on these maps. The most densely populated map consists of 258 loci arranged into nine Lgs covering a

total distance of 820 cM based on a broccoli x cabbage cross.[19] The map of Landry et al,[20] which is based on 201 loci and a cross between a rapid cycling *Brassica* and a cabbage, has a substantially larger total genetic distance (1112 cM). An updated version of this map with the addition of 90 more markers (cDNA, RAPD, STS and SCAR) has further increased the total map distance to 1546 cM for the nine Lgs (Hubert and Landry, unpublished data). The third *B.oleracea* map[21] is a composite map derived from three intraspecific crosses (collard x cauliflower; collard x broccoli; kale x cauliflower) and an interspecific cross (Kohlrabi x *B.insularis*). The map consists of 108 loci (cDNA and isozymes) covering 747 cM organized into 11 Lgs. Pronounced clustering of loci has been observed in the map of Landry et al,[20] but is not evident in the map of Slocum et al.[19]

RFLP maps are available for two of the three cultivated amphidiploid *Brassica* species, *B.juncea* and *B.napus*. The *B.juncea* RFLP map comprises 343 marker loci arranged in 18 major Lgs covering a total distance of 2073 cM (Cheung et al, submitted). This map is based on segregation of polymorphic *B.napus* cDNA markers in an F1-derived doubled haploid (DH) population from a cross between a Russian mustard line and a canola quality *B.juncea* breeding line. Pronounced segmental duplications between Lgs are observed in the map, implying that complex duplications and subsequent rearrangements have occurred after alloploidy.

With *B.napus* being a major oilseed crop, a number of molecular genetic maps are available.[22-24] Landry et al[22] mapped 120 loci using a F2 population from a narrow cross between two spring canola cultivars, with a set of *B.napus* cDNA probes, which, incidentally, was a subset of the probes used for *B.juncea* mapping described above. The map covers 1413 cM with 19 Lgs, and is estimated to cover about 50% of the *B.napus* genome. Ferreira et al[23] mapped 134 loci (a combination of gDNA, cDNA, cloned *B.napus* and *Arabidopsis* genes of known functions) on their *B.napus* map constructed with a DH population from a cross between an annual canola cultivar and a biennial rapeseed cultivar. The map consists of 22 Lgs and 1016 cM. The map of Uzunova et al[24] has more mapped RFLP loci (204) with genomic and cDNA probes and two RAPD markers, organized into 19 Lgs, covering 1441 cM. It is based on segregation of markers in a DH population derived from a cross between two winter rapeseed cultivars. We have also constructed a *B.napus* map based on a combination of mostly cDNA RFLP and RAPD markers, using also a DH population de-

Table 16.1. Summary of genetic linkage maps of Brassica species

Species	Genome	Total genetic distance	Av. marker interval	number of marker loci	Types of probes	Reference
B.rapa	AA	1850 cM	6.6 cM	280	gDNA	Song et al (1991)[15]
B.rapa	AA	1876 m.u.	5.2 m.u.	360	gDNA	Chyi et al (1992)[16]
B.rapa	AA	1785 cM	13.5 cM	139	gDNA,cDNA, *Arabidopsis* genes	Teutonico and Osborn (1994)[17]
B.nigra	BB	855 cM	3.0 cM	288	gDNA, cDNA known function	Langercrantz and Lydiate (1995)[18]
B.oleracea	CC	820 cM	3.5 cM	258	gDNA	Slocum et al (1990)[19]
B.oleracea	CC	1112 cM	5.5 cM	201	cDNA	Landry et al (1992)[20]
B.oleracea	CC	1546 cM	5.3 cM	291	cDNA, RAPD, SCAR, STS, gDNA	Hubert and Landry (unpublished)
B.oleracea	CC	747 cM	6.9 cM	108	cDNA	Kianian and Quiros (1992)[21]
B.juncea	AABB	2073 cM	6.6 cM	342	cDNA	Cheung et al (submitted)
B.napus	AACC	1413 cM	11.8 cM	120	cDNA	Landry et al (1991)[22]
B.napus	AACC	1441 cM	7.1 cM	204	gDNA,cDNA, RAPD	Uzunova et al (1995)[24]
B.napus	AACC	1016 cM	7.7 cM	132	gDNA known function	Ferreira et al (1994)[23]
B.napus	AACC	2125cM	6.3 cM	342	cDNA,RAPD,STS	Cheung et al (submitted)

rived from a cross between a spring canola and a winter canola breeding lines. This map comprises 342 loci arranged into 19 Lgs covering a total genetic distance of 2125 cM (Cheung et al, submitted).

Overall, RFLPs are still the preferred type of markers for genetic mapping in *Brassica*. The mapping populations used are either F2 or F1-derived DH progenies. In general, a higher degree of segregation distortion is observed with loci mapped with DHs than with F2. Extensive segmental duplications are observed in both diploid and amphidiploid maps, which corroborate the hypothesis that diploid *Brassica* species are balanced ancient polyploids derived from an ancestor with lower original basic chromosome number. However, no duplications of whole linkage groups is found in these maps. Strictly speaking, the reported map distances in different maps cannot be compared directly because they depend on the degree of genome coverage by marker loci, and are based on recombination frequencies, which in turn are influenced by the genetic diversity of the parents and environmental effects on meiosis. This may partially explain the poor correlation of genome size and map distance, among the mapped *Brassica* species. For example, the genome size of *B.rapa* is smaller than *B.oleracea*, and yet all the *B.rapa* maps cover much greater total genetic distances than the *B.oleracea* maps, and have similar distances to the maps of the amphidiploids.

16.3. COMPARATIVE MAPPING

With common sets of markers being included in RFLP maps of different *Brassica* species, comparisons between the maps are possible, and can shed further insights into genome organization and evolution of the diploid genomes in the amphidiploids. Lydiate et al[25] had compared the *B.napus* maps derived from two highly polymorphic *B.napus* crosses: a winter oilseed rape x a spring oilseed rape, and a synthetic *B.napus* x winter oilseed rape, and found almost complete collinearity of both maps. They were also able to identify eight of the nine

B.oleracea Lgs and 10 *B.rapa* Lgs in the *B.napus* map based on allele assignments in the cross involved the synthetic *B.napus*,[25] showing that the genome of the synthetic *B.napus* is essentially unrearranged with respect to the *B.oleracea* and *B.rapa* progenitors used to generate the synthetic amphidiploid, as expected for such a cross.

We have also compared our linkage maps of *B.oleracea* and *B.napus*, both of which are fairly densely populated with marker loci (291 loci; 1546 cM for *B.oleracea* and 342 loci; 2125 cM for *B.napus*). The results are summarized in Figure 16.2, where the homoeology between the *B.napus* Lgs and the *B.oleracea* Lgs are shown in relation to the updated *B.oleracea* map (Hubert and Landry, unpublished). In our comparison, six *B.oleracea* Lgs are each homoeologous to two or three *B.napus* Lgs. Only *B.oleracea* Lgs 7 and 8 are homoeologous to a single *B.napus* Lg (Fig. 16.2). However, in most cases, the extent of the *B.oleracea* homoeology only covers part of the *B.napus* Lgs. The observed homoeology in our case is segmental, suggesting that in nature, *B. napus* evolution has been accompanied by more complicated rearrangements than were observed[25] for the synthetic *B.napus*.

The RFLP map of *B.rapa* was compared to the maps of *B.oleracea* and *B.napus* constructed with a common set of markers respectively.[26] Four homoeologous regions, each with at least three common loci in the same order were identified in the *B.rapa* and *B.oleracea* maps. These results support the close evolutionary relationship between these two diploid species, but also indicate that deletions and insertions may have occurred after divergence of the two species. Nine *B.rapa* linkage segments with at least two loci are conserved in the *B.napus* map. This is in agreement with an earlier comparison[26] that revealed 11 conserved segments. It is not uncommon to find a *B.rapa* segment conserved in more than one *B.napus* group, similar to our findings with the *B.oleracea* and *B.napus* comparison. Only a small proportion of

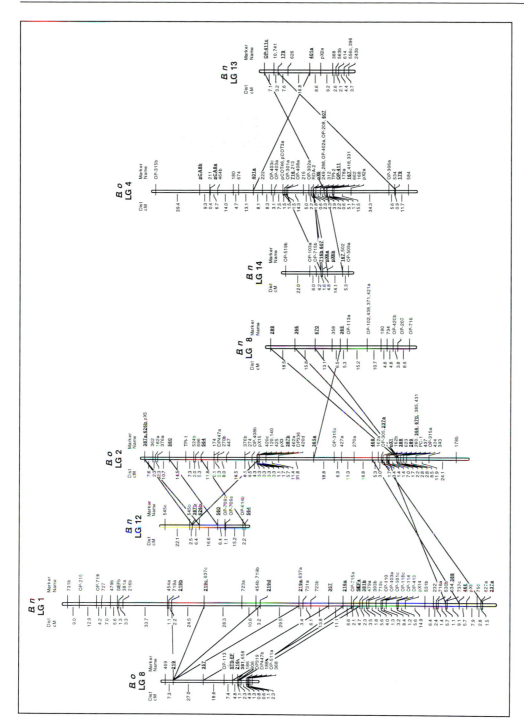

Fig. 16.2. Comparison of the genetic maps of Brassica napus and Brassica oleracea. Major conserved regions between the genomes of B.napus and B.oleracea are shown with reference to the B.oleracea map. Common markers are underlined and joined by solid lines between the corresponding B.oleracea (B.o) and the B.napus (B.n) linkage groups. (Figure is continued on next page.)

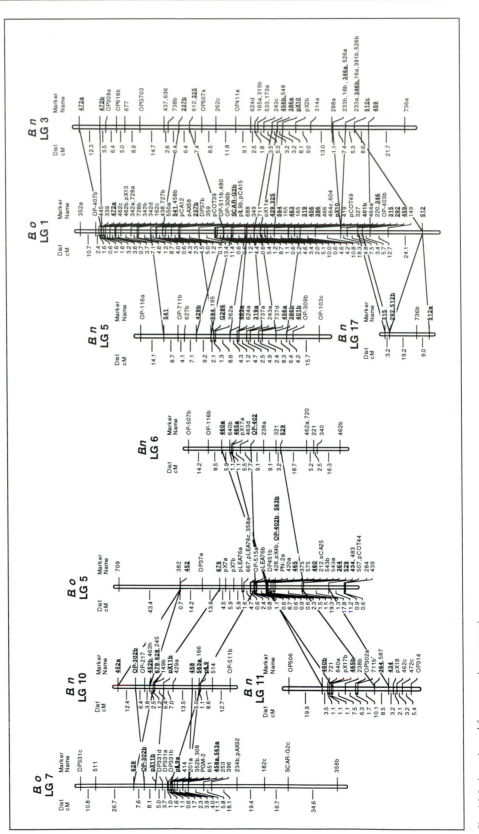

Fig. 16.2. (continued from previous page.)

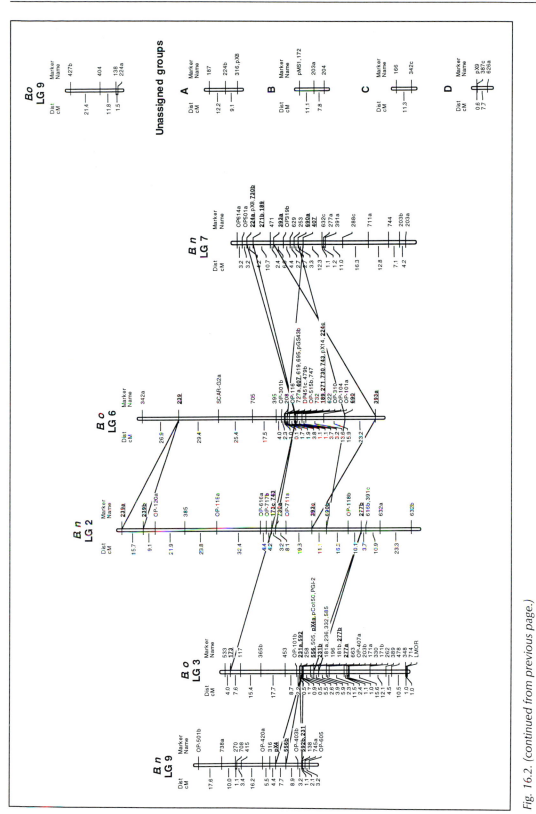

Fig. 16.2. (continued from previous page.)

Arabidopsis cloned genes, which cross-hybridized to *B.rapa* DNA could be mapped onto the *B.rapa* map. These few probes suggested that there are conserved linkage arrangements, but extensive synteny between these two more distantly related species in the *Cruciferae* family was not clear.

Kowalski et al[27] pursued further the comparative mapping of *Arabidopsis* and *B.oleracea* using a larger number of probes from both *B.oleracea* and *Arabidopsis* in both linkage maps. Segments with sizes varying from 3.7 to 49.6 cM are conserved on both maps, accounting for a total of 158.2 cM or 25% of the *Arabidopsis* total map distance, which corresponds to 245 cM or 30% of the *B.oleracea* map distance. About 25 chromosomal rearrangements were estimated to have occurred since the divergence of *B.oleracea* and *Arabidopsis*.

16.4. MAPPING OF PHENOTYPES AND TRAITS

One of the most important applications of genetic maps and DNA markers is to identify markers associated with important agronomic traits that can assist breeders to make more efficient selections in breeding programs. With *Brassica*, the agronomic traits of interest can be divided into four categories: (1) traits of importance in F1 hybrid production; (2) genes for disease resistance; (3) traits controlling the quality and quantity of oil and crushed meals; and (4) genes governing morphological variations.

Creation of F1 hybrids to capture higher yields through heterosis is a major goal in canola breeding. Production of hybrid seeds on a commercial scale depends on simple and efficient methods to force outcrossing in this primarily self-pollinating crop. One of the common methods is the use of self-incompatibility through S-alleles from *B.oleracea* and *B.rapa*, which are responsible for pollen rejection at the stigma surface wherever the pollen and stigma bear identical S-alleles.[28] DNA markers linked to some of the S-alleles have been identified.[29-31] An alternate F1 pro-

duction mechanism involves the use of cytoplasmic male sterility (CMS), which is characterized by the maternally inherited inability to produce pollen.[32] This phenotype can be suppressed by nuclear restorer of fertility (*Rf*) genes. For seed producing Brassicas, a CMS line is pollinated by a restorer line with dominant *Rf* alleles. There are three sources of CMS termed *nap*, *ogu* and *pol* in *B.napus*. For the *ogu*-CMS system, RAPD markers had been identified for the restorer gene (*Rfo*), which was introgressed from a related crucifer, *Raphanus*.[33] For the *pol*- and *nap*-CMS systems, RFLP markers for the restorer *Rfp1*, *Rfp2* and *Rfn* have also been identified (Jean et al, personal communication).

Tight linkage between DNA markers and disease resistance gene(s) is useful to follow gene transfer from one genetic background to another in a breeding program to allow early selection and to avoid difficult multiple screening with pathogens. Fungal diseases like blackleg (caused by *Leptosphaeria maculans*) in *B.napus* and *B.rapa*, white rust (caused by *Albugo candida*) in *B.rapa*, and clubroot (caused by *Plasmodiophora brassicae*) in *B.oleracea* have fueled intense research in mapping the trait in relation to RFLP maps, so that useful DNA markers can be identified. RFLP markers for a French source of blackleg resistance, probably originated from *B.juncea*, have recently been identified by QTL mapping in *B.napus*,[34] and an unknown European source of tolerance for the same disease has also been mapped in *B.napus* (Cheung et al, unpublished results). Genes for clubroot disease resistance have also been mapped in *B.oleracea* and closely associated RFLP markers[20,35] and RAPD markers (Hubert and Landry, unpublished data) have been identified. RFLP markers flanking the gene for white rust resistance have been identified in *B.rapa*.[36] We are currently also mapping white rust resistance in *B.juncea* (Cheung et al, unpublished results).

A large proportion of the cultivated Brassicas are used for oilseeds, and their crushed meal used as animal feed. An important objective of rapeseed breeding has been reduction of glucosinolates and erucic acid in the seeds. Glucosinolates are secondary plant metabolites, of which the hydrolysis products, isothiocyanates, are toxic to domestic animals, and greatly affect the palatability of the crushed meal. Low erucic acid contents are required for human consumption of the oil. These traits are controlled by interacting multiple genes (Quantitative trait loci, QTLs) having interactions with the environment. QTL mapping of the trait allows us to dissect the quantitative phenotype into individual effects, which will lead to better understanding of the genetic basis of the trait. Specific DNA markers linked to these QTLs can be used to follow their transfer through different genetic backgrounds, and as a result allow easier manipulations of these traits. Genes controlling the level of glucosinolates in seeds have been tagged by QTL mapping using RFLP markers in *B.napus*.[24] The toxicity of the glucosinolates is largely determined by the structures of their side chains, which determine the hydrolysis products after tissue damage. McGrath et al[37] had mapped two loci, *Gsl-elong-C* and *Gsl-elong-A* onto a pair of homoeologous Lgs originated from the C and A genomes, respectively. These loci regulate side chain elongation of the amino acid derivatives, which results in the production of the more potent moieties of butyl- and pentyl-glucosinolates. The homologous locus in *Arabidopsis*, *Gsl-elong-Ar* has also been identified. Reduction in 2-hydroxy-3-butenyl glucosinolates, the most abundant glucosinolates (up to 80% of total) in rapeseed, will reduce the major goitrogenic compound in the meal. Parkin et al (1994)[38] had mapped the loci, *Gsl-oh-C* and *Gsl-oh-A*, responsible for the hydroxylation of alkenyl-glucosinolates with the former being a major gene originated from the C genome, and the latter a minor gene from the A genome in *B. napus*.

The gene governing the presence or absence of erucic acid (*Eru*) had been mapped in *B.rapa*.[17] RAPD markers tagged to genes controlling linolenic acid concen-

tration in *B.rapa* had been identified[39] using bulked segregant analysis (BSA).[40]

Development of yellow seeded oilseed cultivars is another popular breeding objective where DNA marker tags will be useful. Yellow seeds have increased oil and protein contents and lower levels of fibers due to thinner seed coat.[41,42] DNA markers flanking the locus controlling yellow seed (*Yls*) have been mapped in *B.rapa*, and can be used to select for yellow seededness in breeding programs. Similarly, RFLP markers for two of the three seed color genes in B.*napus* had been identified using BSA.[43]

The widely varied morphological characteristics in *B. oleracea*, for example between cabbage, broccoli, cauliflower, collard, Brussel sprouts, kohlrabi and kale have stimulated active research in mapping genes underlying the differences. The locus controlling the shape (round vs. fern) of the leaf had been mapped by Landry et al[20] in *B.oleracea*. RFLP markers at an average of 10 cM interval distributed throughout the genome of *B.oleracea* had been used to investigate the genetic control of 22 traits involved in morphological variations in leaf, stem, and flowering measurements by QTL mapping.[44] Morphological variations in *B.rapa* have also been analyzed.[45]

16.5. PHYLOGENETIC STUDIES BASED ON DNA MARKERS

DNA markers, especially RFLP and RAPD, provide easy and efficient means to elucidate phylogenetic relationships. They provide valuable information on: (1) relationships among germplasms; (2) identification of cultivars; and (3) selection of parents for introgression of useful genes, or for hybridization to maximize heterosis.

In *B.napus*, RFLP identification of cultivars and accessions had been reported using both RFLPs[46-48] and RAPDs.[49] Diers and Osborn[47] present the most extensive example involving 83 accessions, which shows clustering into three major groups: (1) the spring cultivars, (2) the winter cultivars, and (3) the rutabagas and oilseed rapes from China and Japan, respectively. The lack of interbreeding between the spring and winter cultivars are clearly reflected in this study, and the information on the genetic diversity in the commonly used *B. napus* accessions will help breeders to choose their desired hybridization combinations. RAPD and RFLP markers were shown to have similar resolution potentials for this type of study. Similar studies have also been carried out on *B. oleracea* with both RFLP and RAPD markers.[50-53] Phylogenetic studies have also been done on other *Brassica* species, for example, *B. juncea*[54] based on RAPD markers and *B. rapa* with isozymes and RFLP markers to investigate the genetic relationships between the crop types and geographical origins.[55]

Interspecific phylogenetic studies on *Brassica* species revealed two distinct lineages: the rapa/oleracea (*Brassica*) lineage and the nigra (*Sinapis*) lineage based on chloroplast RFLP.[3] Pradhan et al[56] used both chloroplast and mitochondrial RFLP and found the two distinct *Brassica* and *Sinapis* lineages and two additional lineages: *Diplotaxis* and *Erucastrum*. There were inconsistencies with existing taxonomic classification based mainly on morphology. Phylogeny of the six cultivated *Brassica* species summarized in the "U triangle" and *Raphanus sativa* were determined using RFLP or/and RAPD markers.[57-58] The latter using both RFLP and RAPD markers showed that even though both types of markers had almost equal resolution power in intraspecific analyses, RFLP markers were more reliable than RAPD markers in interspecific or intraspecific comparisons of more distantly related accessions. This may be explained by the fact that RAPD fragments similar in molecular weight may not necessarily be homologous at the interspecific level.

16.6. REPETITIVE SEQUENCES AND TRANSPOSONS IN *BRASSICA* SPECIES

Repetitive sequences in *Brassica* species have not been systematically studied.

Nonetheless, a number of tandem repeats have been isolated and characterized.[59-64] A major family of 175 bp centromeric satellite sequences is found at about 1.5×10^5 copies per haploid genome in a variety of *Brassica* species and their wild relatives in the Cruciferae family, but not in *Arabidopsis*.[60,62] There are examples of more genome specific repeats: an A-T rich satellite characteristic of the *nigra/sinapis* lineage,[65] and an A-genome specific repeat with a 29 bp repeat motif found at the centromeres of six *B. rapa* chromosomes.[66] The rDNA repeats are found in *B. rapa* and *B. oleracea* at 1300-1590 copies per haploid genome at the nucleolus organizer region (NOR) on the second largest *B. rapa* chromosome. This array of repeats has also been genetically mapped in *B. oleracea*.[21]

More recently, driven by the development of new types of highly polymorphic DNA markers, variable number tandem repeats (VNTR) have been studied in the Brassicas.[67] VNTR have revealed alleles that differ in the number of core sequence repeats which may be long (greater than 100 bp) as in macrosatellites, or as short as four bp in microsatellites or simple sequence repeats (SSR: see chapter 10). Longer PCR-synthetic-tandem-repeat (PCR-STR) probes, which are generated using synthetic oligonucleotides of 3-6 repeats of core sequences as primers in PCR reactions, can reveal highly variable and widely distributed nuclear loci that are transmitted in a Mendelian fashion.[67]

As an example of endogenous transposons, a family of SINE retroposons (SIBN) have been identified in the genome of *B. napus*.[68] The repetitive elements are 170 bp long and dispersed at 500 loci per haploid genome. They have characteristic features of SINE retroposons such as 3' A-rich region, two conserved polymerase III motifs and flanking direct repeats of variable sizes, and a primary and secondary sequence homology to several tRNA species. The use of transposon tagging in *Brassica* species has not been exploited, probably because of the duplicated nature of the genome even in the "diploid" species, which may hamper the efficiency of this approach.

16.7. PROSPECTS FOR PHYSICAL MAPPING AND MAP-BASED CLONING

The development of pulsed field gel electrophoresis (PFGE) to analyze large DNA molecules has extended the possibility of physical mapping in *Brassica*. High molecular weight (HMW) *Brassica* nuclear DNA in agarose plugs or microbeads is restricted with methylation-sensitive restriction enzymes that cut infrequently, therefore generating large DNA molecules. The cleaved DNA is separated using PFGE, and fragments ranging up to millions of bases in sizes can be analyzed. The major challenges for physical mapping by PFGE in Brassicas are: (1) the properties of the probes we used in genetic linkage mapping, which we need to establish the physical linkage, and (2) the segmental duplicated nature of even the diploid *Brassica* genomes. Most of the probes used in genetic mapping, cDNA or genomic DNA, reveal multiple copies of the sequences in the genome. Hence, extra effort is needed to determine which one of the hybridized large DNA fragment on the pulsed field gel (PFG) corresponds to the target locus on the linkage map. If polymorphisms can be readily detected among the PFGE resolved DNA fragments of the parents, DNA samples from a selection of individuals from the progeny can be digested, analyzed on PFG, and hybridized to the probe. Co-segregation of the PFGE pattern with previously obtained segregation data of the mapped marker reveals which megabase DNA fragment corresponds to the locus of interest. Correct location of two mapped loci on a single DNA fragment on the PFG will enable us to deduce the relationship between the maximum physical distance (the size of the DNA fragment) and the genetic distance (the map distance) between the two loci. However, if polymorphism cannot be easily discerned, more complicated procedures have to be employed, involving further digestion of PFG resolved large DNA fragments in the gel slice with the restriction enzymes used to detect the mapped RFLP loci. This digest is then analyzed by normal agarose gel electrophoresis at a right angle to the PFGE direction as described.[69] Despite these complications, we are working on physical mapping by PFGE on a few genetically mapped regions in *B.oleracea*, *B.napus* and *B.juncea* to lay the foundations for future map-based cloning projects. In general, the correlation of genetic distances to physical distances vary greatly with loci and their locations, and the Brassicas are no exceptions.

Physical mapping by in situ hybridization on metaphase and interphase nuclei (see chapter 11) has not been fully exploited in *Brassica* species, probably due to the small sizes of the chromosomes. However, with the more sophisticated and more sensitive fluorescent in situ hybridization (FISH) technique already used successfully in *Arabidopsis* for tandem repeats[70] and in cotton with BAC clones,[71] we can look forward to greater insights into the distribution of various repetitive sequences along the *Brassica* chromosomes, and perhaps ultimately mapping of low-copy sequences corresponding to DNA markers mapped on linkage maps.

Since the *Arabidopsis* genome project can be exploited to benefit *Brassica* reseach,[27] thus a *Brassica* genome effort can hardly be justified. Nevertheless, there are niches where the *Brassica* genes or traits are agronomically important and specific to a particular *Brassica*, and no counterparts can be easily found in *Arabidopsis*, for example, disease resistance genes. These are the areas that map-based cloning in *Brassica* is justified. Towards this end, we have recently constructed a complete bacterial artificial chromosome (BAC) library of *B.juncea* for the isolation of the gene(s) underlying resistance to white rust. To our knowledge, this is the first HMW DNA library in *Brassica*. This ordered library consists of 59,000 clones corresponding to 4.8 genome-equivalent coverage for *B.juncea* (1105 Mb/1C) with estimated average clone size of 90 kb, but clones up to 200 kb or more have been recovered (Cheung et al, unpublished results).

16.8. FUTURE PROSPECTS

The progress and directions of genome analyses in the *Brassica* species have so far been governed largely by the commercial importance of the species, and driven by the applied aspects of the research. This trend is likely to continue in the future decade with maximum exploitation of the fruits of the *Arabidopsis* genome project: the organized contigs and the DNA sequence of the complete *Arabidopsis* genome. Gene tagging and mapping of important agronomic traits using DNA markers in the major oilseed crops, *B.napus* and *B.rapa*, and to a lesser extent on *B.oleracea* will remain important. *B.juncea*, which has been earmarked for development as a major oilseed crop due to its disease resistance to blackleg, and better stress tolerance, will join *B.napus* and *B.rapa* as a target for more active mapping studies. Bulk segregant analysis, which has been widely used to target more markers to regions of interest, will be used to saturate the vicinity of mapped loci of interest to facilitate chromosome landing.[72] Representational difference analysis (RDA)[73] offers a potential alternative for the same purpose. DNA markers associated with important traits will be converted to PCR-based diagnostic markers suitable for automation, and will be used in breeding programs for marker-assisted selection.

It will be beneficial for the *Brassica* research community to have more collaborative and concerted efforts in comparative mapping both within the same species and between related species. By incorporating a common set of markers distributed through the genome, the existing maps of the same species can be aligned and compared to assess the extent of conservation in synteny and the possibility of developing a consensus map for the species. Interspecific comparative mapping has shown that extensive segmental duplications and rearrangements have occurred since speciation. More complete comparison of the maps of the amphidiploid species with those of their progenitors will offer further insights into genome evolution. Incorporation of more *Arabidopsis* DNA markers into the maps of *Brassica* species can help to rationalize map-based cloning of *Brassica* genes in *Arabidopsis*.

In terms of new molecular marker development, AFLP (amplified fragment length polymorphisms)[74] and microsatellites have great potential, and should be incorporated into genetic maps to further increase the densities of markers on the linkage maps and to extend the coverage of the genome.

Fingerprinting and phylogenetic studies that have sparked much research in *Brassica* will continue to flourish for protection of breeders' rights (see chapter 9), for more efficient sampling of germplasm (see chapter 8), and for maximizing heterotic potential in making crosses.

Map-based cloning in Brassicas will most likely be carried out only for alleles that are specific to the *Brassica* species with no obvious counterparts in *Arabidopsis*. PFGE will be used for local long-range physical mapping around these loci of interest. BAC libraries will be constructed and characterized for the isolation of these genes, and contigs covering the loci will be constructed. Further organization of contigs to cover the whole chromosome or linkage group as in the case of *Arabidopsis* will be unlikely due to limited resources. For traits where the counterparts can be identified in *Arabidopsis* or mapped in areas where good local collinearity exist between the *Brassica* and *Arabidopsis* genomes, the cloning will be done in *Arabidopsis* to capitalize on the availability of megabase libraries, physical maps and, in the near future, complete DNA sequences. We have come to an exciting era where molecular and genetic tools are available, and the fruits of fundamental genome research can be reaped for the benefits of agriculture.

REFERENCES

1. Al-Shehbaz IA. The Biosystematics of the genus *Thelypodium* (Cruciferae). Contrib Gray Herb Harv Univ 1973; 204:3-148.

2. Hedge IC. A systematic and geographical survey of the old world cruciferae. In: MacLeod AJ, Jones BMJ, eds. The Biology and Chemistry of the Cruciferae. London: Academic Press, 1976:1-46.

3. Warwick SI, Black LD. Molecular systematics of *Brassica* and allied genera (subtribe Brassicinae, Brassiceae)-chloroplast genome and cytodeme congruence. Theor Appl Genet 1991; 82:81-92.

4. Meyerowitz EM. *Arabidopsis*, a useful weed. Cell 1989; 56:263-269.

5. Dean C, Schmidt R. Plant Genomes: a current molecular discription. Annu Rev Plant Physiol Plant Mol Biol 1995; 46:395-418.

6. U N. Genome analysis in *Brassica* with special reference to the experimental formation of B.napus and peculiar mode of fertilization. Jpn J Bot 1935; 7:389-452.

7. Prakash S, Hinata K. Taxonomy, cytogenetics, and origin of crop *Brassica*, a review. Opera Bot 1980; 55:1-59.

8. Bennett MD, Smith JB. Nuclear DNA amounts in angiosperms. Philos Trans R Soc London Ser 1976; B274:227-74.

9. Arumuganathan K, Earle ED. Nuclear DNA content of some important plant species. Plant Mol Biol Rep 1991; 9:208-18.

10. Stringham GR. Linkage relationships of four seedling mutants in turnip rape. J Hered 1977; 68:391-4.

11. Arus P, Shields CR. Cole crops (*Brassica oleracea* L.). In:Tanksley SD, Orton TJ, eds. Isozymes in Plant Genetics and Breeding. Amsterdam, Oxford and New York: Elsevier, 1983:339-350.

12. Chen BY, Heneen WK, Simonsen V. Genetics of isozyme loci in *Brassica campestris* L. and in the progeny of a trigenomic hybrid between *Brassica napus* and *Brassica campestris* L. Genome 1990; 33:433-440.

13. Quiros CF, Ochoa O, Kianian SF, Douches D. Analysis of *Brassica oleracea* genome by generation of *B.campestris-oleracea* chromosome addition lines: characterization by isozymes and rDNA genes. Theor Appl Genet 1987; 74:758-766.

14. Truco MJ, Arus P. Comparitive study on the isozymes of *Brassica campestris*, *B.oleracea*, and *B.napus*. Eucarpia Cruciferae News 1987; 12:18-19.

15. Song KM, Suzuki JY, Slocum MK et al. A linkage map of *Brassica rapa* (syn.campestris) based on restriction fragment length polymorphism loci. Theor Appl Genet 1991; 82:296-304.

16. Chyi Y-S, Hoenecke ME, Sernyk JL. A genetic linkage map of restriction fragment length polymorphism loci for *Brassica rapa* (syn. campestris). Genome 1992; 35:746-57.

17. Teutonico RA, Osborn TC. Mapping of RFLP and Quantitative trait loci in *Brassica rapa* and comparison to the linkage maps of *B.napus, B.oleracea,* and *Arabidopsis thaliana*. Theor Appl Genet 1994; 89:885-94.

18. Lagercrantz U, Lydiate DJ. RFLP mapping in *Brassica nigra* indicates differing recombination rates in male and female meiosis. Genome 1995; 38:255-64.

19. Slocum MK, Figdore SS, Kennard WC, et al. Linkage arrangement of restriction fragment length polymorphism loci in *Brassica oleracea*. Theor Appl Genet 1990; 80:57-64.

20. Landry BS, Hubert N, Crete R, et al. A genetic map of *Brassica oleracea* based on RFLP markers detected with expressed DNA sequences and mapping resistance genes to race 2 of *Plasmodiophora brassicae* (Woronin). Genome 1992; 35:409-19.

21. Kianian SF, Quiros CF. Generation of a *Brassica oleracea* composite RFLP map: linkage arrangements among various populations and evolutionary implications. Theor Appl Genet 1992; 84:544-54.

22. Landry BS, Hubert N, Etoh T, et al. A genetic map for *Brassica napus* based on restriction fragment length polymorphisms detected with expressed DNA sequences. Genome 1991; 34:543-52.

23. Ferreira ME, Satagopan J, Yandell BS, et al. RFLP mapping of *Brassica napus* using doubled haploid lines. Theor Appl Genet 1994; 89:615-21.

24. Uzunova M, Ecke W, Weissleder K, et al. Mapping of the genome of rapeseed (*Brassica napus* L.) I. Consruction of an RFLP linkage map and localization of QTLs for seed glucosinolate content. Theor Appl

Genet 1995; 90:194-204.

25. Lydiate DJ, Sharpe A, Lagercrantz U,et al. Mapping the *Brassica* genome. Outlook Agri 1993; 22:85-92.

26. Hoenecke M, Chyi Y-S. Comparison of *Brassica napus* and *B.rapa* genomes based on restriction fragment length polymorphism mapping. In: McGregor DI ed. Proc 8th Int Rapeseed Cong GCIRC, Saskatoon 1991:1102-07.

27. Kowalski SP, Lan T-H, Feldmann KA,et al. Comparative mapping of *Arabidopsis thaliana* and *Brassica oleracea* chromosomes reveals islands of conserved organization. Genetics 1994; 138:1-12.

28. Nasrallah JB, Nishio T, Nasrallah ME. The self-incompatibility genes of *Brassica*: Expression and genetic ablation of floral tissues. Annu Rev Plant Physiol 1991; 42:393-422.

29. Brace J, Ockendon DJ, King GL Development of a method for the identification of S alleles in *Brassica oleracea* based on digestion of PCR-amplified DNA with restriction endonucleases. Sex Plant Reprod 1993; 6:133-8.

30. Nishio T, Sakamoto K, Yamaguchi J. PCR-RFLP of S locus for identification of breeding lines in cruciferous vegetables. Plant Cell Rep 1994; 13:546-50.

31. Nishio T, Kusaba M, Watanabe M, et al. Registration of S alleles in *Brassica campestris* L. by restriction fragment sizes of SLGs. Theor Appl Genet 1995; (in press).

32. Hanson MR. Plant Mitochondrial mutations and male sterility. Ann Rev Genet 1991; 25:461-86.

33. Delourme R, Bouchereau A, Hubert N, et al. Identification of RAPD markers linked to a fertility restorer gene for the ogura radish cytoplasmic male sterility of rapeseed (*Brassica napus* L.). Theor Appl Genet 1994; 88:741-48.

34. Dion Y, Gugel RK, Rakow GFW. RFLP mapping of resistance to the blackleg disease [causal agent, *Leptosphaeria maculans* (Desm.) Ces.et de Not.] in canola (*Brassica napus* L.). Theor Appl Genet 1995; 91: 1190-1194.

35. Figdore SS, Ferreira ME, Slocum MK et al. Association of RFLP markers with trait loci

affecting clubroot resistance and morphological characters in *Brassica oleracea* . Euphytica 1993; 69:33-44.

36. Kole C, Teutonico R, Mengistu A, et al. Molecular mapping of a locus controlling resistance *Albugo candida* in *Brassica rapa*. Plant genome III Abstract #P129.

37. Magrath R, Bano F, Morgner M, et al. Genetics of aliphatic glucosinolates. I. Side chain elongation in *Brassica napus* and *Arabidopsis thaliana*. Heredity 1994; 72:290-9.

38. Parkin I, Magrath R, Keith D, et al. Genetics of aliphatic glucosinolates. II. Hydroxylation of alkenyl glucosinolates in *Brassica napus*. Heredity 1994; 72:594-8.

39. Hu J, Quiros C, Arus P, et al. The mapping of a gene determining linolenic acid concentration in rapeseed with DNA-based markers. Theor Appl Genet 1995; 90:258-62.

40. Michelmore RW, Paran I, Kesseli RV. Identification of markers linked to disease resistance genes by bulked segregant analysis: a rapid method to detect markers in specific genomic regions using segregating populations. Proc Natl Acad Sci USA 1991; 88:9828-32.

41. Abraham V, Bhatia CR. Development of strains with yellow seedcoat in Indian mustard (*Brassica juncea* Czern & Coss). Plant Breed 1986; 97:86-88.

42. Downey RK, Robbelen G. *Brassica* species. In: Robbelen G, Downey RK, Ashri A eds. Oil Crops of the World. New York: McGraw-Hill, 1989:339-62.

43. Van Deynze AE, Landry BS, Pauls KP. The identification of restriction fragment length polymorphisms linked to seed colour genes in *Brassica napus*. Genome 1995; 38:534-42.

44. Kennard WC, Slocum MK, Figdore SS, et al. Genetic analysis of morphological variation in *Brassica oleracea* using molecular markers. Theor Appl Genet 1994; 87: 721-32.

45. Song K, Slocum MK, Osborn TC. Molecular marker analysis of genes encoding morphological variation in *Brassica rapa* (syn. campestris). Theor Appl Genet 1995; 90:1-10.

46. Song K, Osborn TC. Polyphyletic origins

of *Brassica napus* : new evidence based on organelle and nuclear RFLP analyses. Genome 1992; 35:992-1001.

47. Diers BW, Osborn TC. Genetic diversity of oilseed *Brassica napus* germ plasm based on restriction fragment length polymorphisms. Theor Appl Genet 1994; 88:662-68.

48. Hallden C, Nilsson NO, Rading IM, et al. The evaluation of RFLP and RAPD markers in a comparison of *Brassica napus* breeding lines. Theor Appl Genet 1994; 88:123-28.

49. Mailer RJ, Scarth R, Fristensky B. Discrimination among cultivars of rapeseed (*Brassica napus* L.) using DNA polymorphisms amplified from arbitrary primers. Theor Appl Genet 1994; 87:697-704.

50. Hu J, Quiros CF. Identification of broccoli and cauliflower cultivars with RAPD markers. Plant Cell Rep 1991; 10:505-11.

51. Kresovich S, Williams JGK, McFerson JR, et al. Characterization of genetic identities and relationships of *Brassica oleracea* L. via a random amplified polymorphic DNA assay. Theor Appl Genet 1992; 85:190-96.

52. Nienhuis J, Slocum MK, DeVos DA, et al. Genetic similarity among *Brassica oleracea* L. genotypes as measured by restriction fragment length polymorphisms. J Amer Soc Hort Sci 1993; 118:298-303.

53. Santos JB, Nienhuis J, Skroch P,et al. Comparison of RAPD and RFLP genetic markers in determining genetic similarity among *Brassica oleracea* L. genotypes. Theor Appl Genet 1994; 87:909-915.

54. Jain A, Bhatia S, Banga SS, et al. Potential use of random amplified polymorphic DNA (RAPD) technique to study the genetic diversity in Indian mustard (*Brassica juncea*) and its relationship to heterosis. Theor Appl Genet 1994; 88:116-122.

55. McGrath JM, Quiros CF. Genetic diversity at isozyme and RFLP loci in *Brassica campestris* as related to crop type and geographical origin. Theor Appl Genet 1992; 83:783-90.

56. Pradhan AK, Prakash S, Mukhopadhyay A, et al. Phylogeny of *Brassica* and allied genera based on variation in chloroplast and mitochondrial DNA patterns: molecular and taxonomic classifications are incongruous. Theor Appl Genet 1992; 85:331-340.

57. Demeke T, Adams RP, Chibbar R. Potential taxonomic use of random amplified polymorphic DNA (RAPD): a case study in *Brassica*. Theor Appl Genet 1992; 84: 819-24.

58. Thormann CE, Ferreira ME, Camargo LEA, et al. Comparison of RFLP and RAPD markers for estimating genetic relationships within and among cruciferous species. Theor Appl Genet 1994; 88:973-80.

59. Grellet F, Delcasso D, Panabières F, et al. Organization and evolution of a higher plant alphoid-like satellite DNA sequence. J Mol Biol 1986; 187:495-507.

60. Hallden C, Bryngelsson T, Sall T, et al. Distribution and evolution of a tandemly repeated DNA sequence in the family *Brassicaceae*. J Mol Evol 1987; 25:318-23.

61. Reddy AS, Srivastava V, Mukherjee G. A tandemly repeated DNA sequence from *Brassica juncea*. Nucleic Acids Res 1990; 17:5849.

62. Harbinder S, Lakshmikumaran M. A repetitive sequence from Diplotaxis erucoids is highly homologous to that of *Brassica campestris* and *B. oleracea*. Plant Mol Biol 1990; 15:155-56.

63. Lakshmikumaran M, Ranade SA. Isolation and characterization of a highly repetitive DNA of *Brassica campestris*. Plant Mol Biol. 1990; 14:447-48.

64. Sibson DR, Hughes SG, Bryant JA, et al. Sequence organization of a simple, highly repetitive DNA elements in *Brassica* species. J Exp Bot 1991; 42:243-49.

65. Gupta V, Jagannathan V, Lakshmikumaran M. A novel A-T rich tandem repeat of *Brassica nigra*. Plant Sci 1990; 68:223-29.

66. Iwabuchi M, Itoh K, Shimamoto K. Molecular and cytological characterization of repetitive DNA sequences in *Brassica*. Theor Appl Genet 1991; 81:349-55.

67. Rogstad SH. Inheritance in turnip of variable-number tandem-repeat genetic markers revealed with synthetic repetitive DNA probes. Theor Appl Genet 1994; 89:824-30.

68. Deragon J-M, Landry BS, Pélissier T, et al. An analysis of retroposition in plants based on a family of SINEs from *Brassica napus*. J Mol Evol 1994; 39:378-86.

69. Funke RP, Kolchinsky A, Gresshoff PM.

Physical mapping of a region in the soybean (*Glycine max*) genome containing duplicated sequences. Plant Mol Biol 1993; 22:437-446.

70. Maluszynska J, Heslop-Harrison JS. Localization of tandemly repeated DNA sequences in *Arabidopsis thaliana*. Plant J 1991; 1:159-66.

71. Hanson RE, Zwick MS, Choi S, et al. Fluorescent in situ hybridization of a bacterial artificial chromosome. Genome 1995; 38:646-651.

72. Tanksley SD, Ganal MW, Martin GB. Chromosome landing: a paradigm for map-based gene cloning in plants with large genomes. Trends Genet 1995; 11:63-68.

73. Lisitsyn N, Lisitsyn N, Wigler M. Cloning the differences between two complex genomes. Science 1993; 259:946-51.

74. Zabeau M, Vos P. European Patent Application. 1992.

CHAPTER 17

GENOME MAPPING IN LEGUMES (FAM. FABACEAE)

Nevin D. Young, Norman F. Weeden and Gary Kochert

17.1. INTRODUCTION

The Fabaceae, better known as legumes, is the second most important family of crop plants after the Poaceae.[1] Legumes comprise a large and diverse family with more than 700 recognized genera and 17,000 species. There are three main subfamilies, Mimosoideae, Caesalpinioideae and Papilionoideae. Of these, the Papilionoideae contains nearly all economically important species and is the only one extensive in temperate regions. Members of this subfamily are distributed worldwide and include everything from small annual herbs to large woody trees. In agriculture, legumes are important because of their ability to form symbiotic relationships with nitrogen-fixing bacteria of the genus *Rhizobium*. Consequently, legume crops are valuable as a source of vegetable protein for people and livestock, and as green manure in the field. This is especially true in less developed countries around the world.

The genetics of many legume species have been studied extensively because of their central role in world agriculture. The garden pea (*Pisum sativum*) was used by Mendel in his pivotal studies of transmission genetics. Classical genetic maps composed of morphological mutants and disease resistance loci have been constructed for most economically important legume species.[2] More recently, researchers have focused on the development of genetic linkage maps composed of DNA markers such as restriction fragment length polymorphisms (RFLP),[3] random amplified polymorphic DNA (RAPD)[4] and simple sequence repeat (SSR) markers.[5] This chapter briefly reviews the status and future prospects for genome mapping in the major cultivated legume species.

Because of their broad botanical diversity, legume crop species are difficult to treat as a single genetic system. Cultivated members can be split according to centers of diversity (temperate versus tropical) or use

Genome Mapping in Plants, edited by Andrew H. Paterson. © 1996 R.G. Landes Company.

by people (forage versus grain).[1] In this chapter, cultivated legumes originating in the tropics or subtropics (including the genera *Glycine*, *Phaseolus*, and *Vigna*) will be discussed together, while cool season species (including the genera *Trifolium*, *Medicago*, *Melilotus*, *Cicer*, *Pisum*, *Vicia* and *Lens*) will be discussed separately. Another economically important legume, *Arachis* (groundnut or peanut), is sufficiently distinct to merit discussion by itself.

17.2. *GLYCINE, PHASEOLUS, AND VIGNA*

17.2.1. OVERVIEW

These three genera, all members of the tribe Phaseoleae, include some of the most important crops in the world. *Glycine max*, the soybean, is grown on more than 26 million hectares in the United State alone.[1] Edible and industrial oils and proteins, livestock feed, printing inks, waterproofing materials, and synthetic fibers are just a few of the products derived from soybean. The center of domestication for soybean is northern China. Soybean has one of the longest recorded histories of cultivation among crop plants, dating back to the Chou Dynasty in 664 B.C.[6] *Phaseolus*, including *P. vulgaris* (common bean), *P. coccineus* (scarlet runner bean), and *P. lunatus* (lima bean) are important food crops that are consumed as either fleshy pods or mature, protein-rich seeds. All three bean species were domesticated in Central and South America. *Vigna*, which was previously grouped together with *Phaseolus*, includes the Asian crop species *V. radiata* (mung bean) and the African species *V. unguiculata*.

Classical taxonomic and cytogenetic evidence strongly indicates a close relationship between *Phaseolus* and *Vigna* and a looser connection to *Glycine*.[7] Cultivated members of *Phaseolus* and *Vigna* are all true diploids with 2n = 2x = 22, while soybean behaves as a diploidized tetraploid[8] with 2n = 40. The chromosomes in all three genera are small and difficult to see under the microscope. Indeed, mung bean

and common bean both have genomes that are quite small in size, between 0.94 and 1.04 pg per diploid genome in mung bean and 1.28 pg in common bean.[9,10] The diploid genome size of soybean is somewhat larger at 2.24 pg.[9] All three are strongly inbreeding and within each genus only very limited success can be achieved in the production of fertile interspecific hybrids. Classical genetic maps (including isozyme markers) exist for soybean and common bean, though there are only 63 loci on the soybean map and 35 markers on the common bean map.[2]

17.2.2. DNA MARKER MAPS

Genetic maps composed of DNA markers have recently been constructed for soybean,[11] common bean,[12] cowpea[13] and mung bean.[14] In general, the maps are based on RFLPs and RAPDs, although there are efforts to develop SSR marker maps in soybean[5] since this type of marker typically exhibits a high level of polymorphism (see chapter 10).

In soybean, molecular genetic maps based on both intraspecific[15] and interspecific (*G. max* x *G. soja*)[16] populations have been constructed. The interspecific map consists of 400 marker loci spanning more than 3000 centiMorgans (cM) on 24 linkage groups.[17] By contrast, the intraspecific map has 132 marker loci spanning approximately 1500 cM on 31 linkage groups.[18] Clearly these maps should be still be considered incomplete since they have not yet condensed to the known haploid number of chromosomes (1n = 20), nor has there been any systematic attempt to tag telomeres with DNA markers (chapter 3). The relationship between the molecular marker maps and maps of classical markers has been investigated.[19] Crosses segregating for seven pigmentation and seven isozyme markers were mapped with 110 RFLPs and eight RAPD markers. In this way, half of the 19 classical linkage groups of soybean have been correlated with DNA marker groups. The diversity of RFLP markers tends to be low in soybean. For example, 54% of the marker loci were monomorphic in a survey with 86 loci on 108 soybean cultivars of the southern U.S.[20] In a separate survey 64 northern soybean cultivars and their ancestors were analyzed with RFLP markers well distributed throughout the soybean genome.[18] The authors identified a core set of RFLP markers with locus diversity values greater than 0.3 (an indication that any two accessions tested would carry different alleles; 0.5 would be the maximum for marker loci with two allelic forms). They recommended this core set of RFLPs as a starting point for researchers interested in soybean mapping.

In common bean, two distinct DNA marker maps have been constructed. In one, 115 RFLPs, seven isozyme and eight RAPD markers map onto 15 linkage groups.[19] The total length of the map is slightly greater than 800 cM with an average of 6.5 cM between markers. The other map, based on a backcross between Mesoamerican and Andean genotypes, includes 224 RFLPs spanning 960 cM.[20] Isozyme as well as DNA marker data indicate that common bean can be split into two well-defined geographical gene pools.[21] The Mesoamerican pool is found from Mexico through Central America down to northern Peru. The other, the Andean, is found only in Peru and Argentina. Levels of DNA polymorphism are relatively high in common bean. Between the Mesoamerican and Andean gene pools, 80% to 90% of DNA clones exhibit polymorphisms, while within either gene pool, the level of polymorphism lies between 50% and 60%.[21] By examining sequence evolution of phaseolin, a major seed storage protein, Kami et al[22] identified an ancestral sequence in a nearly extinct bean accession of Ecuador and Peru that was intermediate between the Mesoamerican and Andean groups. In crosses between members of the two gene pools, Mesoamerican alleles seem to be preferentially inherited.[23]

Maps have been developed for both of the major cultivated *Vigna* species (mung bean and cowpea). In mung bean, the map includes 172 RFLPs assigned to 11 large linkage groups and three small linkage groups of four markers or less.[14] The map spans 1570 cM with an average spacing between markers of 9 cM. In cowpea, the map consists of 83 RFLP and 5 RAPD marker loci spanning 680 cM on 10 linkage groups plus several unlinked marker loci.[13] Linkage relationships among RFLPs common to both *Vigna* maps have been examined in order to compare similarities in genome organization.[24] While there was a high degree of conservation in gene order between the two species, variations in copy number were detected and several rearrangements in linkage orders appeared to have occurred since the divergence of the species. Nonetheless, large conserved blocks nearly the entire length of some linkage groups were also detected.

Comparative genome analysis between members of all three genera has also been performed.[25] Consistent with classical taxonomic assignments, mung bean and common bean exhibited a high degree of linkage conservation and marker order. With two exceptions, all mung bean linkage groups corresponded to only one or two common bean linkage groups (Fig. 17.1). In fact, the mung bean and common bean genomes appear to be nearly as similar to one another as mung bean is to cowpea.[24] On average, linkage blocks more than 36 cM in length were conserved between mung bean and common bean. The authors suggest that these crop species can probably be treated as a "single genetic system" in much the same way as many cultivated grasses are now viewed.[26] By contrast, the soybean genome contained much shorter blocks in common with either mung bean or common bean. On average, conserved linkage blocks were only 12 to 13 cM in length and there were many examples of marker loci contiguous in mung bean and common bean that were unlinked in soybean. Still, the many examples of conserved linkage among the three genera offer valuable opportunities for increasing marker density in genomic regions of interest through comparative mapping.

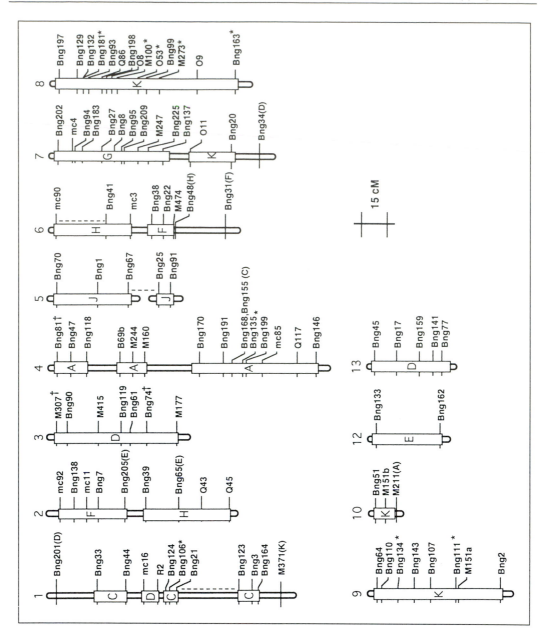

Fig. 17.1. Molecular map of mung bean showing extensive genome conservation with common bean. Mung bean linkage groups are shown as tall rods and conserved blocks from common bean are shown as superimposed rectangles. Numbers at the top of mung bean linkage groups are the same as in ref. 14, with the exception that there is a new linkage group 13 composed solely of common bean markers in the figure. Letters in rectangles and parentheses refer to common bean linkage groups described.[20] The rectangles indicate regions where collinearity has been maintained between mung bean and common bean, with the exception that asterisks indicate marker loci out of order at LOD (log of the odds ratio) greater than or equal to 2.0 and daggers indicate marker loci out of order at less than LOD 2.0. Horizontal tick marks indicate the locations of RFLP markers mapped in common between the two species; Bng markers come from common bean and markers labeled with mc, M, O, R and Q come from mung bean. Dashed lines indicate that the evidence for linkage between marker loci was less than LOD 3.0. Figure reprinted with permission.[25]

17.2.3. GENOME ORGANIZATION

Of the three cultivated species in the Phaseolideae, soybean genome organization has been especially well investigated. Just as DNA markers can be used to analyze genome relationships between related genera, markers can also be used to examine paralogous regions within the same genome (related genomic blocks repeated within a single genome). Duplicated genomic regions are common in soybean and some linkage groups contain as many as 33 loci duplicated elsewhere in the genome (R. Shoemaker et al, unpublished). The size of paralogous regions in soybean ranged from 9 to 127 cM in length. Rather than simply observing two homeologous segments for repeated blocks, segments were found to be duplicated an average of 4.2 times with many examples of duplicated segments nested within others. The authors interpret these results to indicate that one of the original genomes contributing to the polyploid nature of the soybean genome may have undergone an additional round of ancient tetraploidization.

Other studies of soybean genome organization suggest that duplicated sequences are common. For example, Kolchinsky et al[27] identified a highly repetitive sequence based on a 92 base pair (bp) motif. Altogether, this family of repeated sequences is more than seven million base pairs (Mbp) in length, organized into giant blocks over 1 Mbp in size. Significantly, there are at least 10 times fewer copies of this sequence in *V. unguiculata*. Similarly, a long interspersed repetitive DNA element, known as SIRE-1, has been described that terminates in an inverted repeat.[28] This element is repeated between 500 and 800 times per genome and each is approximately 10,600 bp in length. Indeed, some evidence for endogenous transposable elements in soybean has been uncovered, including the *Tgm* family.[29] Detailed study of this sequence family has demonstrated that many members are partial elements with internal and border deletions common.

Phenotypic mapping

Many agriculturally important traits have been mapped in the cultivated members of *Glycine*, *Phaseolus* and *Vigna*. In part, these efforts have been directed toward an understanding of the underlying genetics of the traits. Equally important have been efforts to identify DNA markers useful in indirect selection of target traits without the need for phenotypic scoring and to provide entry points for positional gene cloning.[30] Traits of interest include both genetically simple Mendelian genes as well as complex polygenic characters under the control of multiple quantitative trait loci (QTLs). Examples of simple, monogenic characters mapped in soybean include resistance genes against different races of *Phytophthora infestans*[31] and a locus controlling supernodulation by *Bradyrhizobium*.[32] In mapping studies of resistance to soybean cyst nematode (*Heterodera glycines*), a major partial resistance locus has been identified and resolved into a qualitative locus.[33]

There have been numerous mapping studies of QTLs associated with agronomic traits in soybean. Given the many economic uses of soybean, it is not surprising that QTLs underlying characters like seed oil and protein,[34] maturity,[35] hard seededness,[36] height[35] and a variety of other reproductive and morphological traits[35,37,38] have all been mapped in relation to DNA markers. Interestingly, a single genomic region on linkage group J contains the highest concentration of agronomically important genes, including loci imparting resistance to *Phytophthora*[39] and soybean cyst nematode resistance,[33] responsiveness to *Bradyrhizobium*,[32] and several other quantitative characters. In the process of phenotypic mapping in soybean, profound insights into epistatic interactions between genes were uncovered.[40] Various soybean QTLs were analyzed for their dependence on particular alleles at a second marker locus. The results indicated that interactions between QTLs were frequent, and could account for large portions of phenotypic variance.

Although not as numerous, many important phenotypic traits have been mapped in common bean and mung bean. RAPD markers linked to major genes for rust (*Uromyces appendiculatus*)[41] and bean common mosaic virus[42] resistance have been identified in common bean. In both cases, the markers were less than 2 cM from the gene of interest and should be useful in marker-assisted breeding. Several QTLs for partial resistance to bacterial pustule (*Xanthomonas campestris*) have been mapped in common bean and one is located very near a gene controlling nodule number.[43] In mung bean a major gene for resistance to the insect pest, *Callosobruchus* (pod weevil), has been mapped with tightly linked DNA markers.[44] Resistance QTLs for powdery mildew (*Erysiphe polygoni*) of mung bean have also been mapped.[45] Interestingly, the array of QTLs important in powdery mildew resistance changed during the course of the infection cycle.

One phenotypic trait mapped in both mung bean and cowpea is of special interest. Seed weight is a quantitative character showing nearly normal distribution in F2 populations in both species. QTL mapping of this trait uncovered the same genomic interval as the major locus in both species,[46] despite the fact the two species evolved and were domesticated on separate continents. Recently, the same marker loci associated with seed weight in mung bean and cowpea have also been shown to be associated with this character in the garden pea (G. Timmerman-Vaughn, Christchurch, New Zealand, personal communication).

17.2.4. PHYSICAL MAPPING

There have been relatively few physical mapping studies in *Glycine*, *Phaseolus*, or *Vigna*. Still, some clusters of loci tightly linked by linkage mapping have been examined with pulsed field gel analysis in soybean. In one study of markers near a supernodulation locus,[47] the relationship between genetic and physical distance was estimated to be 550 kbp per cM, very close to the calculated average for the entire

genome. In other studies, this relationship was found to be as low as 150 kbp per cM near a soybean cyst nematode resistance locus that is found near the end of a linkage group.[48] Needless to say, physical mapping studies are complicated in soybean by the frequent presence of duplicated sequences. Still, this complication could be resolved by the authors through the use of two dimensional electrophoresis[47] or segregational analysis of high molecular weight fragments.[48] One interesting result in these studies was the observation that rare cutting restriction enzyme sites seem to be conserved between homeologous regions.[47]

17.3. TEMPERATE LEGUMES

17.3.1. OVERVIEW

The species covered in this section belong to tribes which, in contrast to the Phaseoleae, are primarily temperate in origin.[7] Classification as 'temperate' does not mean that these genera are found only in temperate latitudes or that other legumes do not grow under temperate conditions, but that the ancestors of these genera apparently evolved in temperate regions of the world. The primary focus of the section will be on species in *Viceae* (pea, lentil, faba bean), *Cicereae* (chickpea), and *Trifolieae* (alfalfa and sweetclover), which together represent a cohesive and monophyletic group easily identified by the absence of the inverted repeat on the chloroplast genome.[49,50]

Again in contrast to the Phaseoleae, the Viceae, Cicereae and Trifolieae all have relatively low basal chromosome numbers (characteristically n = 7, 8). In the Viceae and Cicereae polyploidy is rare, though alfalfa is a well-known autotetraploid and a number of clover and certain other taxa of the Trifolieae are polyploid. Surprisingly, the lower chromosome number is not associated with a reduced DNA content relative to Phaseoleae. Pea (*Pisum sativum*) and lentil (*Lens culinaris*) both have n = 7 but have relatively large genomes of approximately 8.5 pg DNA/2C, over 10-fold larger than that of mung bean or common bean.[9,51] Faba bean (*Vicia faba*), with n = 6, has about 25.0 pg DNA/2C. Diploid species in the Cicereae and Trifolieae (usually n = 8) have much smaller genomes, only slightly larger than those in the Phaseoleae: chickpea (*Cicer arietinum*, 1.5 pg/2C), sweetclover (*Melilotus alba*, 2.3 pg/2C), and even tetraploid alfalfa (*Medicago sativa*, 3.1 pg/2C). Thus, within the temperate legumes, the Viceae clearly has had a significant increase in DNA content per chromosome. As we shall see below, this increase in DNA content appears to be correlated with an increase in recombination frequency per chromosome and total length of the linkage map.

17.3.2. GENETIC MAPS

Of the temperate legumes, pea has the greatest number of mapped genes, particularly with regard to morphological mutations, isozyme loci, and RFLPs. Linkage maps from several programs have been published.[52-55] DNA markers have been such powerful and abundant tools for genetic mapping studies that relatively little emphasis has been made in some cases to place classical (e.g., morphological markers listed in Blixt[52]) on the new maps. Thus, despite having over 1000 DNA markers 'mapped' on linkage groups in various laboratories, it often has been difficult to identify homologous linkage groups on maps from different laboratories. Fortunately, this situation is improving with the identification of standard DNA probes[56] and greater communication between programs. The most recent 'consensus' map[57] had over 300 genes located relative to each other, but that number has now more than doubled and does not include the many RAPDs and amplified fragment length polymorphisms (AFLPs) that have been added to previous maps.

Specific regions on the pea map containing genes of particular interest have been saturated with DNA markers to nearly one marker per cM.[58] However, the overall map remains something of an enigma, being well-saturated and reproducible in certain regions, while apparently devoid of markers in other areas. The classical map of Blixt, which presented the known morphological markers arranged into seven distinct and fairly well-saturated linkage groups (maximum distance between adjacent markers about 20 cM) has turned out to be an overly optimistic interpretation of the linkage data of Lamprecht (see summary in ref. 59) and other previous pea geneticists. Chromosome 7 of the Blixt map has become part of chromosome 5, and chromosomes 1 and 2 each now have been split into two independent linkage groups.[59] Translocations are well known in pea,[61,62] but both Blixt and Lamprecht were aware of many of these translocations and scrupulously tested their parental lines for the presence of the standard karyotype. It is unlikely that the differences between the classical and the molecular maps can be attributed to the use of lines with major differences in karyotype.

The addition of molecular markers has increased the length of the pea linkage map to nearly 2000 cM[54] and has provided much greater saturation (approaching 2 cM saturation in many regions), but surprisingly has failed to coalesce the map into seven major linkage groups. None of the maps developed in individual laboratories has fewer than 12 linkage groups, albeit some are very small, and the consensus map still has nine major linkage groups. Although a consensus map with seven linkage groups is expected within the next few years, it is remarkable that pea has proven so reticent to reveal its complete genetic fingerprint, despite its popularity and high level of genetic diversity.

Maps of other members of the Viceae are less complete. Lentil, as opposed to pea, has very few morphological markers and no linkage map was available before the development of molecular markers. A relatively extensive linkage map is now available for lentil[63,64] consisting of about 75 morphological, isozyme and RFLP markers arranged in 10 linkage groups with a total length of over 700 cM. The arrangement of genes on the map has been compared to that in pea using isozyme and cDNA markers.[65] Eight regions making up about 40% of the lentil map were identified as containing the same arrangement of linked markers as that found in pea.

A similar conservation of linkage groups has been determined in chickpea.[66] Unfortunately, the crosses used in chickpea displayed fewer RFLPs than those in pea or lentil and most of the comparison had to be performed using isozyme loci. However, many morphological polymorphisms also have been mapped in chickpea and the map appears to be relatively complete with a total length of about 600 cM. A preliminary map for faba bean has been published by Torres et al.[67] This map consisted primarily of RAPD markers and the linkage relationships cannot be easily compared with those in the other species. There were no obvious regions of linkage conservation.

Alfalfa and sweetclover are two members of the Trifolieae with extensive linkage groups. Alfalfa, despite its importance as a crop plant, has been slow to develop a classical genetic map because of its high level of outcrossing and autotetraploid nature. Molecular markers have made it possible, however, to construct high density linkage maps. All the alfalfa maps have been constructed using diploid species or diploid derivatives of tetraploid alfalfa to simplify linkage calculations. Three alfalfa maps have been independently constructed. Brummer et al[68] used a cross between a diploid *Medicago sativa* ssp. *falcata* accession and a diploid derivative of *M. sativa* (a cultivated alfalfa at the diploid level or CADL) line. A single F1 plant was selfed, and mapping was performed in the F2 by selecting those loci that were heterozygous in the F1 hybrid. More than 150 RFLP (cDNAs) have been mapped in this population, and a number are being added to the map (N. Diwan et al, unpublished). This map is notable for its very low rate of recombination; the entire map totals only about 700 cM. Many loci were also detected that exhibited severe segregation distortion, usually in favor of a great excess of heterozygotes. This may relate to

the high degree of inbreeding depression seen in cultivated alfalfa, which was the source of the CADL line used in this cross. Kiss et al[69] used a similar approach to develop an alfalfa linkage map. In this case the initial cross was between a diploid *M. sativa* ssp. *quasifalcata* and a diploid *M. sativa* ssp. *coerulea*. Many of the 89 RFLP, RAPD, isozyme and morphological markers mapped showed segregation distortion, and again the map was relatively short (< 700 cM) indicating a low level of recombination. Echt et al[70] developed a mapping population of 87 progeny using a backcross developed from two CADL populations. Both RFLP and RAPD markers were used and 34% of the markers were shown to exhibit significant segregation distortion and levels of recombination were low.

A linkage map for sweetclover was assembled using the interspecific cross *M. alba* x *M. polonica*.[71] Relatively few isozyme polymorphisms or RFLPs were detected between the parents, and the map was primarily developed from 150 segregating RAPDs. Nine linkage groups were identified with a total length of 252 cM, perhaps the shortest linkage map yet reported for a higher plant. The shortness of the map may be partially a result of repression of recombination in and interspecific cross, but the length is comparable with the short linkage map of alfalfa.

17.3.3. GENOME ORGANIZATION

In one of the earliest studies of plant genome organization, Thompson et al[72] demonstrated that pea possessed a much greater percentage (85%) of repetitive sequences than mung bean. These sequences were widely scattered throughout the pea genome. Later work suggests that these sequences consist of many different types of repeats, as virtually every arbitrary primer-amplified sequence tested seems to contain repetitive sequences, yet few of these amplified DNA fragments possess the same repeat (N.F. Weeden and G.M. Timmerman-Vaughan, unpublished). Currently, there is little available information on lentil repetitive sequences and only a few gene families have been investigated in other members of the Viceae.[73-74]

Several studies have investigated the effect of changes in genome size on a variety of related parameters. Significant differences in genome size within *Lathyrus* did not appear to have much affect on the distribution or divergence of repetitive DNA sequences[75] and the large faba bean genome was shown to be highly tolerant of duplications of chromosomal segments, but not of deletions.[76] One particularly interesting result from comparative mapping studies is that although there is about a two-fold difference in length of the linkage maps of pea and chickpea, the recombination frequency within conserved linkage blocks appears to be approximately the same.[77] Although this finding is based on very few comparisons, it suggests that the difference between pea and chickpea in map length may be primarily caused by DNA inserted between 'islands' of conserved linkages.

17.3.4. Phenotype Mapping

One of the most exciting areas of genetic research in temperate legumes has been the application of the maps to tag genes and investigate traits with a complex genetic basis. The development of bulked segregant analysis using arbitrary primers[77] has made the identification of DNA markers linked to specific genes a routine task. Most of the disease resistance genes in pea have now been tagged with DNA markers.[78-80] QTLs for seed size and color retention have been identified in pea by Timmerman-Vaughan et al.[81]

Pea has been particularly amenable to genetic studies on nodule formation and metabolism and ranks among the best model systems for exploring this complex trait. Over 40 mutants that are in some way defective in nodulation capacity have been isolated, and although allelism tests are still being performed, at least 30 different genes have been defined, including those encoding the nodulins leghemo-

globin, ENOD2, ENOD3, ENOD5, ENOD7, and ENOD40.[82-84] These genes are widely distributed over the pea linkage map, although one region on linkage group I appears to have several genes, three nodulin loci (*Lb, Enod7* and *Enod40*) as well as four mutants (*sym2, sym5, sym19* and *nod-3*) within 30 cM of each other.[58,85] This region is particularly intriguing because genes involved in the same metabolic pathway or otherwise functionally related are rarely closely linked in higher eukaryotes. In those cases where linkage between functionally related genes does occur (seed protein genes, ribosomal repeats, disease resistance genes), the genes usually form very tight clusters and are believed to be related by gene duplication. The nodule-associated genes in this region do not appear to be related by gene duplication and, thus, may form a unique type of loose cluster not previously described in plants.

Only a few phenotypic traits have been mapped in alfalfa. These include a dominant purple flower color factor (C2), several isozymes, a male sterility factor, a dwarf mutation, a "stick leaf" mutant, an anthocyanin mutant and a xanthophyll mutant.[69,70]

Map-based cloning

The large size of the genomes in the Viceae has intimidated most plant molecular geneticists from seriously attempting map-based cloning in these taxa. Although no specific determination of the relationship between physical and genetic maps have been made in this tribe, the relationship probably averages 5-10 Mbp per cM based on comparisons with Phaseolus and other taxa where such determinations have been made. Chickpea, sweetclover, and alfalfa appear to be much better organisms for genomic walks; however, the low levels of recombination that are seen in the Trifoleae so far analyzed will make map-based cloning difficult. For example, one cM would equal about 1.5 Mbp in the mapping population analyzed by Kiss et al.[69]

17.4. *ARACHIS*

17.4.1. OVERVIEW

The genus *Arachis* is a member of the legume tribe Aeschynumeneae, subtribe Stylosanthinae, and it shares morphological characters such as a staminal tube, alternately attached basal and dorsal anthers, pinnate leaves and flowers in small heads or spikes with related genera such as Stylosanthes and Arthrocarpum. *Arachis* is, however, different from its closest relatives in that it produces aerial flowers but subterranean fruit. This character clearly delineates *Arachis* from closely-related genera[86] and has interesting implications in terms of susceptibility to soil-borne diseases, such as *Aspergillus* infection, which can lead to aflatoxin contamination.

The genus *Arachis* contains about 70 taxa that are grouped into several sections based on morphological characters and crossability.[87] Five species are important commercially as forage or seed crops. By far the most important of these is *Arachis hypogaea*, domesticated peanut, which is one of the principal food legumes of the world. Domesticated peanut is an allotetraploid (2N=4X=40) of South American origin and about 20 million hectares are planted to domesticated peanut worldwide. Most experimental work on peanut has been confined to domesticated peanut and its relatives in section Arachis, which contains about 20 diploid wild species and a single wild tetraploid species. These related species are important potential sources of genes for pest and disease resistance that could be introgressed into domesticated peanut.

When Spanish and Portuguese explorers arrived in the New World, domesticated peanut was being grown over a large area of South America, in Mexico, and on some of the Caribbean islands.[86] Recent cytogenetic and molecular marker studies indicate that domesticated peanut originated in northern Argentina or southern Bolivia and resulted from a hybridization event between the diploid wild species *A. duranensis* and *A. ipaensis*[88,89] (G. Kochert

et al, unpublished). *A. duranensis* was apparently the female parent, since its chloroplast DNA is more similar to that of *A. hypogaea* than is that of *A. ipaensis* (G. Kochert et al, unpublished).

Domesticated peanut exhibits a considerable amount of genetic variation for morphological traits such as upright versus prostrate habit, seed coat color, number of seeds per pod, seed size and patterns of flower production on the stems.[87,90] In addition there is variation for resistance to insects and diseases among the various cultivars and landraces of domesticated peanut, although only partial resistance has been found for pests and pathogens. On the basis of morphological characters, *A. hypogaea* has been divided into two subspecies (*A. hypogaea* ssp. *hypogaea* and *A. hypogaea* ssp. *fastigiata*, with two and four varieties, respectively), and four "market types" (Virginia and Runner in ssp. *hypogaea* and Valencia and Spanish in ssp. *fastigiataa*).[87]

In contrast to morphological variation, molecular marker surveys have detected very little genetic variation among improved cultivars or landraces of *A. hypogaea*. This low level of polymorphism is exhibited for isozymes,[91] RFLPs,[88,92,93] and RAPDs.[94] Microsatellite (SSR) analysis is just beginning. However, it is clear that even with these markers, which exhibit a large number of alleles and high levels of diversity in other crop plants, only a very low level of genetic variation is detected in domesticated peanut (S. Kresovich and G. Kochert, unpublished). This apparent contradiction can be reconciled under the following scenario. Domesticated peanut arose only once in the relatively recent past and all landraces and cultivars of domesticated peanut are descendants of that one event. Polyploidization isolated domesticated peanut genetically in the sense that ploidy differences prevented further gene flow with related diploid species. Native cultivators recognized the potential of the polyploid plants and, during domestication, selected for the small number of genes which now differ among the varieties of domesticated peanut and which are responsible for the morphological variation observed. These would have been genes for traits useful in any agronomic system, such as larger seed size, as well as genes for traits useful in specialized agricultural ecosystems, such as an upright habit for cultivation on riverine sandbars.

Domesticated peanut chromosomes are small, and represent difficult objects for cytogenetic study. One chromosome pair (the A chromosomes) differ markedly in size between the two constituent genomes present in domesticated peanut; the genome derived from *A. duranensis* contains the conspicuously smaller (A) pair, but the genome derived from *A. ipaensis* does not.[89,95] The other chromosomes of the basic set are hard to distinguish cytogenetically. The genome size of domesticated peanut is approximately 3.6 Mbp per diploid genome,[9] but is not yet known how this DNA is distributed between the *"duranensis"* and *"ipaensis"* genomes of domesticated peanut. No classical genetic map of conventional traits or isozymes exists for domesticated peanut, although several studies of segregation ratios for individual genes have been reported.[96]

17.4.2. DNA Marker Maps

The low level of polymorphism present among domesticated peanut accessions makes molecular map construction impractical with RFLPs or RAPDs. Microsatellite markers are being developed, but it is not yet clear whether enough variability will be present to allow cost-effective map construction and analysis of segregating populations of domesticated peanut. However, peanut molecular maps have been constructed using segregating populations developed from an interspecific cross between two diploid wild species related to domesticated peanut.[97] The two species used were *A. stenosperma* and *A. cardenasii*. F1 plants from this cross are fertile and produce F2 populations which exhibit only a small degree of segregation distortion. An F2 mapping population of 90 individuals was used to produce one of the molecular maps.

This map presently contains about 135 RFLP loci, the great majority of which were detected using cDNA probes from libraries developed from root or hypocotyl tissue derived from domesticated peanut seedlings. A few SSRs, RAPDs and clones of known genes have also been mapped (Fig. 17.2). All of the mapped cDNA clones have been partially sequenced, and comparisons to gene databases have identified about 15% of the formerly anonymous cDNA clones (S. Kresovich and G. Kochert, unpublished). The map spans about 1500 cM and consists of 11 linkage groups. The interspecific F1 was also backcrossed to *A. stenosperma* and this population was used to construct a map of RAPD markers. This map was constructed mainly to allow placement of RAPD markers used to locate introgressed chromosome segments in interspecific crosses.[98] The F2 map and the backcross map have been merged by mapping a number of RFLP markers onto the RAPD map (G. Garcia et al, unpublished).

A large number of cDNA probes from the peanut libraries have been mapped in soybean and an effort is underway to construct a comparative map (G. Kochert, unpublished). Peanut cDNA probes are also being screened against a variety of legume DNAs to try to locate common mapping probes which would give strong hybridization signals and relatively simple hybridization patterns.

17.4.3. Genome Organization

Almost nothing is known about genome organization in *Arachis* species. A large proportion of cDNA clones randomly chosen during map construction produce complicated hybridization patterns and represent repeated sequences (G. Kochert, unpublished), which seems to indicate that extensive genome duplication has occurred. Thus far, map construction has concentrated on clones that produce simple patterns, but future analysis of some of the complex patterns will yield more information about genome organization.

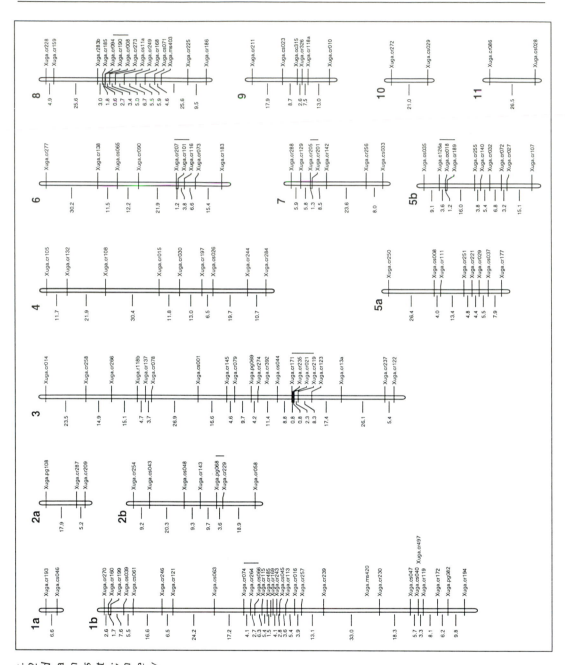

Fig. 17.2. Molecular map of peanut. The map was produced using an F2 population of 90 individuals derived from a cross of Arachis stenosperma x A. cardenasii. Symbols: cr (as in XUGA.cr010) denotes a locus mapped with a cDNA from a root library; cs = a shoot cDNA library; ms = a microsatellite. Clones with no mark connecting them to the map have only been approximately mapped to the interval denoted.

17.4.4. PHENOTYPIC MAPPING

No phenotypic traits were mapped in domesticated peanut prior to the development of molecular maps and unless more polymorphic markers are found, it is unlikely that it will be possible to map traits segregating in crosses between different accessions of domesticated peanut. However, it is possible to use the existing molecular maps to detect chromosome segments introgressed into domesticated peanut from wild species (as long as such wild species are different from the two species which hybridized to form domesticated peanut) and to map the position of these traits on the molecular maps derived from interspecific crosses. This work is now beginning. For example, Garcia et al (unpublished) have been able to map a nematode resistance gene introgressed into domesticated peanut from *A. cardenasii*. This appears to be a dominant gene that confers near immunity to the root-knot nematode.

17.5. CONCLUSION

From its beginning as a model system in Mendel's original experiments, the legume family has provided many profound insights into the nature of plant genetics. Recent applications of DNA mapping technology to legumes, especially economically valuable species, promises a future with many more useful discoveries. Legumes include species with small genomes that are suitable for positional cloning,[9] unique phenotypic traits like nitrogen fixing nodulation[32] and broad comparative mapping studies that indicate substantial conservation at the genome level.[25,65] Given that billions of people depend upon legumes for food, fiber, oil, and industrial products, aggressive efforts to utilize genome mapping technology in legumes may fundamentally improve the quality of life for people worldwide.

ACKNOWLEDGMENTS

We thank Dr. J. Groth for helpful suggestions regarding the manuscript. This paper is published as contribution No. 22,232 of the Minnesota Agricultural Experiment Station on research conducted under Project 015, supported by G.A.R. funds.

REFERENCES

1. Langer RHM, Hill GD. Agricultural Plants. 2nd ed., Cambridge, U.K.: Cambridge University Press, 1991.
2. O'Brien SJ. Genetic Maps Locus Maps of Complex Genomes. 6th ed., Cold Spring Harbour, NY: Cold Spring Harbour Laboratory Press, 1993.
3. Botstein D, White RL, Skolnick M, Davis RW. Construction of a genetic linkage map in man using restriction fragment length polymorphisms. Amer Journ Hum Genet 1980; 32:314-331.
4. Williams J, Kubelik A, Livak K, Rafalski J, Tingey S. DNA polymorphisms amplified by arbitrary primers are useful as genetic markers. Nucl Acids Res 1990; 18:6531-6535.
5. Akkaya MS, Bhagwat AA, Cregan PB. Length polymorphisms of simple sequence repeat DNA in soybean. Genetics 1992; 132:1131-1139.
6. Harlan JR. Crops & Man. 2nd ed., Madison, WI: American Society of Agronomy, 1992.
7. Polhill RM, Raven PH, ed. Advances in Legume Sytematics. Kew, England: Royal Botanic Gardens, 1981.
8. Simmonds NW, ed. Evolution of Crop Plants. New York: Longman, Inc., 1976.
9. Arumuganathan K, Earle E. Nuclear DNA content of some important plant species. Plant Molec Biol Report 1991; 9:208-218.
10. Murray MG, Palmer JD, Cuellar RE, Thompson WF. Deoxyribonucleic acid sequence organization in the mung bean genome. Biochemistry 1979; 18:5259-5264.
11. Shoemaker RC, Olson TC. Molecular linkage map of soybean (Glycine *max* L. Merr.) (2N = 40). In: Genetic Maps, O'Brien SJ, ed., Cold Spring Harbour, NY: Cold Spring Harbour Laboratory Press 1993:6.131-6.138.
12. Gepts P, Linkage map of common bean (*Phaseolus vulgaris*, L.) 2N = 22. In: Genetic Maps, O'Brien SJ, ed., Cold Spring Harbour,

NY: Cold Spring Harbour Laboratory Press 1993:6.101-6.109.

13. Fatokun CA, Danesh D, Menancio-Hautea D, Young ND. A linkage map for cowpea [*Vigna unguiculata* (L.) Walp.] based on DNA markers (2N=22), In: Genetic Maps, O'Brien SJ, ed., Cold Spring Harbour, NY: Cold Spring Harbour Laboratory Press 1993:6.256-6.258.

14. Menancio-Hautea D, Kumar L, Danesh D, Young N. A genome map for mungbean (*Vigna radiata* (L.) Wilczek) based on DNA genetic markers (2N=2X=22). In: Genetic Maps, O'Brien SJ, ed., Cold Spring Harbour, NY: Cold Spring Harbour Laboratory Press 1993:6.259-6.261.

15. Lark KG, Weisemann JM, Matthews BF, Palmer R, Chase K, Macalma T. A genetic map of soybean (*Glycine max* L.) using an intraspecific cross of two cultivars: 'Minsoy' and 'Noir 1'. Theor Appl Genet 1993; 86:901-906.

16. Shoemaker RC, Specht JE. Integration of the soybean molecular and classical genetic linkage groups. Crop Sci 1995; 35:436-446.

17. Skorupska HT, Shoemaker RC, Warner A, et al. Restriction fragment length polymorphism in soybean germplasm of the southern USA. Crop Sci 1993; 33:1169-1176.

18. Lorenzen LL, Boutin S, Young N, Specht JE, Shoemaker RC. Soybean pedigree analysis using map-based molecular markers 1. Tracking RFLP markers in cultivars. Crop Sci 1995; 35:1326-1336.

19. Nodari RO, Tsai SM, Gilbertson RL, Gepts P. Toward an integrated linkage map of common bean. 2. Development of an RFLP-based linkage map. Theor Appl Genet 1993; 85:513-520.

20. Vallejos CE, Sakiyama NS, Chase CD. A molecular marker-based linkage map of *Phaseolus vulgaris* L. Genetics 1992; 131: 733-740.

21. Becerra Valasquez VL, Gepts P. RFLP diversity of common bean (*Phasolus vulgaris*) in its centre of origin. Genome 1994; 37:256-263.

22. Kami J, Valasquez V, Debouck DG, Gepts P. Identification of presumed ancestral DNA sequences of phasolin in *Phaseolus vulgaris*. Proc Natl Acad Sci U.S.A. 1995; 92:1101-4.

23. Paredes OM, Gepts P. Segregation and re-combination in inter-gene pool crosses of *Phaseolus vulgaris* L. Journ Hered 1995; 86:95-106.

24. Menancio-Hautea D, Fatokun C, Kumar L, Danesh D, Young N. Comparative genome analysis of mungbean (*Vigna radiata* (L.) Wilczek) and cowpea (*V. unguiculata* (L.) Walpers) using RFLP mapping data. Theor Appl Genet 1993; 86:797-810.

25. Boutin S, Young N, Olston T, Yu ZH, Shoemaker R, Vallejos C. Genome conservation among three legume genera detected with DNA markers. Genome 1995; 38: 928-37.

26. Helentjaris T. Implications for conserved genomic structure among plant species. Proc Natl Acad Sci USA 1993; 90:353-363.

27. Kolchinsky A and Gresshoff PM. A major satellite DNA of soybean is a 92-base pairs tandem repeat. Theor Appl Genet 1995; 90:621-626.

28. Laten HM, Morris RO. SIRE-1, a long interspersed repetitive DNA element from soybean with weak sequence similarity to retrotransposons: initial characterization and partial sequence. Gene 1993; 134:153-159.

29. Rhodes PR, Vodkin LO. Organization of the Tgm family of transposible elements in soybean. Genetics 1988; 120:597-604.

30. Tanksley SD, Young ND, Paterson AH, Bonierbale MW. RFLP mapping in plant breeding: new tools for an old science. Bio/Technology 1989; 7:257-264.

31. Diers BW, Mansur L, Imsande J, Shoemaker RC. Mapping *Phytophthora* resistance loci in soybean with restriction fragment length polymorphism markers. Crop Sci 1991; 32:377-383.

32. Landau-Ellis D, Angermü ller S, Shoemaker R, Gresshoff PM. The genetic locus controlling supernodulation in soybean (*Glycine max* L.) co-segregates tightly with a cloned molecular marker. Molec Gen Genet 1991; 228:221-226.

33. Concibido VC, Denny RL, Boutin SR, Hautea R, Orf JH, Young ND. DNA marker analysis of loci underlying resistance to soybean cyst nematode (*Heterodera glycines* Ichinohe). Crop Sci 1994; 34:240-246.

34. Lark KG, Orf J, Mansur LM. Epistatic ex-

pression of quantitative trait loci (QTL) in soybean [*Glycine max* (L.) Merr.] determined by QTL association with RFLP alleles. Theor. Appl. Genet. 1994; 88:486-489.

35. Mansur LM, Orf J, Lark KG. Determining the linkage of quantitative trait loci to RFLP markers using extreme phenotypes of recombinant inbreds of soybean (*Glycine max* L. Merr.). Theor Appl Genet 1993; 86:914-18.

36. Keim P, Diers BW, Shoemaker RC. Genetic analysis of soybean hard seededness with molecular markers. Theor Appl Genet 1990; 79:465-469.

37. Keim P, Diers BW, Olson TC, Shoemaker RC. RFLP mapping in soybean: association netween marker loci and variation in quantitative traits. Genetics 1990; 126:735-742.

38. Mansur LM, Lark KG, Kross H, Oliveira A. Interval mapping of quantitative trait loci for reproductive, morphological, and seed traits of soybean (*Glycine max* L.). Theor Appl Genet 1993; 86:907-913.

39. Polzin KM, Lohnes DG, Nickell CD, Shoemaker RC. Integration of R*ps2*, *Rmd*, and *Rj2* into linkage group J of the soybean molecular map. Journ Hered 1994; 85:300-3033.

40. Lark KG, Chase K, Adler F, Mansur LM, Orf JH. Interactions between quantitative trait loci in soybean in which trait variation at one locus is conditional upon a specific allele at another. Proc Natl Acad Sci U.S.A. 1995; 82.4656-4660.

41. Haley SD, Miklas PN, Stavely JR, Byrum J, Kelly JD. Identification of RAPD markers linked to a major rust resistance gene block in bean. Theor Appl Genet 1993; 86:505-512.

42. Haley SD, Afanador L, Kelly JD. Identification and application of a random amplified polymorphic DNA marker for the I gene. Phytopathology 1994; 84:157-160.

43. Nodari RO, Tsai SM, Guzman P, Gilbertson RL, Gepts P. Toward an integrated linkage map of common bean. 3. Mapping genetic factors controlling host-bacteria interactions. Genetics 1993; 134:341-350.

44. Young ND, Kumar L, Menancio-Hautea DI, Danesh D, Talekar NS, Shanumgasundarum S, Kim DH. RFLP mapping of a major

bruchid resistance gene in mungbean (*Vigna radiata*, L. Wilczek). Theor Appl Genet 1992; 84:839-844.

45. Young N, Danesh D, Menancio-Hautea D, Kumar L. Mapping oligogenic resistance to powdery mildew in mungbean with RFLPs. Theor Appl Genet 1993; 87:243-249.

46. Fatokun CA, Menancio-Hautea D, Danesh D, Young ND. Evidence for orthologous seed weight genes in cowpea and mungbean based on RFLPs. Genetics 1992; 132:841-846.

47. Funke RP, Kolchinsky A, Gresshoff PM. Physical mapping of a region in the soybean (*Glycine max*) genome containing duplicated sequences. Plant Molec Biol 1993; 22:437-466.

48. Danesh D, Young N, Concibido V, Penuela S, Orf J. Physical mapping of DNA markers near a major soybean cyst nematode resistance locus. Soybean Genet Newsl 1995; In press.

49. Palmer JD, Osorio B, Aldrich J, Thompson WF. Chloroplast DNA evolution among legumes: loss of a large inverted repeat occurred prior to other sequence rearrangements. Curr Genet 1987; 11:275-286.

50. Lavin M, Doyle JJ, Palmer JD. Evolutionary significance of the loss of the chloroplast-DNA inverted repeat in the Leguminosae subfamily Papilionoideae. Evolution 1990; 44:390-402.

51. Bennett MD, Smith JB. Nuclear DNA amounts in angiosperms. Phil Trans Royal Soc Lond B 1976; 274:227-273.

52. Blixt S. The pea. In: King RS, ed. Handbook of Genetics, Vol 2. New York: Plenum Press, 1975:181-221.

53. Weeden, N.F. and B. Wolko. 1990. Linkage map for the garden pea (*Pisum sativum*) based on molecular markers. In: O'Brien SJ, ed. Genetic Maps, 5th ed. Cold Spring Harbor: Cold Spring Harbor Laboratory Press. 1990:6.106-6.112.

54. Ellis THN, Turner L, Hellens RP, Lee D, Harker CL, Enard C, Domoney C, Davies DR. Linkage maps in pea. Genetics 1992; 130:649-663.

55. Dirlewanger E, Isaac PG, Ranade S, et al. Restriction fragment length polymorphism analysis of loci associated with disease

resitance genes and developmental traits in *Pisum sativum* L. Theor Appl Genet 1994; 88:17-27.

56. Weeden NF, Swiecicki WK, Timmerman GM, Ambrose M. Guidelines for future mapping studies in *Pisum*. *Pisum* Genetics 1993; 25:13-14.

57. Weeden NF, Swiecicki WK, Ambrose M, Timmerman GM. Linkage groups of pea. *Pisum* Genetics 1993; 25:4.

58. Kozik A, Matvienko M, Scheres B, Bisseling T, Van Kammen A, Ellis THN, LaRue TA, Weeden NF. The pea early nodulin gene *Enod7* maps in the region of linkage group I containing *sym2* and leghemoglobin. Plant Mol Biol (in press)

59. Lamprecht H. Monographie der Gattung Pisum. Graz: Steiermarkische Landesdruckerei, 1974.

60. Kosterin OE 1993. Genes *a* and *d* may not be in the same linkage group. *Pisum* Genet 1993; 25:23-26.

61. Lamm R. Cytogenetical studies on translocations in *Pisum*. Hereditas 1951; 37: 356-372.

62. Folkeson D. Assignment of linkage segments to the satellite chromosomes 4 and 7 in *Pisum sativum*. Hereditas 112:257-263;1990

63. Havey MJ, Muehlbauer FJ. Linkages between restriction fragment length, isozyme, and morphological markers in lentil. Theor Appl Genet 1989; 77:395-401.

64. Simon CJ, M. Tahir M, Muehlbauer FJ. Linkage map of lentil (*Lens culinaris*). In: O'Brien SJ, ed. Genetic Maps, 6th ed. Cold Spring Harbor: Cold Spring Harbor Laboratory Press. 1993:6.97-6.100.

65. Weeden NF, Muehlbauer FJ, Ladinzinsky G. Extensive conservation of linkage relationships exists between pea and lentil genetic maps. Journ Hered 1992; 83:123-129.

66. Kazan K, Muehlbauer FJ, Weeden NF, Ladizinsky G. Inheritance and linage relationships of morphological and isozyme loci in chickpea (*Cicer arietinum* L.). Theor Appl Genet 1993; 86:417-426.

67. Torres AM, Weeden NF, Martin A. Linkage among isozyme, RFLP and RAPD markers in *Vicia faba*. Theor Appl Genet 1993; 85:937-945.

68. Brummer EC, Bouton JH, Kochert G.

Development of an RFLP map in diploid alfalfa. Theor Appl Genet 1993; 86:329-332.

69. Kiss GB, Csanadi G, Kalman K, Kalo P, Okresz L. Construction of a basic genetic map for alfalfa using RFLP, RAPD, isozyme and morphological markers. Mol Gen Genet 1993; 238:129-137.

70. Echt, CS, Kidwell KK, Knapp SJ, Osborn TC, McCoy TJ. Linkage mapping in diploid alfalfa (*Medicago sativa*). Genome 1994; 37:61-71.

71. Cargnoni TL, Weeden NF, Gritton ET. A DNA marker correlated with tolerance to *Aphanomyces* root rot is tightly linked to Er-1. *Pisum* Genetics 1994; 26:11-12.

72. Thompson WF, Murray MG, Cuellar RE. Contrasting patterns of DNA sequence organization in plants. In: Leaver CJ ed. Genome Organization and Expression in Plants. London: Plenum Press. 1980:1-15.

73. Heim U, Schubert R, H. Bä umlein H, Wobus U. The legumin gene family: structure and evolutionary implications of *Vicia faba* B-type genes and pseudogenes. Plant Mol Biol 1980; 653-663.

74. Kato A, Nakajima T, Yamashita J, Yakura K, Tanifuji S. The structure of the large space region of the rDNA in *Vicia faba* and *Pisum sativum*. Plant Mol Biol 1990; 14:983-993.

75. Kuriyan PN, Narayan RKJ. The distribution and divergence during evolution of families of repetitive DNA sequences in *Lathyrus* species. J Mol Evol 1988; 27:303-310.

76. Schubert I, Rieger R, Michaelis A. On the toleration of duplications and deletions by the *Vicia faba* genome. Theor Appl Genet 1988; 76:64-70.

77. Michelmore RW, Paran I, Kesseli RV. Identification of markers linked to disease resistance genes by bulked segregant analysis: a rapid method to detect markers in specific genomic regions by using bulked segregant populations. Proc Natl Acad Sci USA 1991; 88:9828-9832.

78. Timmerman GM, Frew TJ, Miller AL, Weeden NF, Jermyn WA.Linkage mapping of *sbm-1*, a gene conferring resistance to pea seed-borne mosaic virus, using molecular

markers in *Pisum sativum*. Theor Appl Genet 1993; 85:609-615.

79. Timmerman GM, Frew TJ, N.F. Weeden NF, Miller AF, Jermyn WA. 1994. Linkage analysis of *er-1*, a recessive *Pisum sativum* gene for resistance to powdery mildew fungus (*Erysiphe pisi*). Theor Appl Genet 88:1050-1055.

80. Yu J, Gu WK, Provvidenti R, Weeden WF. Identifying and mapping two DNA markers linked to the gene conferring resistance to pea enation mosaic virus. Journ Am Soc Hort Sci 1995; 120:730-733.

81. Timmerman-Vaughan GM, McCallum J, Frew T, Weeden NF. QTL mapping seed traits in pea. Plant Genome III, Final Program & Abstracts. January 15-19, San Diego, CA. p.81;1995.

82. Franssen HJ, Nap JP, Bisseling T. Nodulins in root nodule development. In: Stacey G, Burris RH, Evans HJ, ed., Biological nitrogen Fixation. London: Chapman and Hall. 1992:598-624.

83. LaRue TA, Weeden NF. The symbiosis genes of pea. *Pisum* Genetics 1992; 24:5-12.

84. Tsyganov VE, Borsiov AY, Rozov SM, Tikhonovich IA. New symbiotic mutants of pea obtained after mutagenesis of line SGE. *Pisum* Genet. 1995; 26:36-37.

85. Weeden, NF, Kneen BE, LaRue TA. Genetic analysis of sym genes and other nodule-related genes in *Pisum sativum*. In: Nitrogen Fixation: Achievements and Objectives. P. Gresshoff P, Roth X, Stacey X, Newton X, ed., New York: Chapman and Hall. 1991:323-330.

86. Hammons RO. Origin and early history of the peanut. In: HE Pattee HE, Young CT, ed., Peanut Science and Technology. Yoakum, TX: Amer Peanut Res Ed Soc 1982; 1-20.

87. Krapovickas A, Gregory WC. Taxonomia del genero *Arachis* (Leguminosae). Bonplandia 1994; 8:1-186.

88. Kochert G, Halward T, Branch WD, Simpson CE. RFLP variability in peanut (*Arachis hypogaea*) cultivars and wild species. Theor Appl Genet 1991; 81:565-570

89. Fernandez A, Krapovickas A. Cromosomas y evolucion en *Arachis* (Leguminosae). Bonplandia 1994; 8:187-220.

90. Gregory WC, Krapovickas A, Gregory MP. Structure, variation, evolution, and classification in *Arachis*. In: Summerfield RJ, Bunting AH, ed. Advances in Legume Science, Vol. 1. London: British Museum of Natural History. 1980:469-481.

91. Lacks GD, Stalker HT. Isozyme analyses of *Arachis* species and interspecific hybrids. Peanut Sci 1991; 20:76-81.

92. Halward TM, Stalker HT, Larue EA, Kochert G. Genetic variation detectable with molecular markers among unadapted germ-plasm resources of cultivated peanut and related wild species. Genome 1991; 34:1013-1020.

93. Paik-Ro OG, Smith, RL, Knauft, DA. Restriction fragment length polymorphism evaluation of six peanut species with the *Arachis* section. Theor Appl Genet 1992; 84:201-208

94. Halward T, Stalker T, Larue E, Kochert G. Use of single-primer DNA amplifications in genetic studies of peanut (*Arachis hypogaea* L.). Plant Mol Biol 1992; 18:315-325

95. Stalker HT, Moss JP. Speciation, cytogenetics, and utilization of *Arachis* species. Adv Agronomy 1987; 41:1-40

96. Wynne JC, Coffelt TA. Genetics of *Arachis hypogaea* L. In: HE Pattee, Young CT, ed. Peanut Science and Technology. Yoakum, TX: Amererican Peanut Research and Education Society. 1982:50-94.

97. Halward T, Stalker HT, Kochert G. Development of an RFLP linkage map in diploid peanut species. Theor Appl Genet 1993; 87:379-384.

98. Garcia GM, Stalker HT, Kochert G. Introgression analysis of an interspecific hybrid population in peanuts (*Arachis hypogaea* L.) using RFLP and RAPD markers. Genome 1995; 38:166-176.

CHAPTER 18

STATUS OF GENOME MAPPING IN THE MALVACEAE (COTTONS)

Andrew H. Paterson, David M. Stelly, Jonathan F. Wendel and Xinping Zhao

18.1. BRIEF DESCRIPTION OF TAXON

The Malvaceae include about 100 genera and 1500 species, native to tropical and temperate regions worldwide.[1] Several genera have been domesticated as ornamentals, including *Hibiscus*, *Alcea* (hollyhock), and *Abutilon* (flowering maple). In addition, *Hibiscus cannabinus* (kenaf) is cultivated as a fiber crop, its rapidly-growing stems providing a suitable alternative to trees for newsprint and other purposes, and *Hibiscus sabdariffa* (roselle) is used to make drinks and jellies, as well as fiber.[2]

Other Malvaceae, particularly kenaf, are of considerable interest as potentially important crops—however, the genus *Gossypium* L., which includes cotton, the world's leading natural fiber crop, has long been a focus of genetic, systematic and breeding research. *Gossypium* comprises about 50 diploid and tetraploid species indigenous to Africa, Central and South America, Asia, Australia, the Galapagos Islands, and the Hawaiian Islands.[3,4]

Cultivated forms of cotton derive from four species, namely *G. hirsutum* L. (n = 2x = 26), *G. barbadense* L. (n = 2x = 26), *G. arboreum* L. (n = x = 13), and *G. herbaceum* L. (n = x = 13). These four species provide the world's leading natural fiber, and are also a major oilseed crop. Cotton was among the first species to which the Mendelian principles were applied,[5] and has a long history of improvement through breeding, with sustained long-term yield gains of 7-10 kg lint/ha/yr.[6] The annual world cotton crop of about 65 million bales (of 218 kg/bale), has a value of about US$15-20 billion/year.

Diploid species of the genus *Gossypium* are all n = 13, and fall into 7 different "genome types," designated A-G based on chromosome pairing relationships.[7,8] A total of five tetraploid (n = 2x = 26) species are generally recognized. All tetraploid species exhibit disomic chromosome pairing, with A and D "subgenomes" seldom if ever undergoing inter-genomic reciprocal chromosomal exchange.[9] Chromosome pairing in interspecific crosses between diploid and tetraploid cottons suggests that tetraploids contain two distinct genomes, which resemble the extant A genome of *G. herbaceum* (n = 13) and D genome of *G. raimondii* Ulbrich (n = 13), respectively. The A- and D-genome species are suggested to have a common ancestor about 6-11 million years ago.[10] The putative A x D polyploidization event occurred in the New World, about 1.1-1.9 million years ago, and required transoceanic migration of the maternal A-genome ancestor,[8,9] which is indigenous to the Old World.[3] Polyploidization was followed by radiation and divergence, with distinct n = 26 AD genome species now indigenous to Central America (*G. hirsutum*), South America (*G. barbadense*, *G. mustelinum* Miers ex Watt), the Hawaiian Islands (*G. tomentosum* Nuttall ex Seemann), and the Galapagos Islands (*G. darwinii* Watt).[3] Some authors refer to an additional tetraploid species, *G. lancelolatum*; however we consider this to be a form of *G. hirsutum*.[12]

18.2. STATUS OF GENETIC MAP

A detailed molecular map of the cotton genome has been published. Reinisch et al[13] used a cross between the two predominantly cultivated species of cotton, *Gossypium hirsutum* and *G. barbadense*, to assemble 705 RFLP loci into 41 linkage groups and 4675 cM, with average spacing between markers of about 7 cM. Since the gametes of these cotton species each contain 26 chromosomes, it was clear that additional DNA markers were needed to bring the map to completion. Estimates of recombination in cotton based upon meiotic configuration analyses indicate a minimum overall map length of 4660 cM,[14] suggesting that most regions of the genome are already covered by the present map, although gaps in some linkage groups remain to be filled. Given that the 26 chromosomes of cotton are presently represented by 41 linkage groups, and that we

can detect linkage at up to 30 cM, filling the gaps between existing linkage groups should add at least 450 cM (=30 cM x 15 gaps) to the map, for a minimum overall length of 5125 cM. Once the map reaches a density of about one marker per 5 cM, or a total of about 1025 markers, fewer than 1% of intervals between markers should measure > 25 cM, and the map should "link up" into 26 linkage groups corresponding to the 26 gametic chromosomes of cotton.

18.2.1. ASSIGNMENT OF LINKAGE GROUPS TO CHROMOSOMES

The polyploid nature of cultivated cotton renders it more tolerant of chromosomal deficiencies than many organisms. The ability of the cotton plant, and its megagametophytes, to survive even when deficient for entire chromosomes, has afforded a facile means to determine the correspondence between "linkage groups" and most chromosomes. A subset of mapped DNA probes were hybridized to genomic digests of a previously-developed series of monosomic and monotelodisomic interspecific substitution stocks, with a single *G. barbadense* chromosome substituted for one *G. hirsutum* chromosome, or chromosome arm (Fig. 18.1A). The hybrid nature of these stocks affords a modified form of deficiency mapping, in that high rates of interspecific DNA polymorphism obviate reliance upon dosage analysis. Loci in the affected chromosome are hemizygous for the *G. barbadense* allele, and lack the *G. hirsutum* allele. By contrast, loci on non-affected chromosomes exhibit the *G. hirsutum* allele. This method is conceptually similar to use of chromosome-deficient lines to determine chromosomal location of DNA probes,[15,16] except that assignment of probes to chromosomes is based on detection of an RFLP, thus is less subject to errors. Based on evidence from a minimum of three genetically-linked loci which correspond to the same aneuploid substitution stock, we have tentatively determined which linkage groups correspond to *chrs. 1, 2, 4, 6, 9, 10, 17, 22, and 25*. The

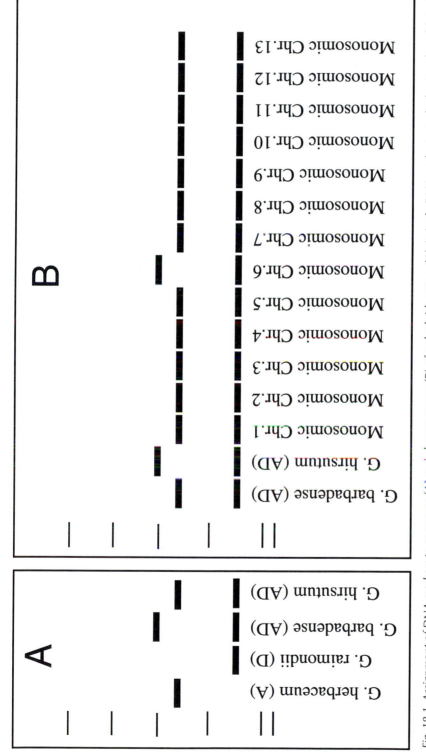

Fig. 18.1. *Assignment of DNA markers to genomes (A) and chromosomes (B) of polyploid cotton. (A) A single DNA probe is applied to Southern-blotted, EcoRI-digested DNA from each of four genotypes (as labeled), representing the A, D, and tetraploid (AD) genomes. The single genomic DNA segment corresponding to this probe in the A genome, migrates at a different rate than the D genome DNA segment. Both tetraploid cottons contain the D genome segment, but only G. hirsutum contains the A genome segment. G. barbadense contains a segment not found in either of the diploids, which presumably represents a mutant allele. It is inferred that the segment which is monomorphic in the two tetraploids is located on the chromosome which traces back to the D genome ancestor, and that the segment which is polymorphic in the tetraploids is located on the chromosome which traces back to the A genome ancestor.*

Although only a single representative of the A and D genomes were shown here, in most instances it is preferable to analyze several divergent representatives of these genomes, to be sure that the data represents a consensus of the respective genomes, rather than simply a recent polymorphism in one accession (such as the one found in G. barbadense).

identity of *chrs. 5, 14, 15, 18,* and *20* is suggested by single loci, which are neither corroborated nor contradicted by any other locus on the linkage group.[13] Identification of several additional chromosomes is in progress.

18.2.2. DETERMINATION OF HOMOEOLOGY FROM MAP POSITIONS OF DUPLICATED LOCI

Mapping of multiple RFLPs derived from individual cotton DNA probes has revealed the consequences of genome-wide duplication in tetraploid (n = 2x = 26) cotton. We tacitly assume that most such associations between linkage groups are a result of the New World A x D hybridization event responsible for the occurrence of tetraploid (n = 2x = 26) cottons.

Based on criteria described in detail,[13] we have tentatively identified (at least parts of) 11 of the 13 expected homoeologous pairs in n = 26 cottons based upon 62 duplicated DNA probes, spanning 1668 cM or 35.6% of the genome. Among the 265 mapped duplicated loci, 124 (46.8%) are accounted for by homoeology. Additional loci may be accounted for as the remaining small linkage groups coalesce. The occurrence of proximal duplications, and duplicated loci inconsistent with homoeologous relationships, together with independent evidence such as retro-transposition,[17] reflect the activity of other sequence duplication mechanisms in cotton and further support the need for determining homoeology based on several loci along a linkage group.

Among the homoeologous relationships found by RFLP mapping were three pairs of tentatively-identified chromosomes, chrs. 1 and 15, chrs. 5 and 20, and chrs. 6 and 25. Homoeology between chrs. 1 and 15 is supported by prior evidence that duplicated mutations in three morphological traits map to this pair of chromosomes,[18] with one inversion in order. Homoeology between chrs. 6 and 25 has been suggested previously,[8] based upon similar phenotypes of plants monosomic for each of these chro-

mosomes. Although several mutant phenotypes have been assigned to chr. 5, no evidence is available regarding homoeology of chromosomes 5 and 20. Independent studies have demonstrated homoeology of chromosomes 7 and 16, and chromosomes 12 and 26 based on morphological markers,[18] and chromosomes 9 and 23 based on in situ hybridization.[19]

Most of the homoeologous chromosomes in AD cottons are distinguished by one or more inversions. In only two cases are putative homoeologs homosequential over the entire region of the chromosome for which homoeology can be inferred, although large blocks of sequence conservation are evident in most cases. Four pairs clearly differ by at least one inversion, and two additional pairs differ by at least two inversions. At least one pair shows a translocation, while several additional pairs may represent either translocations or simply small linkage groups which have not yet linked up. A study in progress (C.L. Brubaker, A.H.P. and J.F. Wendel, in preparation) will describe mapping of a subset of the tetraploid DNA probes in F_2 populations of *G. arboreum* x *G. herbaceum* (A-genome) and *G. raimondii* x *G. trilobum* (D-genome: Fig. 18.2). This should permit discrimination of structural mutations that occurred during divergence of the n = 13 A- and D-genomes from mutations that occurred after formation of the n = 26 AD-genome.

18.2.3. DEDUCING THE GENOMIC ORIGIN OF LINKAGE GROUPS IN ALLOTETRAPLOID COTTON

Some DNA probes detected genomic fragments in tetraploid cottons that were shared with either A- or D-genome ancestors, but not both (Fig. 18.1B). In a subset of these cases, polymorphism between *G. hirsutum* and *G. barbadense* permitted mapping of one or both of the homoeologous genomic fragments. Such "alloalleles" (e.g., orthologous diploid-genome-specific DNA fragments) have been suggested to be a means of deducing the diploid origin

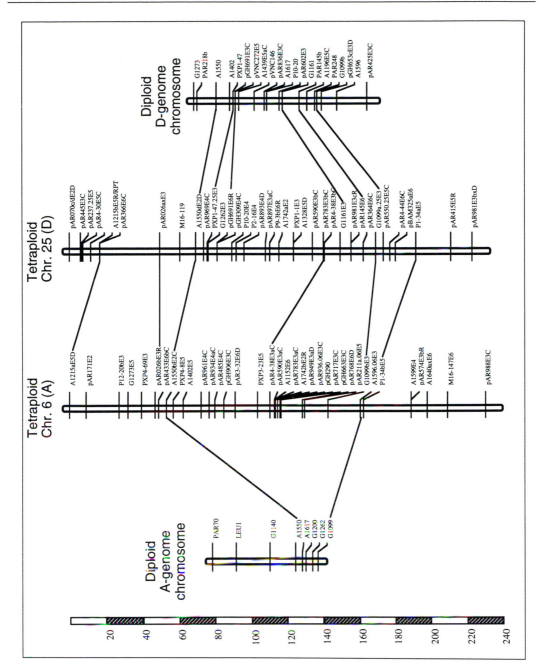

Fig. 18.2. Alignment of chromosomes of cultivated polyploid cotton, and its probable diploid progenitors. Cultivated cotton is a polyploid (n = 2x = 26), resulting from an interspecific hybridization between "A" and "D" genome diploid (n = x = 13) ancestors, about 1.1-1.9 million years ago.[10,11] Most low-copy DNA probes detect at least two restriction fragments in cultivated cotton, and in a subset of cases, both restriction fragments can be mapped as RFLPs. Study of the arrangement of such "duplicated" RFLP loci across the genome has been the basis for identifying 11 of the expected 13 pairs of "homoeologous" chromosomes in polyploid cotton.[13] By means shown in Fig 18.1a and described in text, the subgenomic (A versus D) origin of individual chromosomes can be determined. Finally, by mapping of common DNA markers in different crosses, the chromosomes of diploid cotton relatives can be aligned with those of present-day cultivated polyploid forms. The figure shows alignment of genetic maps for one homoeologous series from crosses between G. herbaceum x G. arboreum (A genome), G. trilobum x G. raimondii (D genome) and G. hirsutum x G. barbadense (AD genome).

of particular linkage groups (or chromosome segments) in tetraploid cotton.[16]

Based on alloallelic loci, the genomic origins of 33 of the present 41 linkage groups, including all of the known chromosomes, have been determined. One linkage group (L.G. U01) showed conflicting information (2 A, 2 D), and seven small linkage groups (totaling 224.8 cM and 34 loci) harbored no alloallelic loci. To evaluate the reliability of our subgenomic inferences, we calculated the frequency at which alloallelic information coincided with the classical assignment of cotton chromosomes to genomes based upon pairing relationships in diploid x tetraploid hybrids (A = *chrs. 1-13*; D = *chrs. 14-26*). Based on the 14 chromosomes we have identified, the majority of alloallelic data for a chromosome agreed with classical subgenomic assignment of chromosomes in all 14 (100%) cases. We conclude that alloallelic information is a reliable, although not infallible, indicator of genomic origin of a chromosome or linkage group.

Six of the 10 loci where alloallelic information disagreed with classical subgenomic assignment of chromosomes occurred as linked pairs on three linkage groups. In each of these three cases (chr. 14, L.G. A03, L.G. U01), the two deviant loci are consecutive along the linkage group (with reference to alloallelic probes). This suggests the possibility that exchanges of chromatin between homoeologous chromosomes may have occurred during the evolution of tetraploid cottons.

18.3. UNDERSTANDING OF GENOME ORGANIZATION

Prior evidence from DNA reassociation kinetics[20-22] suggest that 50-65% of the tetraploid cotton genome is composed of low copy DNA and the remaining 35-50% is repetitive DNA. Evidence of interspersed elements, some of which putatively resulted from ancient retrotransposition, is also well-documented.[17] No active endogenous transposons have been confirmed to date.

We have isolated 103 non-cross-hybridizing families of repetitive DNA elements, which are estimated to comprise 29-35% of the genome, and 60-70% of the repetitive fraction of tetraploid cotton.[23] Therefore, we believe that we have cloned and characterized segments of most of the abundant repetitive DNA elements in the tetraploid cotton genome. Most repeat families (83/103) are interspersed in the cotton genome and have copy numbers that range from 10^3-10^4, thus fall into the moderate-abundance class (10^2-10^5 copies[24]). Tandemly repetitive DNA, the satellite DNA, usually represents the most highly repetitive sequence family in eukaryotic genomes[25,26]—but was relatively rare in cotton. Out of the cloned 103 repeat families, only 20 (19%) families were tandem repeats. Five were estimated to have a copy number in the range of 10^4, and the rest had a lower copy number. None of the tandem repeats meet the traditional threshold of 10^6 copies[25] for classical "highly-repetitive" DNA. While the approach employed to cloning of repetitive DNA might have missed short "satellite" DNAs, the paucity of tandem repeats in cotton is supported by evidence from DNA/DNA reassociation kinetics, which showed the rapidly-reassociating fraction to be small,[21,22] and buoyant density centrifugation experiments which showed no detectable satellites.[20]

Little variation in abundance of repetitive sequences was found among the tetraploid species of *Gossypium*.[23] This suggests that little, if any, net amplification or deletion of individual repetitive sequences in the tetraploid genomes has occurred during the divergence of tetraploid species from a common ancestor 1.1-1.9 million years ago.[10] Southern hybridization analysis based on one to two enzyme digests showed that most repeat families failed to detect RFLPs among tetraploid species, providing further evidence to support the above conclusion.

Sequence analysis of four cotton repetitive elements suggests that many are AT-rich, comprised in part of shorter direct and/or inverted repeats, and contain open reading frames. Although representing only a small subset of the total repertoire of repetitive element families in cotton, each of these features are, in principle, consistent with the possibility that replicative transposition may have played a significant role in cotton genome evolution.[17] Sequence analysis of the entire repertoire of cloned cotton repetitive element families will afford more detailed characterization of the structure and function of these elements (X. Zhao, Y. Si, and A.H. Paterson, in preparation).

Three interspersed repeat families, pXP004, pXP020, and pXP072, were estimated to have copy numbers of 100,000 (each) in tetraploid cotton, almost as abundant as the Alu family in the human genome.[27] Based on in situ hybridization (R. Hanson, C. Crane, D. Stelly, J. Wendel, XZ, and AHP, unpublished), the repetitive DNA families including these three elements are distributed differently across the cotton chromosomes from each other, and from the most prominent retroposon families in cotton.[17] Repetitive DNA families identified in this study provide new tools for physical mapping of chromosomes, fingerprinting of megabase DNA clones,[28] or "painting" of cotton chromosomes by in situ hybridization.[29]

Although limited RFLP or copy number variation in the repetitive elements was found among different tetraploid species, our survey of genome specificity showed that these repetitive elements exhibit varying patterns of genome distribution across the diploid genomes. Either RFLPs or polymorphisms in abundance have been detected to different degrees among cotton species with A, B, C, D, E, F, G, and AD genomes by the repeat families (Zhao et al, in preparation). Repetitive DNA elements of cotton are potentially useful for investigation of species and genome relationships in the genus *Gossypium* L.

18.4. MAPPING OF AGRICULTURALLY- AND DEVELOPMENTALLY-IMPORTANT GENES

Because the molecular map of cotton has only recently been developed, associa-

tion of phenotypes with molecular markers is just beginning. One means by which trait mapping will be accelerated is through integration with the classical map of cotton, based on visible markers. Variation in ploidy among *Gossypium* spp., together with tolerance of aneuploidy in tetraploid species of *Gossypium*, has facilitated use of cytogenetic techniques to explore the inheritance of visible markers in cotton. Among 198 morphological mutants described in cotton, 61 mutant loci have been assembled into 16 linkage groups, through the collective results of many investigators. Using nullisomic, monosomic, and monotelodisomic stocks, 11 of these linkage groups have been associated with chromosomes.[18] Integration of these phenotypes with the molecular map is just beginning.

Numerous introgression events in cotton may be detectable using existing DNA markers. Chromatin from *G. hirsutum* has been introgressed into cultivated *G. barbadense*, with cultivated *G. barbadense* genotypes comprised of an average of 9% *G. hirsutum* chromatin. More than 50% of the introgressed chromatin is found in five specific chromosomal regions, which are largely common to genotypes from each of the leading *G. barbadense* breeding programs throughout the world.[30] An example of one such region is shown (Fig. 18.3). One hypothesis suggests that cultivated *G. barbadense* is of relatively recent origin, and is the result of an interspecific hybrid, followed by selection for fiber attributes. The fact that the *G. hirsutum* introgression has persisted through many breeding cycles, despite opportunities for it to be lost, suggests that non-linear interaction of *G. hirsutum* and *G. barbadense* alleles may confer unique traits to cultivated forms of *G. barbadense*, which are maintained by artificial selection on breeder's plots.[30] In addition, DNA markers hold promise for mapping many traits in cotton which have been introgressed from exotic germplasm, such as *Verticillium* wilt resistance, bacterial blight resistance,[31] nectariless leaves,[32] restoration of cytoplasmic male-sterility,[33,34] and improved fiber quality.[35,36]

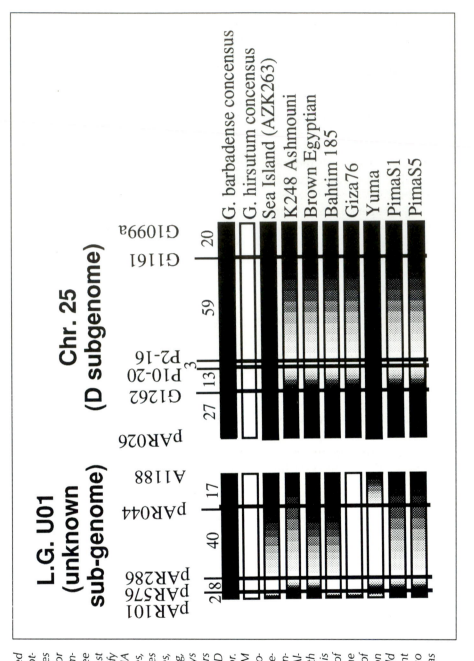

Fig. 18.3. Homoeology of duplicated microsatellite loci in polyploid cotton. Sequence-tagged microsatellites (STMs) are a relatively new tool for genetic mapping, which is becoming more widely used in plants (see chapter 10). In cotton, a modest fraction of STMs are able to amplify DNA at two or more different DNA marker loci. In most of these cases, the loci are at corresponding places on homoeologous chromosomes, indicating that the rapidly-evolving, polymorphic microsatellite arrays have persisted for 6-11 million years since the divergence of the A and D genomes from a common ancestor. It is noted that addition of the STM locus LGA16 to the map of chromosomes 1 and 15 reveals a rearrangement in gene order which distinguishes the two chromosomes. Although this appears inconsistent with the other duplicated markers, it is consistent with the map order of duplicated visible markers on the two chromosomes.[18] Assignment of the visible markers to locations on the molecular map (see text) should confirm that the rearrangement which distinguishes these two homoeologous chromosomes has been identified.

QTL mapping of cotton is considered a high priority, in view of the general observation that most measures of cotton quality and productivity are polygenic. Emphasis is being placed on phenotypes of direct relevance to agricultural productivity and/or quality of cotton. Efforts are in progress to map mutations which eliminate extrafloral nectaries and reduce the density of trichomes (each associated with reduced insect damage; A.H. Paterson et al, in preparation), impart resistance to various races of important pathogens such as "bacterial blight" (R. Wright, A.H. Paterson, P.M. Thaxton, K.M. El-Zik, in preparation), restore cytoplasmic male-sterility,[37] improve several aspects of fiber quality (A. H. Paterson et al, in preparation), and improve productivity of cotton under water deficit (Y. Saranga, D. Yakir, A.H.P, unpublished).

As in many largely self-pollinated crops, the gene pools of each of the cultivated cotton species show only modest levels of DNA polymorphism. Routine application of DNA markers to cotton breeding may benefit from technologies such as microsatellite-based markers, which are being superimposed on the existing RFLP map.[38] Mono- and dinucleotide microsatellites are found in the cotton genome. Our estimation shows that on average $(GA)_n$ and $(CA)_n$ microsatellites occur at a frequency of every 919 kb (about 2 cM) in the cotton genome, which is lower than in other plant species characterized. It is possible that cotton simply contains fewer microsatellite arrays than other plants. However, these estimates are conservative, since we used conditions which should only detect arrays of 10 repeat units or more (the arrays which have been sequenced from this screen ranged from 9-29 units). The longer arrays detected under our hybridization and wash conditions tend to be more polymorphic than shorter arrays.[39]

Most cotton sequence-tagged microsatellites (STMs) mapping to two or more loci show homoeologous polymorphisms (Fig. 18.4), suggesting that microsatellite arrays have persisted since divergence of the A and D genomes, an estimated 6-11 million years.[10,11]

Microsatellite-based markers in cotton show moderately higher levels of allelic variation than RFLPs, however, the incremental advantage of STMs in cotton is clearly smaller than in most animals, and perhaps also in other plants. In an interspecific cotton mapping population,[13] a total of 16 STMs segregated at an average of 1.25 loci per STM, versus an average of 0.6 loci per low-copy DNA probe evaluated (Reinisch et al) with 4-6 restriction enzymes.[13] Study of *G. hirsutum* cultivars from across the U.S. reveals an average of 1.64 alleles per STM primer pair, versus 1.28 alleles per low-copy DNA probe. Since this evaluation was based on STMs containing relatively long (and consequently polymorphic[39]) microsatellite arrays, "average" STMs in cotton may be somewhat less polymorphic.

18.5. PROSPECTS FOR MAP-BASED CLONING

The physical size of a centiMorgan in cotton is within reach of map-based cloning, but the lengthy genetic map will require a large number of DNA markers, in order to be confident of having a marker sufficiently close to most genes for "chromosome walking." With a genome size of 2246 Mb,[40] the average physical size of a centiMorgan in cotton is about 400 kb, only moderately larger than that of *Arabidopsis* (ca. 290 kb), and smaller than that of tomato (ca. 750 kb), both species in which map-based gene cloning has been accomplished (see chapter 12). However, even with the assumption that we can infer the homoeologous site for most DNA markers (see below), the genetic map of 5000 cM will require ca. 3000 DNA probes to map at average 1 cM density, and the physical genome of 2246 Mb will require ca. 75,000 YACs/BACs of average size 150 kb for 5x coverage. The suggestion above that repetitive DNA is recombinationally inert will further improve prospects for map-based cloning in cotton,

Fig. 18.4. Introgression from G. hirsutum *into* G. barbadense. *Numerous introgression events have occurred during breeding of cotton, to provide genetic variation which was absent from a particular gene pool. An excellent example is the development of cultivated forms of* G. barbadense *("Sea Island," "Egyptian," or "Pima," cottons) which are actually complex hybrids between* G. barbadense *and* G. hirsutum. G. hirsutum *chromatin appears to be retained in several locations of the genome in* G. barbadense, *as exemplified by the two locations shown. In the majority of these chromosomal regions, such as LG U01, introgression is postulated to have been important to development of Sea Island cottons, the ancestor of most cultivated* G. barbadense. *In the case of chromosome 25, introgression postdates the development of Sea Island cottons, being found predominantly in the more recently-developed Egyptian and Pima cottons. Note that Yuma, an early "Pima-type" cultivar, has Sea Island cotton directly in its pedigree, which may account for the lack of introgression on chr. 25.*

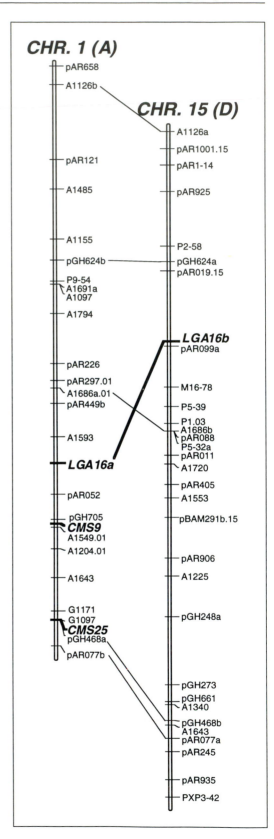

as a sizable fraction of the genome may contribute little or nothing to the 5000 cM genetic length. Finally, we note that the estimate of physical genome size we have employed (2246 Mb[40]) falls at the median of several estimates[22,41,42] which vary by ± 30%, introducing a corresponding range into our figures.

Map-based cloning in polyploids such as cotton introduces a new technical challenge not encountered in diploid (or highly diploidized) organisms, i.e., that virtually all "single-copy" DNA probes occur at two or more unlinked loci. This makes it difficult to assign megabase DNA clones to their site of origin. One possible approach to this problem is the utilization of diploids in physical mapping and map-based cloning; for example, exploiting the relatively small genome of D-genome diploid cottons (approximately 65% of the size of the A genome, and 39% of the tetraploid[22]). However, since D-genome cottons produce minimal fiber, and are not cultivated, they are of limited utility for dissecting the fundamental basis of cotton productivity. The relatively larger size of the A genome,[22] despite low-copy sequence complexity similar to that of the D genome,[22] suggests that the subgenome-specific repetitive elements may provide a means of characterizing the genomic identity of individual megabase DNA clones (YACs, BACs, P1, or others) from tetraploid cottons (X. Zhao, R. Wing, and AHP, submitted). Such an approach may prove generally applicable to map-based cloning in other major crops, many of which are disomic polyploids.

Potential complications associated with map-based cloning in cotton (and other disomic polyploids) are partly compensated for by some unique advantages of polyploid genomes. Utilization of genetic information from homoeologous relationships will accelerate development of high-density maps of DNA markers in polyploid genomes. About 21.7% of DNA probes segregated for RFLPs at two or more loci, accounting for 265 (37.5%) of the mapped loci. These duplicated loci provide a skeleton for inferring the approximate homoeologous locus of DNA probes monomorphic in one subgenome but polymorphic in the other, increasing the marker density of the map. As additional markers are mapped it should be possible to catalog homoeologous relationships over most of the genome, greatly increasing the density of DNA markers in maps of disomic polyploids such as cotton.

Techniques for isolation and cloning of megabase DNA in cotton have been established,[43] and rapid progress is being made toward construction of BAC libraries for cotton.

18.6. LIKELY DIRECTIONS OF FUTURE RESEARCH

In view of the economic importance of cotton, it seems likely that future research in cotton genetics will increasingly focus on traits related to agricultural productivity and quality. The detailed molecular map holds the opportunity for identification of DNA markers for rapid assay of segregants, even using nondestructive assay of ungerminated seed.[44] Already, characterization of selected gene pools within cultivated cotton is reaching an advanced state,[30] however, a much more comprehensive DNA fingerprint of the cultivated gene pool is needed. In addition, the map offers an opportunity to undertake molecular dissection of complex measures of quality and/or productivity, such as fiber attributes. Detailed characterization of genome organization, together with ongoing development of megabase DNA tools, holds promise that map-based cloning of mutations unique to cotton will be a reality in the foreseeable future.

Cotton may also make a unique contribution to better understanding of genome evolution. As a naturally-occurring polyploid, with at least seven diploid genomes occurring in the genus and distributed over four continents (America, Africa, Asia, Australia), cotton is a facile system for study of genome divergence, both at the level of individual DNA sequences and entire chromosomes. Regarding the former, homoeologous exchange of repetitive DNA sequences in cotton has recently been demonstrated,[45] and the extensive array of repetitive elements which have been cloned in cotton[23] represent a valuable body of tools for further studies. Further, considerable progress has been made toward comparative analysis of chromosome organization in polyploid cotton and its diploid progenitors (C.L. Brubaker, A.H.P., J.F.W., in preparation), and the future is likely to hold a better understanding of molecular-level relationships between the various genomes of this polymorphic taxon.

REFERENCES

1. Sauer J. Historical Geography of Crop Plants: A Select Roster. Boca Raton, FL: CRC Press, 1993.

2. Wilson FD and MY Menzel. Kenaf (Hibiscus cannabinus), roselle (Hibiscus sabdariffa). Econ Bot 1964; 18:80-91.

3. Fryxell PA. The Natural History of the Cotton Tribe. College Station, TX: Texas A&M University Press, 1979.

4. Fryxell PA. A revised taxonomic interpretation of *Gossypium* L. (Malvaceae). Rheedea 1992; 2:108-165.

5. Balls WL. Studies in Egyptian cotton. Cairo, Egypt: Yearbook Khediv Agric Soc, 1906:29-89.

6. Meredith WR, RR Bridge. Genetic contributions to yield changes in upland cotton. In: Fehr WR, ed. Genetic Contributions to Yield Gains of Five Major Crop Plants. Madison WI: Crop Science Society of America, 1984:75-86.

7. Beasley JO. Meiotic chromosome behavior in species hybrids, haploids, and induced polyploids of *Gossypium*. Genetics 1942; 27:25-54.

8. Endrizzi JE, and G Ramsay. Monosomes and telosomes for 18 of the 26 chromosomes of *Gossypium hirsutum*. Can J Genet Cytol 1979; 21:531-536.

9. Kimber G. Basis of the diploid-like meiotic behavior of polyploid cotton. Nature 1961; 191:98-99.

10. Wendel JF. New World cottons contain Old World cytoplasm. Proc Natl Acad Sci USA 1989; 86:4132-4136.

11. Wendel JF, and VA Albert. Phylogenetics of the Cotton genus (*Gossypium*): Character-state weighted parsimony analysis of chloroplast-DNA restriction site data and its systematic and biogeographic implications. Syst Bot 1992; 17:115-143.

12. Brubaker CL and JF Wendel. Molecular evidence bearing on the specific status of *Gossypium lanceolatum* Todaro. Genet Resrcs & Crop Evol 1993; 40:165-170.

13. Reinisch AJ, J-M Dong, C Brubaker, D Stelly, J Wendel, AH Paterson. A detailed RFLP map of cotton (Gossypium hirsutum x G. barbadense): Chromosome organization and evolution in a disomic polyploid genome. Genetics 1994; 138:829-847.

14. Reyes-Valdes M and DM Stelly. A maximum likelihood algorithm for genome mapping of cytogenetic loci from meiotic configuration data. Proc Natl Acad Sci USA 1995; 9824-9828.

15. Helentjaris T, DF Weber, S Wright. Use of monosomics to map cloned DNA fragments in maize. Proc Natl Acad Sci USA 1988; 83:6035-6039.

16. Galau GA, HW Bass, and DW Hughes. Restriction fragment length polymorphisms in diploid and allotetraploid *Gossypium*: Assigning the late-embryogenesis-abundant (Lea) alloalleles in *G. hirsutum*. Mol Gen Genet 1988; 211:305-314.

17. Vanderwiel PS, DF Voytas, JF Wendel. *Copia*-like retrotransposable element evolution in diploid and polyploid cotton (*Gossypium* L.) J Mol Evol 1993; 36:429-447.

18. Endrizzi JE, EL Turcotte, RJ Kohel. Qualitative genetics, cytology, and cytogenetics. In: RJ Kohel and CF Lewis, eds. Cotton. Madison, WI: ASA/CSSA/SSSA Publishers, 1984:81-129.

19. Crane CF, HJ Price, DM Stelly, DG Czeschin, Jr, and TD McKnight. Identification of a homoeologous chromosome pair by in situ hybridization to ribosomal RNA loci in meiotic chromosomes of cotton (*Gossypium hirsutum* L.). Genome 1994; (in press).

20. Walbot V, and Dure, LS. Developmental biochemistry of cotton seed embryogenesis

and germination. VII. Characterization of the cotton genome. J Mol Biol 1976; 101:503-536.

21. Wilson JT, Katterman, FRH, and Endrizzi, JE. Analysis of repetitive DNA in three species of *Gossypium*. Biochem Genet 1976; 14:1071-1075.

22. Geever RF, Katterman, F, and Endrizzi, JE. DNA hybridization analyses of *Gossypium* allotetraploid and two closely related diploid species. Theor Appl Genet 1989; 77:553-559.

23. Zhao X, R Wing, and AH Paterson. Cloning and characterization of the majority of repetitive element families in cotton (*Gossypium spp.*). Genome 1995; (in press).

24. Bouchard RA. Moderately repetitive DNA in evolution. Int Rev Cytol 1982; 76:113-93.

25. Singer MF. Highly repetitive sequences in mammalian genomes. Int Rev Cytol, 1982; 76:67-112.

26. Flavell RB. Repetitive sequences and genome architecture. In: Ciferri O and Dure L, eds. Structure and Function of Plant Genomes. New York: Plenum Press, 1982:1-14.

27. Deininger PL, Jolly DJ, Rubin CM, Friedmann T, and Schmid CW. Base sequence studies of 300 nucleotide renatured repetitive human DNA clones. J Mol Biol 1981; 151:17-33.

28. Nelson DL, Ledbetter S, Corbo L, Victoria DH, and Caskey CT. *Alu* polymerase chain reaction: A method for rapid isolation of human-specific sequences from complex DNA sources. Proc Natl Acad Sci USA 1989; 86:6686-6690.

29. Stallings RL, Torney DC, Hildebrand CE, Longmire JL., Deaven, LL, Jeff, JH, Dogget, NA, and Moyzis, RK. Physical mapping of human chromosomes by repetitive sequence fingerprinting. Proc Natl Acad Sci USA 1990; 87:6219-6222.

30. Wang G, J Dong, and AH Paterson. Genome composition of cultivated Gossypium barbadense reveals both historical and recent introgressions from G. hirsutum. Theor Appl Genet 1995; (in press).

31. Staten G. Breeding Acala 1517 cottons, 1926-1970. New Mexico State Univ. College of Agric. and Home Econ. Memoir Series No. 4, 1971.

32. Tyler FJ. The nectaries of cotton. US Dept Agr Plant Ind Bull No 131, 1908:45-54.

33. Meyer V. Male sterility from *Gossypium harknessii*. J Hered 1975; 66:23-27.

34. Weaver DB. and JB Weaver. Inheritance of pollen fertility restoration in cytoplasmic male-sterile upland cotton. Crop Sci 1977; 17:497-499.

35. Culp TW, and DC Harrell. Breeding quality cotton at the Pee Dee Experiment Station, Florence S.C. USDA Publ. ARS-5-30. Washington, DC: US Gov't Printing Offc, 1974.

36. Culp TW, DC Harrell, and T Kerr. Some genetic implications in the transfer of high-fiber strength genes to upland cotton. Crop Sci 1979; 19:481-484.

37. Lan T-H, C Cook, AH Paterson. Molecular mapping of a gene for restoration of cytoplasmic male-sterility in cotton. San Diego, CA: Proc Plant Genome II, 1992.

38. Zhao X, X Ding, and AH Paterson. Sequence-tagged microsatellites detect polymorphic homoeologous loci in cotton (*Gossypium* spp.). Submitted.

39. Weber JL. Genomics 1990; 7:524-530.

40. Arumuganathan, K and ED Earle. Nuclear DNA contents of some important plant species. Plt Mol Biol Rptr 1991; 9:208-218.

41. Michaelson MJ, HJ Price, JR Ellison, JS Johnston. Comparison of plant DNA contents determined by Feulgen microspectrophotometry and laser flow cytometry. Am J Bot 1991; 78:183-188.

42. Gomez M, JS Johnston, JR Ellison, HJ Price. Nuclear 2c DNA content of *Gossypium hirsutum* L. accessions determined by flow cytometry. Biol Zent bl 1993; 112:351-357.

43. Zhao X, H Zhang, RA Wing, AH Paterson. An efficient and simple method for isolation of intact megabase-sized DNA from cotton. Plt Mol Biol Rptr, 1994; 12:126-131.

44. Wang G, R Wing, and AH Paterson. PCR amplification of DNA extracted from single seeds, facilitating DNA-marker assisted selection. Nucl Acids Res 1993; 21:2527.

45. Wendel JF, A Schnabel, T Seelanan. Bidirectional interlocus concerted evolution following allopolyploid speciation in cotton (Gossypium). Proc Natl Acad Sci USA 1995; 92:280-284.

=CHAPTER 19=

GENOME MAPPING IN TEMPERATE GRAINS AND GRASSES IN THE FAMILY GRAMINEAE (POACEAE)

James C. Nelson, Jorge Dubcovsky, Susan R. McCouch and Mark E. Sorrells

Of particular interest for genome mapping in recent years have been the staple cereal crops and their relatives. Here we discuss grasses in the tribe Triticeae, which include wheat and other *Triticum* species, rye (*Secale*), barley (*Hordeum*) and many other genera. Also covered will be oat (*Avena* spp.; tribe Aveneae) and rice (*Oryza* spp.; tribe Oryzeae of subfamily Oryzoideae).

19.1. TAXONOMY AND CYTOLOGY

19.1.1. TRITICEAE AND AVENEAE

The basic chromosome number in the cereals is x = 7 and all are considered to have their centers of origin in southwestern Asia. Wheat (*T. aestivum* L. em. Thell), an allohexaploid containing three distinct sets of homoeologous chromosomes designated as the A, B, and D genomes, is thought to have arisen in the Middle East as early as 7000 BC via chance pollination of cultivated *T. dicoccum* (genome formula AABB) by the wild diploid *T. tauschii* (= *Aegilops squarrosa*) (DD).[1] The AABB and AAGG tetraploid wheats, in turn, trace their A genomes to *T. urartu* (AA),[2] while the ancestral donor of the B genome remains unknown. *T. monococcum* is a diploid closely related to *T. urartu*, with a genome designated A[m]. Triticeae grasses possess numerous other genomes classified by their meiotic pairing affinities. Some (e.g., rye (*Secale cereale*), *T. speltoides*, *T. tripsacoides*) are outcrossing species but the main cereal crops are self-pollinating.

Genome Mapping in Plants, edited by Andrew H. Paterson. © 1996 R.G. Landes Company.

Barley (*Hordeum vulgare* L.) is a diploid crop domesticated in the Near East and now used principally for feed and malting. It has rivaled wheat as a subject of genetic mapping studies if not cytogenetic manipulations.

Rye, a diploid crop of less economic importance than the other cereals, is proposed to have arisen as a weed in association with them. Genomic mapping in the many noncultivated species in the Triticeae, except for *T. tauschii* and *T. monococcum*, has received little attention as yet. The taxonomy of the Triticeae, reviewed in reference 3, is complicated by the many ancestral intergeneric hybridizations apparent from genome analysis.

In the Aveneae, oat (*Avena sativa* L.) is one of the major world cereal crops. Oat, like wheat, appears in all ploidy levels up to the cultivated hexaploid (AACCDD) and the wild tetraploids *A. magna* and *A. murphyi* have recently been domesticated.[4]

19.1.2. *ORYZA*

Within the genus *Oryza*, the section Sativa consists of cultivated species of *O. sativa* L. (*indica* and *japonica*) and allied wild species, and the cultivated species, *O. glaberrima* Steud. and its allied wild species.[5] *O. sativa* L., a diploid with x = 12, is the most common form of cultivated rice. It is cultivated on all continents except Antarctica, while *O. glaberrima* is cultivated only in some parts of West Africa. Members of the section Sativa are either annual or perennial, and their genomes, all of which are highly homoeologous with A, are classified as either A, A^b, or A^g. Other sections of the genus consist of both diploid and tetraploid species possessing the B, C, BC, or CD genomes. None of these genomes combines well with the A genomes of the section Sativa.

19.2. STATUS OF THE GENETIC MAPS

Wheat is endowed with a rich set of tools for genetic analysis in the form of the aneuploid stocks in cv Chinese Spring (CS) constructed largely by the late E.R. Sears.[6]

Around 100 genes had been placed on the classical wheat map by 1980 via monosomic or substitution analyses or variants of these. Comparative mapping in the Triticeae was then advanced by Hart[7] who used the CS stocks to place numerous isozyme loci on homoeologous chromosomes in the wheat genome and also used disomic barley addition lines[8] in wheat to identify homoeologous barley chromosomes by isozyme allele patterns. The CS aneuploids remain of great value for chromosomal localization of RFLP markers.[9,10] It is now routine in wheat RFLP work to localize markers to chromosome and arm, without the necessity of a mapping cross, by simply hybridizing DNA probes to a panel of nullitetrasomic and ditelosomic CS stocks. Given polymorphic DNA markers, linkage mapping in allopolyploids is not as difficult as it is often supposed to be by analogy with the laborious task, in classical mapping experiments, of genotyping a plant that may contain from zero to three doses of an orthologous gene on three homoeologous chromosomes. The larger genome size is compensated by the frequency with which two or three markers segregate from a single RFLP probe, with the accompanying possibility of mapping in two or three homoeologous chromosomes at once.

RFLP mapping in wheat began at the IPSR laboratory in Norwich, England, from which have come maps of wheat homoeologous groups 7,[11] 2, 3,[12,13] and 5[14] and others listed in reference 15; this map now comprises > 800 markers.[16] The Tsunewaki laboratory in Kyoto, Japan produced a 204-marker RFLP map.[17] Of late a third hexaploid wheat map of > 1100 RFLP markers has been developed by U.S. and French collaborators.[18-22] Because of the relatively low polymorphism among cultivated wheats, the major maps have been constructed in progeny from the cross of a bread wheat with a synthetic hexaploid. Such a synthetic arises from a cross between a *T. durum* or *T. spelta* (AABB) wheat and *T. tauschii*, the D-genome ancestor of cultivated wheat. Polymorphism and hence marker coverage has been generally higher in the A and B genomes than in the D genome. As expected, molecular maps of homoeologous chromosomes are very similar with respect to marker order and distance, aside from a set of rearrangements involving chromosomes 4A, 5A, and 7B.[21,23-26] The major maps have yet to be united into a consensus map by RFLP analysis with common probes, although a new physical map based on hybridizing probes to deletion stocks (see below) offers an excellent framework for this. Nor have chromosome ends been defined with telomeric sequences. The most recent hexaploid wheat RFLP linkage map is shown in Figure 19.1.

Barley has large chromosomes, a rich classical genetic map, and a nearly complete set of wheat single-chromosome or arm addition lines. Barley mapping has employed readily generated doubled-haploid populations from male or female F_1 gametes from variety crosses[27-30] or crosses between *H. vulgare* and *H. spontaneum*[12,28,31] and produced maps better populated than those of the other Triticeae species and much better united with markers from common probes. Not surprisingly, in view of the homoeology earlier demonstrated for wheat and barley, the maps of barley chromosomes 1H through 7H are virtually identical at the present DNA-level resolution with those of the unrearranged chromosomes of wheat (but see refs. 32, 33).

RFLP mapping of the rye genome has also received attention from the Norwich group and associates,[34,35] who demonstrated[36] both the extensive homoeology and segmental rearrangement earlier characterized via pairing studies[23] with respect to the wheat map. Notable in rye is a series of translocations described in detail,[35,36] some with similarity to known reciprocal translocations in wheat and at least one[37] differentiating *S. cereale* from its wild relative *S. montanum*. Isozyme, cytological, and/or morphological maps of rye chromosomes have also been produced; see references 38 (60 markers) and 39-43. A summary of rye mapping results appears in reference 44.

The maps of *T. tauschii*, *T. monococcum*, and diploid oat illustrate the strategy, chosen for polyploid species to reduce complexity and increase polymorphism, of mapping the genomes of diploid progenitors. The *T. tauschii* D genome was mapped[45,46] in an F_2 population with a set of *T. tauschii* genomic probes (Fig. 19.2); and other maps exist.[47,48] The cross-mapping of *T. tauschii* genome markers in wheat and barley[18-20,49] has confirmed the close similarity between the Triticeae maps.

In *T. monococcum*, a diploid wheat with abundant RFLP polymorphism, a 333-marker map (Fig. 19.3) was constructed in 74 F_2/F_3 plants with RFLPs, isozymes, seed storage proteins, and morphological markers.[32] Since half of the markers were common to barley maps, it could be shown that the two genomes differ by a reciprocal translocation and two paracentric inversions. The marker order identical to that in the wheat B and D genomes places these rearrangements either in the evolution of barley or in the evolutionary lineage of the *Triticum* prior to the radiation of *Triticum* species. The NORs (nucleolar organizing regions) of *T. monococcum*, wheat, and barley are located in different regions of otherwise colinear maps, indicating their migration during radiation of Triticeae species.[50]

Mapping in tetraploid durum wheat (*T. turgidum*) has covered homoeologous group 6[51] in recombinant substitution lines developed from crosses between the durum cv 'Langdon' and *T. dicoccoides* disomic chromosome-6A and -6B substitution lines in 'Langdon' backgrounds. Chromosome arm 1BS was mapped in similar material.[50] There is no reason to expect significant differences between the RFLP maps of the durum A and B genomes and their counterparts in the diploid and hexaploid Triticeae. *T. turgidum* possesses the 4AL–5AL–7BS translocation and probably the 4A pericentric inversion.[52]

In oat, the diploid map (from a cross between A-genome diploids)[53, 54] (Fig. 19.4) and hexaploid map[55] (Fig. 19.5) were based on the same libraries of oat and barley

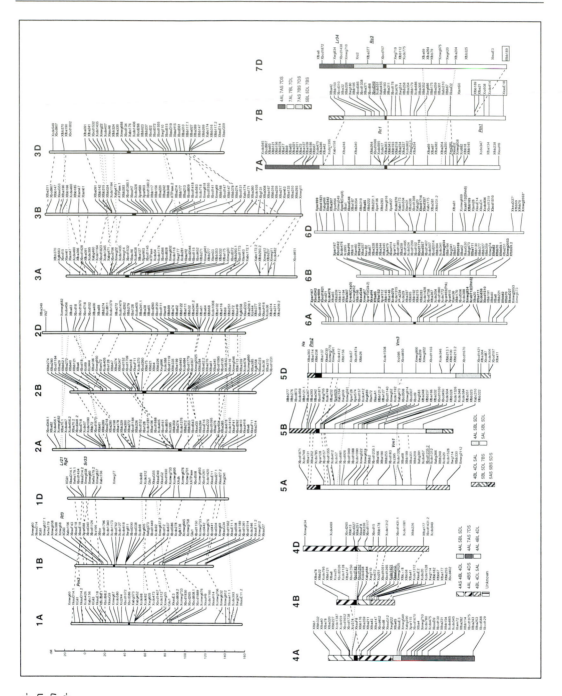

Fig. 19.1. RFLP map of hexa-ploid wheat (Triticum aestivum L.) Adapted from figures supplied by the authors of references 18–22.

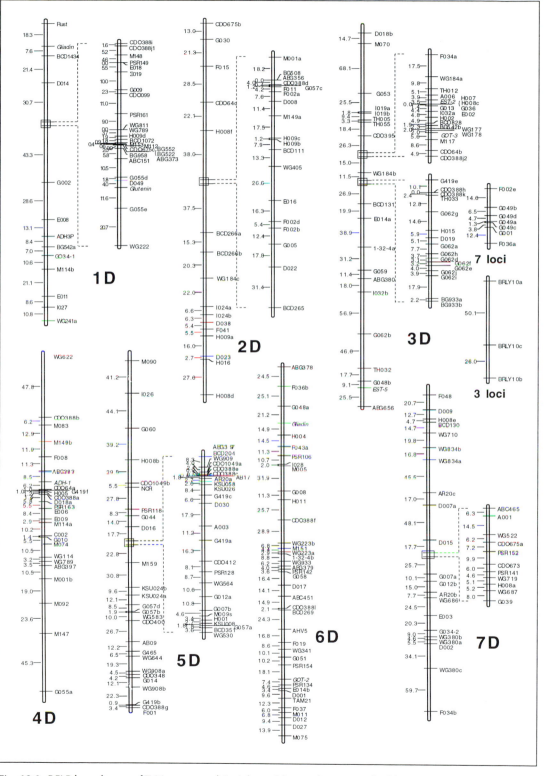

Fig. 19.2. RFLP-based map of Triticum tauschii. Adapted from a figure supplied by KS Gill.[45, 46]

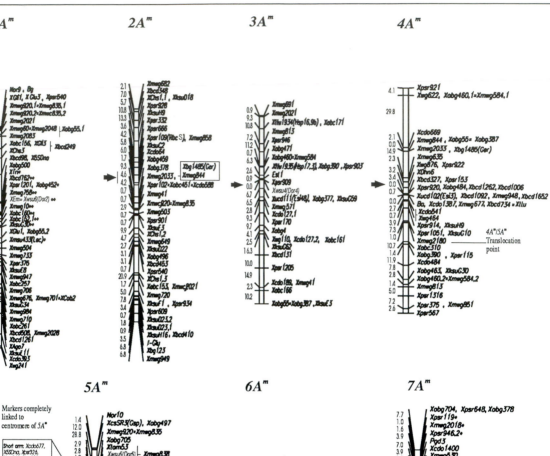

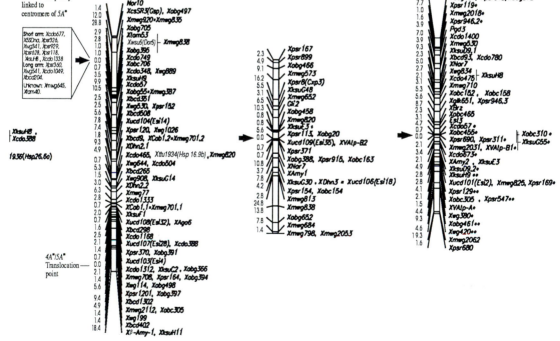

Fig. 19.3. RFLP-based map of Triticum monococcum. *Adapted from a figure supplied by J Dubcovsky.*[32]

*Fig. 19.4. (opposite) RFLP-based map of diploid oat (*Avena atlantica x A. hirtula*). Adapted from a figure supplied by A Van Deynze.*[53,54]

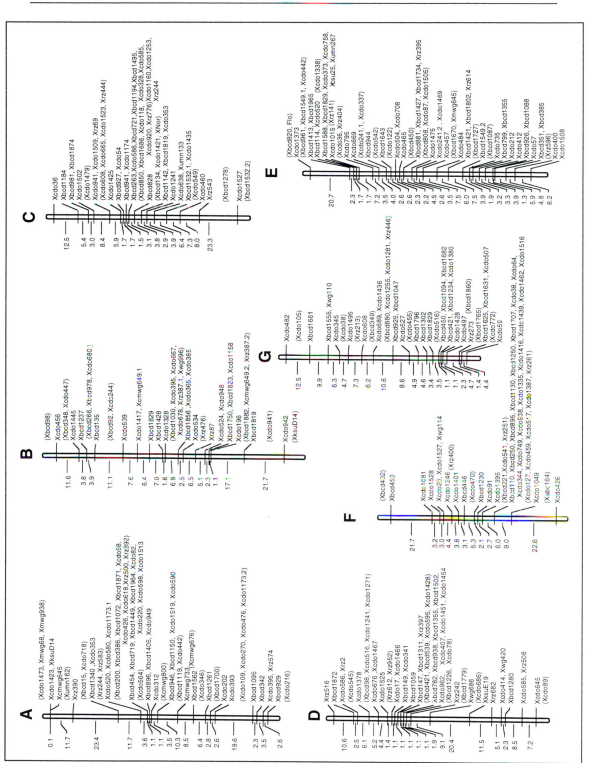

cDNAs used for two major wheat and barley maps cited above. Owing to the lack of aneuploid stocks, genetic studies in cultivated oat have lagged behind those in wheat and only recently has monosomic analysis[56] been used to characterize the homoeologous groups of hexaploid oat and assign the 38 linkage groups among the 21 chromosomes. Results from RFLP analysis[55-57] and from genomic in situ hybridization[58] have confirmed earlier studies showing extensive chromosomal interchanges among the A, C, and D genomes that compose hexaploid oat. While some homoeologous relationships are revealed by these approaches, oat chromosomes have clearly undergone too much rearrangement to be placed neatly into homoeologous groups like those of wheat.[54]

In rice, two molecular linkage maps provide a saturation of about 1 marker per 0.9 cM. One, of 722 markers,[59] is based on an interspecific backcross population from the cross *O. sativa/O. longistaminata/O. sativa* (Fig. 19.6). The other, of 1,383 markers,[60] is based on an F$_2$ population from the cross Nipponbare (japonica) x Kasalath (indica). Efforts to integrate these two molecular maps are underway. A classical linkage map developed over the last 50 years with data from many different crosses contains some 170 phenotypic trait markers and trisomic analysis has placed an additional 150 morphological markers on chromosomes.[61]

19.3. GENOME ORGANIZATION IN CEREALS

The haploid size of the barley genome, as of any of the individual genomes of hexaploid wheat or oat, is around five billion bp of DNA. Up to 92% of genomic DNA in the grasses consists of noncoding repeated sequences, as shown by renaturation experiments.[62] Digestion studies with cytosine-methylation-sensitive restriction enzymes such as *Not*I[63] indicate that pericentromeric regions of chromosomes yield longer fragments, corresponding to higher methylation, than do distal regions. Moore et al[63] combined evidence[64] of a nonrandom association between unmethylated CpG dimers and gene coding regions with concordant findings from mammalian genome research to present a model of cereal genome organization. In this model proximal chromosomal regions are rich with long tandem repetitive sequences. Their density and length decrease with distance from the centromere until in distal regions both unmethylated restriction sites and genes occur in high density. The extreme inhibition of recombination near centromeres and the extra-high density of low-copy sequences in distal regions are obvious from genetic maps and from physical-mapping studies.[66-68]

The proportions of shared repeated-sequence fractions between cereal genomes, derived from DNA–DNA annealing experiments,[69] support the morphologically based phylogenetic tree in which oat, then barley, and finally, rye diverged from the wheat lineage.

Several repeated sequences (e.g., BIS, WIS) with retrotransposon structure have been found in the Triticeae dispersed over all chromosomes.[70-72] The isolation of one (WIS 2-1A) as an insertion sequence in a single wheat variety pointed at a recent transposition event, but its nonfunctional reverse-transcriptase domain meant that activation could not be autonomous. However, autonomous excision has been verified for an Ac element introduced into wheat via protoplast transfection with wheat dwarf virus.[73]

In rice, Southern analysis of large numbers of clones has given information about the distribution of repeated sequences. Based on evaluation of moderate-stringency hybridization of 210 *Pst*I clones, about 58% showed one or two copies, 20% multiple copies, and 22% highly repeated sequences.[74] By comparison, of 2,950 random rice cDNAs analyzed at moderate stringency by Southern analysis, some 30% were single or low-copy.[60] Fewer than 20% of clones from a *Not*I linking library or yeast artificial chromosome (YAC) end-clones showed single or low-copy RFLPs that could be mapped.[60] Apparently at least

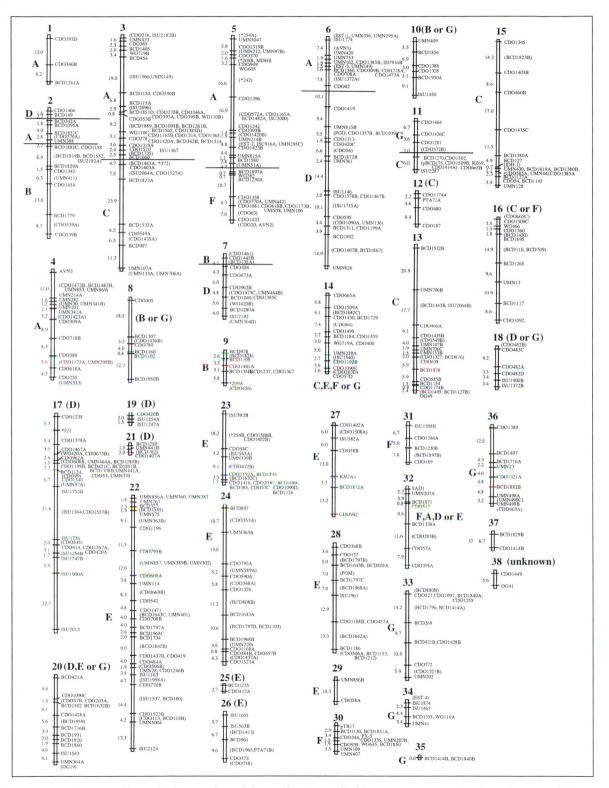

Fig. 19.5. RFLP map of hexaploid oat. Adapted from a figure supplied by A Van Deynze. Numbers designate linkage groups; letters A–F indicate putative homoeology with diploid oat chromosomes (cf. Fig. 19.4).[55]

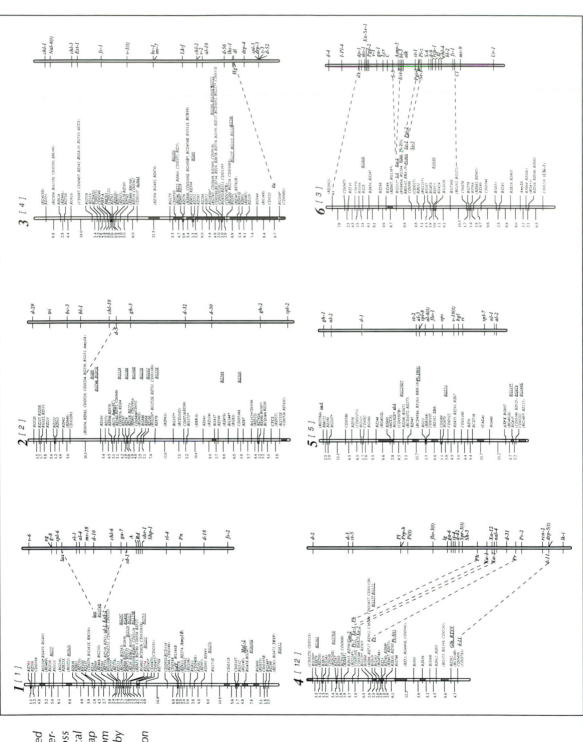

Fig. 19.6. RFLP-based map of rice inter-specific backcross (left) with classical morphological map (right). Adapted from a figure supplied by SR McCouch.[59] (Figure continued on facing page)

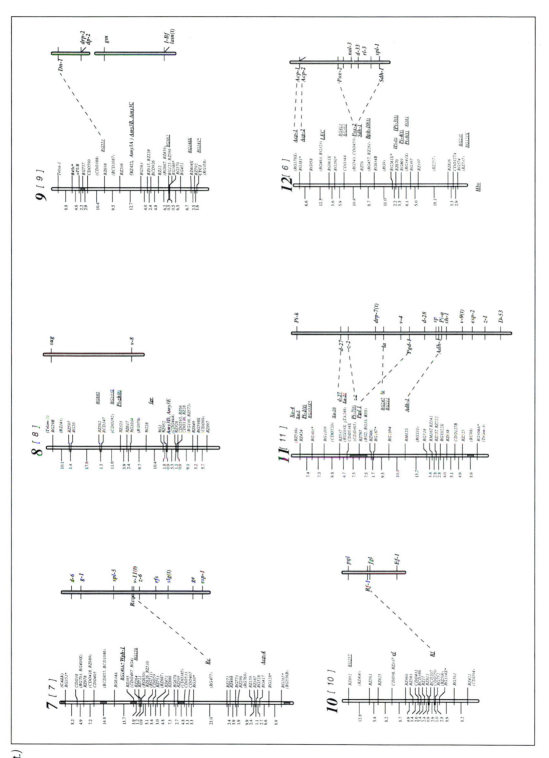

Fig. 19.6. (cont.)

50% of the rice genome consists of moderate to highly repetitive DNA. Several multiple-locus DNA markers were mapped by hybridization with a YAC library.[75] Members of repeated cDNA families were tightly clustered in the rice genome.

Rice presents evidence of active retrotransposition[76] and of defective autonomous elements[77] and other retroposon sequences.[78,79] Transposons are promising tools for tagging rice genes. Integration and transposition of maize transposable elements Ac and Ds is possible via protoplast electroporation[80-82] or viral transfection.[83]

19.4. COMPARATIVE MAPPING IN THE GRASSES

Comparative mapping aims to reveal the genomic similarities between related species so that genetic information about one species may be extended to another and phylogenetic inferences can be made. Ahn et al[84] presented a picture of segmental homoeology of the rice and wheat maps, complemented by a similar picture for rice and maize.[85] These results have been expanded with further studies in these species[86-88] and in oat.[54] Schemes of chromosomal evolution from common ancestors based on map differences have been proposed for rye[36] and maize.[87]

Genomic similarity is conserved not only for biochemical markers[7] and neutral RFLP loci, but also for genes. A good example is that of time to flowering. RFLP mapping[21,89] shows that the well-known *Vrn* genes on the group-5 chromosomes of wheat reside at the same genomic locations as *Sh2* and *Sp1*, major genes controlling earliness in barley[90,91] and rye,[92] as well as a hitherto unnamed locus on hexaploid oat group 24.[54] Somewhat less precise marker evidence links major earliness genes on the group-2 chromosome homoeologs of barley and wheat.[21,93-95] There are hundreds of known genes in the classical maps of cereal genomes for which no such counterparts have yet been sought, partly because of the still only rough alignment between the classical and molecular maps.

19.5. ALIGNING GENETIC AND MORPHOLOGICAL MAPS AND SEARCHING FOR QUANTITATIVE TRAIT LOCI (QTLs)

The task of uniting the classical and molecular genetic maps in Triticeae and oat is being carried out piecemeal because mapping crosses have been chosen more to maximize molecular polymorphism than for the segregation of classical genetic markers. The North American Barley Genome Map (NABGM) (Kleinhofs et al)[29,30] and the Norwich wheat map are richer in isozyme markers and thus more readily aligned with the classical map than are other major oat, wheat, and barley maps which have relied solely on RFLP markers. The NABGM doubled haploids derived from Steptoe x Morex have been extensively evaluated for quantitative characters of agronomic and industrial importance.[96] The inbred lines from the synthetic x 'Opata 85' wheat cross have permitted mapping at least nine named genes[19-21] and await evaluation for many disease resistance and quantitative traits. Further gene tagging in cereals and their relatives has been achieved in special-purpose populations. Among many examples are that of cereal-cyst nematode[97] and Hessian fly[98] resistance in *T. tauschii*; powdery-mildew resistance genes in barley and wheat,[99-103] and stress-induced genes in *T. monococcum*.[104] With the dense maps and rich probe libraries now available in the grasses, any desired gene may be readily tagged with an appropriate cross. Classical aneuploid mapping techniques in the Triticeae remain vital, as shown by the mapping of salt-tolerance[105] and aluminum-tolerance[106] genes in disomic substitution lines in Chinese Spring.

In rice, over 80 known-function genes have been mapped with molecular markers[59] and factors have been mapped for blast resistance,[107,108] sheath blight resistance,[109] root characters associated with drought avoidance[110] and penetration ability,[111] osmotic adjustment,[112] yield components,[113] height and heading date,[114] and seedling vigor.[115] The integration of the molecular and classical maps is well advanced.[116-118]

19.6. STATUS OF PHYSICAL MAP

Physical map construction in the Triticeae and oats has not relied on the conventional approach in which megabase-sized genomic fragments are cloned into yeast or bacterial artificial chromosomes (YACs, BACs) and then used in conjunction with smaller cosmid clones and RFLP probes to "tile" the genome. This method is most effective in genomes with a lower proportion of repeated sequences. The only such library reported for a Triticeae species consists of 18,000 YACs of 100 to 1000 kb in size, containing collectively about half of the barley genome.[119]

Methods of physical mapping exploit the special genetic resources in grass species. Chromosomes of the wheat B genome give strong C-banding patterns. In selected crosses and translocation stocks this cytological feature has been used to characterize recombination rates in various areas of B-genome chromosomes 1B and 5B.[66] It was estimated that a centiMorgan (cM) of genetic distance corresponded to some 1.5 mb in distal chromosome regions but 234 mb or more near the centromere owing to the typical suppression of recombination in this region. A similar estimate was obtained via another approach[120] in which large DNA fragments generated by rare-cutting enzymes were separated by pulsed-field gel electrophoresis (PFGE) and probed with α-amylase-1 gene clones on wheat chromosomes 6AL and 6BL. Recombination between markers on a single fragment of known size yielded a 1 mb/1 cM correspondence between genetic and physical distance.

In rye, where interstitial C-bands are scarce, a physical map of 40 translocation breakpoints was constructed by synaptonemal-complex analysis at meiotic pachytene.[121]

In a wheat background, a gametocidal chromosome from *Triticum cylindricum* produces wheat stocks with apparently random terminal deletions in all chromosomes.[122]

Single-deletion stocks are characterized by dye-banding and physical measurements and their DNA probed with RFLP probes, resulting in an ordered map of genetic loci without need of a mapping cross. However, the extent of chromosomal rearrangement in these stocks may be greater than cytologically apparent,[33,123] and reliance on RFLP probes to reveal interstitial deletions and inversions thus makes linkage and physical mapping techniques interdependent. Deletion maps now exist for all wheat chromosomes.[25,122-127] The density of deletion events, about 1 per 2.7 Mbp, in the distal regions of chromosomes suggests the application of PFGE, in combination with a dense genetic map in *T. tauschii*, for physical mapping of the many genes in these regions.

An analogous if somewhat more laborious physical-mapping technique has been reported[128] for barley, where PCR was used to amplify sequences from microisolated single chromosomes of reciprocal translocation stocks.

Two strategies for correlating the physical and genetic positions of telomere-associated sequences in terminal chromosomal regions have been demonstrated. Primers were designed for specific PCR amplification of barley telomere-associated sequences,[129] and PFGE and in situ hybridization revealed that a tandemly repeated satellite sequence lies physically close to a barley telomeric repeat sequence.[130] The variability revealed in both of these approaches allowed the respective markers to be genetically mapped in barley mapping populations.

In situ hybridization (ISH)-based physical mapping of single-copy sequences such as those from RFLP probes, a technique used in human mapping[131] has recently been reported for cereal chromosomes.[132] Two fluorescent dyes may be used to reveal the relative chromosomal positions of two sequences.[133] Genomic ISH (GISH), the hybridization of labeled full genomic DNA of one species to the chromosomes of another for the purpose of identifying introgressed chromatin from the first, has been used to detect alien chromatin for example, in wheat[134] and A- and C-genome contributions to the oat AACCDD genome.[58] Recently[135] a faster technique of preannealing was substituted for the blocking step usual in GISH to eliminate hybridization to sequences common to the background and target genomes.

The rice genome, at 450 Mb,[136] is among the smallest of any monocot known and the DNA:cM ratio, based on current maps, is thus approximately 250-300 kb/cM. In rice, several large insert libraries, each of 5-6 genome equivalents, are available in the form of YACs,[137] bacterial artificial chromosomes (BACs),[138,139] and cosmids,[140] and a physical map based on the alignment of YAC contigs is under construction.[60]

19.7. DNA MARKER DEVELOPMENT

Marker development remains a challenge in the grasses. The use of RAPD markers in wheat has been evaluated.[141] While of value in species with smaller genomes and less repeated-sequence DNA, this technique seldom amplified markers from homoeologous chromosomes and thus could not exploit a major advantage of RFLP mapping in wheat; furthermore, it produced dominant markers, which cannot be reliably genotyped in early-generation segregants, and was apt to amplify repeated sequences rather than single-copy loci. A more promising application of PCR to wheat mapping involves sequence-tagged sites (STS) detected by PCR primers synthesized to complement low-copy RFLP clone sequences.[142] While markers in this study generally mapped to the chromosomes to which the source RFLPs mapped, the detection of polymorphism usually required digestion of the PCR product with restriction enzymes. STSs were similarly developed for barley mapping.[143] The expense of sequencing and the difficulty of assuring that the primer pairs amplify the polymorphic region detected by the original clone suggest that while useful in applications such as gene tagging,[144] this

technique may not find general use in the grasses. One possible source of markers is chromosome-specific libraries generated via PCR amplification following microdissection[145-147] or flow-sorting.[148] They could be further enriched by subtractive hybridization with DNA from appropriate nullisomic stocks. Clone libraries enriched for small fragments from genomic digests with the methylation-sensitive enzyme HpaII contained 80% low-copy clones with application both to RFLP mapping in grasses and to physical mapping in wheat.[65]

Polymorphism can be uncovered by representational difference analysis.[149] This PCR-adapted genomic subtraction procedure yielded several markers specific for rye segments in wheat backgrounds.[150] For some PCR-amplified polymorphisms, gel electrophoresis across a denaturing[151] or temperature[152] gradient allows separations not possible in ordinary gels.

The rye genome possesses several families of well-studied dispersed repeated sequences, one of which, R173, was shown[153] to have the potential to supply numerous well-spaced markers for mapping. PCR primers were designed complementary to sequences within and flanking R173 elements and several were mapped in rye. It was proposed that a limited sequencing effort with the existing R173 clone library could generate hundreds of other markers in the rye genome. Other rye-specific PCR primers[154,155] would be useful for detecting rye chromatin in an alien, e.g., wheat, genomic background. Techniques similar to these should be applicable with BIS and other retrotransposon sequences in wheat and barley. Recently a PstI rye genomic clone library with 44% low-copy sequences was developed for rye mapping.[156]

Microsatellite (also called simple-sequence-repeat or SSR) markers have been tested in wheat,[157-159] barley,[160,161] and rice.[162-164] The labor and cost of sequencing and primer synthesis make development of SSR libraries expensive. However, SSR markers are highly polymorphic and well distributed across the genomes of cereals tested to date. Unlike most PCR markers, they are codominant. Unlike RFLPs, they can be assayed overnight on micro amounts of DNA with nonradioactive detection. They can be multiplexed (as many as 300 assays can be separated on a single gel) and show stable inheritance across decades of rice pedigrees (SR McCouch, unpublished data). In the grasses as in any species, efficient DNA-based genotyping methods should be multiplex, microscale, and semiautomated in order to minimize handling, time, and materials costs.

19.8. PROSPECTS FOR FUTURE RESEARCH AND APPLICATIONS

The findings of homoeology between grass-species maps have excited interest in a gene-mapping and gene-isolation approach proceeding from the simplest genomes to the most complex ones. The base genome proposed[18,63,165] is that of rice, which features more than 800 cDNAs that have been sequenced for comparison with archived sequences to determine their likely gene functions.[60] These STSs could be used as probes to identify, map, and sequence related genes in the other grass species. With the recent construction of dense barley and wheat RFLP maps and the characterization of interspecies map homoeology at an ever finer scale, this gene-directed approach seems likely to be applied next for integrating the genetic maps of neutral markers and major genes with the physical maps of grass-species chromosomes. The focus of attention on rice STSs avoids direct cloning efforts in the larger genomes with their abundant repetitive DNA.

Phylogenetic investigations in the Gramineae should benefit from comparative sequence analysis of genes thus isolated. Gene sequences may also be used to design PCR primer sets for detection of polymorphism in adjacent genomic regions.[166,167]

For cereal breeders the main promise of genome mapping is in the identification of genes with desirable agronomic effect that can be easily moved into culti-

vars. At present, realizing this promise for any but major genes of large economic importance is impractical. Even where an easily assayable marker has been found for a gene segregating in a mapping population, not only must it show polymorphism in the breeding population but the phase of its linkage with the gene must be shown to be the same in the breeding material as in the mapping material. This latter is easy only with genes of major effect. There seem to be only a few ways to address these obstacles. One is to restrict breeding attention to germplasm already used for mapping (e.g., the barley Steptoe x Morex cross), which may yield useful information about the degree of improvement to be achieved via the pyramiding of QTLs, but will not produce widely adapted cultivars. Another is to assume that the linkage phase between a marker and a nearby QTL is constant, i.e., that natural or unwitting artificial selection have maintained a desired coupling linkage. The low polymorphism in the self-pollinated cereals argues for the conservation of sizable portions of the genome, plausibly including a segment containing a QTL lying between two markers possessing the same allele in both a mapping parent and a breeding parent. Genotyping studies of ancestral lines could then allow breeders and researchers to infer gene complements in germplasms of interest without resort to conventional linkage or QTL mapping. Allozyme and rDNA studies in cereals show correlations between specific alleles and adaptedness and productivity.[168] "Pedigree-based mapping" with RFLPs has been demonstrated in soybean[169] and a body of data with similar potential application is available for oat.[170,171] Highly polymorphic markers from microsatellites or 5S rDNA sequences[160,172,173] may have much to tell us about the relationship of gene flow to genome organization.

ACKNOWLEDGMENTS

Dr. Gary Hart is thanked for corrections and suggestions.

REFERENCES

1. Miller TE. Systematics and evolution. In: Lupton FGH, ed. Wheat Breeding: Its Scientific Basis. London: Chapman and Hall, 1987:1-28.

2. Dvorak J, di Terlizzi P, Zhang HB et al. The evolution of polyploid wheats: identification of the A genome donor species. Genome 1993; 36:21-31.

3. West JG, McIntyre CL, Appels R. Evolution and systematic relationships in the *Triticeae* (*Poaceae*). Pl Syst Evol 1988; 160:1-28.

4. Ladizinsky G. Domestication via hybridization of the wild tetraploid oats *Avena magna* and *A. murphyi*. Theor Appl Genet 1995; 91:639-646.

5. Matsuo T, Hoshikawa K. Science of the Rice Plant. Volume 1: Morphology. Tokyo, Japan: Ministry of Agriculture, Forestry and Fisheries, Food and Agriculture Policy Research Center, 1993.

6. Sears ER. Nullisomic–tetrasomic combinations in hexaploid wheat. In Riley R, Lewis KR, eds. Chromosome manipulations and plant genetics. London: Oliver and Boyd, 1966:29-45.

7. Hart GE. Determination of relationships in the tribe Triticeae by genetic and biochemical studies of enzymatic and nonenzymatic proteins. In: Heyne E, ed. Wheat and Wheat Improvement. Madison, Wis: American Society of Agronomy, 1987:199-214.

8. Islam AKMR, Shepherd KW. Incorporation of barley chromosomes into wheat. In: Bajaj YPS, ed. Biotechnology in Agriculture and Forestry; Wheat. Vol. 13. Berlin, Heidelberg: Springer-Verlag,1990:128-151.

9. Sharp PJ, Chao S, Desai S et al. The isolation, characterization and application in the Triticeae of a set of wheat RFLP probes identifying each homoeologous chromosome arm. Theor Appl Genet 1989; 78:342-348.

10. Anderson JA, Ogihara Y, Sorrells ME et al. Development of a chromosomal arm map for wheat based on RFLP markers. Theor Appl Genet 1992; 83:1035-1043.

11. Chao S, Sharp PJ, Worland AJ et al. RFLP-based genetic maps of wheat homoeologous group 7 chromosomes. Theor Appl Genet 1989; 78:495-504.

12. Devos KM, Atkinson MD, Chinoy CN et al. RFLP-based genetic map of the homoeologous group 2 chromosomes of wheat, rye, and barley. Theor Appl Genet 1993; 85:784-792.

13. Devos KM, Atkinson MD, Chinoy CN et al. RFLP-based genetic map of the homoeologous group 3 chromosomes of wheat and rye. Theor Appl Genet 1992; 83:931-939.

14. Xie DX, Devos KM, Moore G et al. RFLP-based genetic maps of the homoeologous group 5 chromosomes of bread wheat (*Triticum aestivum* L.). Theor Appl Genet 1993; 87:70-74.

15. Hart GE, Gale MD, McIntosh RA. Linkage maps of *Triticum aestivum* (Hexaploid wheat, 2n=42, genomes A, B & D) and *T. tauschii* (2n=14, genome D). In O'Brien SJ, ed. Genetic Maps, 6th Ed. Cold Spring Harbor: Cold Spring Harbor Press, 1993: 6.204-6.219.

16. Devos KM, Gale MD. The genetic maps of wheat and their potential in wheat breeding. Outlook Agric 1993; 22:93-99.

17. Liu YG, Tsunewaki K. Restriction fragment length polymorphism (RFLP) analysis in wheat. II. Linkage maps of the RFLP sites in common wheat. Jpn J Genet 1991; 66:617-633.

18. Van Deynze AE, Dubcovsky J, Gill KS et al. Molecular-genetic maps for chromosome 1 in *Triticeae* species and their relation to chromosomes in rice and oats. Genome 1995; 38:47-59.

19. Nelson JC, Van Deynze AE, Autrique E et al. Molecular mapping of wheat. Homoeologous group 2. Genome 1995; 38:116-24.

20. Nelson JC, Van Deynze AE, Autrique E et al. Molecular mapping of wheat. Homoeologous group 3. Genome 1995; 38:125-133.

21. Nelson JC, Sorrells ME, Van Deynze AE et al. Molecular mapping of wheat. Major genes and rearrangements in homoeologous groups 4, 5, and 7. Genetics 1995; 141:721-731.

22. Marino CL, Nelson JC, Lu YH et al. RFLP-based linkage maps of the homoeologous group 6 chromosomes of hexaploid wheat (*Triticum aestivum* L. em. Thell). Genome 1995; in press.

23. Naranjo T, Roca P, Goicoechea PG et al.

Arm homoeology of wheat and rye chromosomes. Genome 1987; 29:873-882.

24. Liu CJ, Atkinson MD, Chinoy CN et al. Nonhomoeologous translocations between group 4, 5, and 7 chromosomes within wheat and rye. Theor Appl Genet 1992; 83:305-312.

25. Mickelson-Young L, Endo TR, Gill BS. A cytogenetic ladder-map of the wheat homoeologous group-4 chromosomes. Theor Appl Genet 1995; 90:1007-1011.

26. Devos KM, Dubcovsky J, Dubcovsky J et al. Structural evolution of wheat chromosomes 4A, 5A, and 7B and its impact on recombination. Theor Appl Genet 1995; 91:282-288.

27. Heun M, Kennedy AE, Anderson JA et al. Construction of a restriction fragment length polymorphism map for barley (*Hordeum vulgare*). Genome 1991; 34:437-447.

28. Graner A, Jahoor A, Schondelmaier J et al. Construction of an RFLP map of barley. Theor Appl Genet 1991; 83:250-256.

29. Kleinhofs A, Kilian A, Saghai Maroof M et al. A molecular, isozyme, and morphological map of the barley (*Hordeum vulgare*) genome. Theor Appl Genet 1993; 86:705-712.

30. Kleinhofs A, Kilian A, Kudrna D et al. The NABGMP Steptoe x Morex mapping progress report. Barley Genet Newslett 1993; 23:79-83.

31. Sherman JD, Fenwick AL, Namuth DM et al. A barley RFLP map: alignment of three barley maps and comparisons to Gramineae species. Theor Appl Genet 1995; 91:681-90.

32. Dubcovsky J, Luo MC, Zhong GY et al. Genetic map of diploid wheat,*Triticum monococcum*. L., and its comparison with maps of *Hordeum vulgare* L. Submitted.

33. Hohmann U, Endo TR, Herrmann RG. Characterization of deletions in common wheat induced by an *Aegilops cylindrica* chromosome: detection of multiple chromosome rearrangements. Theor Appl Genet 1995; 91:611-617.

34. Wang ML, Atkinson MD, Chinoy CN et al. RFLP-based genetic map of rye (*Secale cereale* L.) chromosome 1R. Theor Appl Genet 1991; 82:174-178

35. Rognli OA, Devos KM, Chinoy CN et al. RFLP mapping of rye chromosome 7R reveals a highly translocated chromosome relative to wheat. Genome 1992; 35:1026-1031.

36. Devos KM, Atkinson MD, Chinoy CN et al. Chromosomal rearrangements in the rye genome relative to that of wheat. Theor Appl Genet 1993; 85:673-680.

37. Riley R. The cytogenetics of the differences between some *Secale* species. J Agric Sci 1955; 46:277-283.

38. Philipp U, Wehling P, Wricke G. A linkage map of rye. Theor Appl Genet 1994; 88:243-248.

39. Baum M, Appels R. The cytogenetic and molecular architecture of chromosome 1R, one of the most widely utilized sources of alien chromatin in wheat varieties. Chromosoma 1991; 101:1-10.

40. Alonso-Blanco C, Goicoechea PG, Roca A et al. A cytogenetic map on the entire length of rye chromosome 1R, including one translocation breakpoint, three isozyme loci and four C-bands. Theor Appl Genet 1993; 85:735-744.

41. Benito C, Gallego FJ, Zaragoza C et al. Biochemical evidence of a translocation between 6RL/7RL chromosome arms in rye (*Secale cereale* L.). A genetic map of 6R chromosome. Theor Appl Genet 1991; 82:27-32.

42. Benito C, Zaragoza C, Gallego FJ et al. A map of rye chromosome 2R using isozyme and morphological markers. Theor Appl Genet 1991; 82:112-116.

43. Benito C, Llorente F, Henriques-Gil N et al. A map of rye chromosome 4R with cytological and isozyme markers. Theor Appl Genet 1994; 87:941-946.

44. Melz G, Schlegel R, Thiele V. Genetic linkage map of rye (*Secale cereale* L.). Theor Appl Genet 1992; 85:33-45.

45. Gill KS, Lubbers EL, Gill BS et al. A genetic linkage map of *Triticum tauschii* (DD) and its relationship to the D genome of bread wheat (AABBDD). Genome 1991; 34:362-374.

46. Gill BS, Gill KS, Raupp WJ et al. Genetic and physical mapping in *Triticum tauschii* and *T. aestivum*. In Hoisington D, McNab A, eds. Progress in genome mapping of wheat and related species, 1993. Proceedings of the 3rd Public Workshop of the International Triticeae Mapping Intiative.

Mexico, D. F.: CIMMYT, Sept. 22-26, 1992:10-17.

47. Lagudah E, Appels R, Brown AHD et al. The molecular-genetic analysis of *Triticum tauschii*, the D-genome donor to hexaploid wheat. Genome 1991; 34:375-386.

48. Hohmann U, Lagudah ES. C-banding polymorphism and linkage of nonhomoeologous RFLP loci in the D genome progenitor of wheat. Genome 1993; 36:235-243.

49. Namuth DM, Lapitan NLV, Gill KS et al. Comparative RFLP mapping of *Hordeum vulgare* and *Triticum tauschii*. Theor Appl Genet 1994; 89:865-872.

50. Dubcovsky J, Dvorak J. Ribosomal RNA multigene loci: nomads of the Triticeae genomes. Genetics 1995; 140:1367-1377.

51. Chen Z, Devey M, Tuleen NA et al. Use of recombinant substitution lines in the construction of RFLP-based genetic maps of chromosomes 6A and 6B of tetraploid wheat (*Triticum turgidum* L.) Theor Appl Genet 1994; 89:703-712.

52. Naranjo T. Chromosome structure of durum wheat. Theor Appl Genet 1990; 79:397-400.

53. O'Donoughue LS, Wang Z, Röder M et al. An RFLP-based linkage map of oats based on a cross between two diploid taxa (*Avena atlantica* x *A. hirtula*). Genome 1992; 35:765-771.

54. Van Deynze AE, Nelson JC, O'Donoughue LS et al. Comparative mapping in grasses. Oat relationships. Mol Gen Genet 1995; 249;349-356.

55. O'Donoughue LS, Kianian SF, Rayapati PJ et al. A molecular map of cultivated oat. Genome 1995; 38:368-380.

56. Rooney WL, Jellen EN, Phillips RL et al. Identification of homoeologous chromosomes in hexaploid oat (*A. byzantina* cv Kanota) using monosomics and RFLP analysis. Theor Appl Genet 1994; 89:329-335.

57. Jellen EN, Rooney WL, Phillips RL et al. Characterization of the hexaploid oat *Avena byzantina* cv. Kanota monosomic series using C-banding and RFLPs. Genome 1993; 36:962-970.

58. Chen Q, Armstrong K. Genomic in situ hybridization in *Avena sativa*. Genome 1994; 37:607-612.

59. Causse M, Fulton TM, Cho YG et al. Saturated molecular map of the rice genome based on an interspecific backcross population. Genetics 1994; 138:1251-1274.

60. Kurata N, Nagamura Y, Yamamoto K et al. A 300 kilobase interval genetic map of rice including 883 expressed sequences. Nature Genet 1994; 8:365-372.

61. Kinoshita T. Report of the committee on gene symbolization and linkage map. Rice Genet Newslett 1990; 7:22-50.

62. Flavell RB, Bennett MD, Smith JB et al. Genome size and the proportion of repeated sequence DNA in plants. Biochem Genet 1974; 12:257-269.

63. Moore G, Gale MD, Kurata N et al. Molecular analysis of small grain cereal genomes: current status and prospects. Bio/Technology 1993; 11:584-589.

64. Moore G, Abbo S, Cheung W et al. Key features of cereal genome organization as revealed by the use of cytosine methylation-sensitive restriction endonucleases. Genomics 1993; 15:472-482.

65. Cheung WY, Moore G, Money TA et al. *Hpa*II library indicates "methylation-free islands" in wheat and barley. Theor Appl Genet 1992; 84:739-746.

66. Lukaszewski AJ, Curtis CA. Physical distribution of recombination in B-genome chromosomes of tetraploid wheat. Theor Appl Genet 1993; 86:121-127.

67. Hohmann U, Endo TR, Gill KS et al. Comparison of genetic and physical maps of group 7 chromosomes from *Triticum aestivum* L. Mol Gen Genet 1994; 245:644-653.

68. Hohmann U, Graner A, Endo TR. Comparison of wheat physical maps with barley linkage maps for group 7 chromosomes. Theor Appl Genet 1995; 91:618-626.

69. Flavell RB, O'Dell M, Rimpau J et al. Repeated sequence DNA relationships in four cereal genomes. Chromosoma 1977; 63:205-222.

70. Harberd NP, Flavell RB, Thompson RD. Identification of a transposon-like insertion in a *Glu-1* allele of wheat. Mol Gen Genet 1987; 209:326-332.

71. Moore G, Lucas H, Batty N et al. A family of retrotransposons and associated genomic

variation in wheat. Genomics 1991; 10:461-468.

72. Monte JV, Flavell RB, Gustafson JP. WIS 2-1A: an ancient retrotransposon in the *Triticeae* tribe. Theor Appl Genet 1995; 91:367-373.

73. Laufs J, Wirtz U, Matzeit V et al. Wheat dwarf virus Ac/Ds vectors: expression and excision of transposable elements introduced into various cereals by a viral replicon. Proc Natl Acad Sci USA 1990; 87:7752-7756.

74. McCouch SR, Kochert G, Yu ZH et al. Molecular mapping of rice chromosomes. Theor Appl Genet 1988; 76:815-829.

75. Tanoue H, Umehara,Y, Wang, ZX et al. Poster 230. Plant Genome III. San Diego, Calif. Jan. 15-19, 1995.

76. Hirochika H, Takeda T, Fukuchi A et al. Retrotransposon families in rice. Mol Gen Genet 1992; 233:209-216.

77. Tenzen T, Matsuda Y, Ohtsubo H et al. Transposition of Tnr1 in rice genomes to 5'-Pu-TAPy-3' sites, duplicating the TA sequence. Mol Gen Genet 1994; 245: 441-448.

78. Hirano HY, Mochizuki K, Umeda M et al. Retrotransposition of a plant SINE into the wx locus during evolution of rice. J Mol Evol 1994; 38:132-137.

79. Ohtsubo H, Ohtsubo E. Involvement of transposition in dispersion of tandem repeat sequences (TrsA) in rice genomes. Mol Gen Genet 1994; 245:449-455.

80. Murai N, Li Z, Kawagoe Y et al. Transposition of the maize activator element in transgenic rice plants. Nucleic Acids Res 1991; 19:617-622.

81. Izawa T, Miyazaki C, Yamamoto M et al. Introduction and transposition of the maize transposable element Ac in rice (*Oryza sativa* L.). Mol Gen Genet 1991; 227:391-396.

82. Shimamoto K, Miyazaki C, Hashimoto H et al. Trans-activation and stable integration of the maize transposable element Ds cotransfected with the Ac transposase gene in transgenic rice plants. Mol Gen Genet 1993; 239:354-360.

83. Sugimoto K, Otsuki Y, Saji S et al. Transposition of the maize Ds element from a viral vector to the rice genome. Plant J 1994; 5:863-871.

84. Ahn S, Anderson JA, Sorrells ME et al. Homoeologous relationships of rice, wheat, and maize chromosomes. Mol Gen Genet 1993; 241:483-490.

85. Ahn S, Tanksley SD. Comparative linkage maps of the rice and maize genomes. Proc Natl Acad Sci USA 1993; 90:7980-7984.

86. Kurata N, Moore G, Nagamura Y et al. Conservation of genome structure between rice and wheat. Bio/Technology 1994; 12:276-278.

87. Devos KM, Chao S, Li QY et al. Relationship between chromosome 9 of maize and wheat homeologous group 7 chromosomes. Genetics 1994; 138:1287-1292.

88. Van Deynze AE, Nelson JC, Harrington SN et al. Comparative mapping in grasses. Wheat relationships. Mol Gen Genet 1995; 248:744-754.

89. Galiba G, Quarrie SA, Sutka J et al. RFLP mapping of the vernalization (*Vrn1*) and frost resistance (*Fr1*) genes on chromosome 5A of wheat. Theor Appl Genet 1995; 90:1174-1179.

90. Pan A, Hayes PM. Chen F et al. Genetic analysis of the components of winter-hardiness in barley (*Hordeum vulgare* L). Theor Appl Genet 1994; 89:900-910.

91. Backes G, Graner A, Foroughi-Wehr B et al. Localization of quantitative trait loci (QTL) for agronomic important characters by the use of a RFLP map in barley (*Hordeum vulgare* L.). Theor Appl Genet 1995; 90:294-302.

92. Plaschke J, Borner A, Xie DX et al. RFLP mapping of genes affecting plant height and growth habit in rye. Theor Appl Genet 1993; 85:1049-1054.

93. Worland AJ, Law CN. Genetic analysis of chromosome 2D of wheat. I. The location of genes affecting height, day length insensitivity, hybrid dwarfism and yellow rust resistance. Z Pflanzenzücht 1986; 96: 331-345.

94. Barua UM, Chalmers KJ, Thomas WTB et al. Molecular mapping of genes determining height, time to heading, and growth habit in barley (*Hordeum vulgare*). Genome 1993; 36:1080-1087.

95. Laurie DA, Pratchett N, Bezant JH et al. Genetic analysis of a photoperiod response

gene on the short arm of chromosome 2(2H) of *Hordeum vulgare* (barley). Heredity 1994; 72:619-627.

96. Hayes PM, Liu BH, Knapp SJ et al. Quantitative trait locus effects and environmental interaction in a sample of North American barley germ plasm. Theor Appl Genet 1993; 87:392-401.

97. Eastwood RF, Lagudah ES, Appels R. A directed search for DNA sequences tightly linked to cereal cyst nematode resistance genes in *Triticum tauschii*. Genome 1994; 37:311-319.

98. Ma ZQ, Gill BS, Sorrells ME et al. RFLP markers linked to two Hessian fly-resistance genes in wheat (*Triticum aestivum* L.) from *Triticum tauschii* (coss.) Schmal. Theor Appl Genet 1993; 85:750-754.

99. Schüller C, Backes G, Fischbeck G et al. RFLP markers to identify the alleles on the *Mla* locus conferring powdery mildew resistance in barley. Theor Appl Genet 1992; 84:330-338.

100. Maroof MAS, Zhang Q, Biyashev RM. Molecular marker analysis of powdery mildew resistance in barley. Theor Appl Genet 1994; 88:733-740.

101. Hartl L, Weiss FJ, Jahoor A. Use of RFLP markers for the identification of alleles conferring powdery mildew resistance in wheat (*Triticum aestivum* L.). Theor Appl Genet 1993; 86:959-963.

102. Ma ZQ, Sorrells ME, Tanksley SD. RFLP markers linked to powdery mildew resistance genes *Pm1*, *Pm2*, *Pm3*, and *Pm4* in wheat. Genome 1994; 37:871-875.

103. Schachermayr G, Siedler H, Gale MD et al. Identification and localization of molecular markers linked to the *Lr9* leaf rust resistance gene of wheat. Theor Appl Genet 1994; 88:110-115.

104. Dubcovsky J, Dvorak J. Linkage relationships among 25 clones from stress-induced genes in wheat. Theor Appl Genet 1995; 91:795-801.

105. Omielan JA, Epstein E, Dvorak J. Salt tolerance and ionic relations of wheat as affected by individual chromosomes of salt-tolerant *Lophopyrum elongatum*. Genome 1991; 34:961-974.

106. Aniol, AM. Physiological aspects of alu-

minium tolerance associated with the long arm of chromosome 2D of the wheat (*Triticum aestivum* L.) genome. Theor Appl Genet 1995; 91:510-516.

107. Yu ZH, Mackill DJ, Bonman JM et al. Tagging genes for blast resistance in rice via linkage to RFLP markers. Theor Appl Genet 1991; 81:471-476.

108. Wang GL, Mackill DJ, Bonman JM et al. RFLP mapping of genes conferring complete and partial resistance to blast in a durably resistant rice cultivar. Genetics 1994; 136:1421-1434.

109. Li Z, Pinson SRM, Marchetti MA et al. 1995. Characterization of quantitative trait loci (QTLs) in cultivated rice contributing to field resistance to sheath blight (*Rhizoctonia solani*). Theor Appl Genet 1995; 91:382-388

110. Champoux MC, Wang G, Sarkarung S et al. Locating genes associated with root morphology and drought avoidance in rice via linkage to molecular markers. Theor Appl Genet 1995; 90:969-981.

111. Ray J, Yu L, McCouch SR et al. Mapping quantitative trait loci associated with root penetration ability in rice (*Oryza sativa* L.). Theor Appl Genet 1995; in press.

112. Lilley JM, Ludlow MM, McCouch SR et al. Locating QTL for osmotic adjustment and dehydration tolerance in rice. Submitted.

113. Xiao J, Li J, Yuan L et al. Dominance is the major genetic basis of heterosis in rice as revealed by QTL analysis using molecular markers. Genetics 1995; 140:745-754.

114. Li Z, Pinson SRM, Stansel JW et al. Identification of two major genes and quantitative trait loci (QTLs) for heading date and plant height in cultivated rice. (*Oryza sativa* L). Theor Appl Genet 1994; 91:374-381.

115. Redoña ED, Mackill DJ. Molecular mapping of quantitative trait loci in japonica rice. Genome 1995; in press.

116. Xiao J, Fulton T, McCouch S et al. Progress in integration of the molecular maps of rice. Rice Genet Newsl 1992; 9:124-128.

117. Ideta O, Yoshimura A, Matsumoto T et al. Integration of conventional and RFLP linkage maps in rice, I. Chromosomes 1, 2, 3 and 4. Rice Genet Newslett (1992); 9:128.

118. Yu ZH, McCouch SR, Kinoshita T et al.

Association of morphological and RFLP markers in rice (*Oryza sativa* L.). Genome 1995; 38:566-574

119. Kleine M, Michalek W, Graner A et al. Construction of a barley (*Hordeum vulgare* L.) YAC library and isolation of a *Hor1*-specific clone. Mol Gen Genet 1993; 240:265-272.

120. Cheung WY, Chao S, Gale MD. Long-range physical mapping of the alpha-amylase-1 (a-Amy-1) loci on homoeologous group 6 chromosomes of wheat. Mol Gen Genet 1991; 229:373-379

121. Alvarez E, Alonso-Blanco C, Roca A et al. Physical mapping of translocation breakpoints in rye by means of synaptonemal complex analysis. Theor Appl Genet 1994; 89:33-41.

122. Werner JE, Endo TR, Gill BS. Toward a cytogenetically based physical map of the wheat genome. Proc Natl Acad Sci USA 1992; 89:11307-11311.

123. Ogihara Y, Hasegawa K, Tsujimoto H. Fine cytological mapping of the long arm of chromosome 5A in common wheat by use of a series of deletion lines induced by gametocidal (*Gc*) genes of *Aegilops speltoides*. Mol Gen Genet 1994; 244:253-259.

124. Kota RS, Gill KS, Gill BS et al. A cytogenetically based physical map of chromosome 1B in common wheat. Genome 1993; 36:548-554.

125. Gill KS, Gill BS, Endo TR. A chromosome region-specific mapping strategy reveals gene-rich telomeric ends in wheat. Chromosoma 1993; 102:374-381.

126. Delaney D, Nasuda S, Endo TR et al. Cytologically based physical maps of the group-2 chromosomes of wheat. Theor Appl Genet 1995; 91:568-573.

127. Delaney D, Nasuda S, Endo TR et al. Cytologically based physical maps of the group-3 chromosomes of wheat. Theor Appl Genet 1995; 91:780-782.

128. Sorokin A, Marthe F, Houben A et al. Polymerase chain reaction mediated localization of RFLP clones to microisolated translocation chromosomes of barley. Genome 1994; 37:550-555.

129. Kilian A, Kleinhofs A. Cloning and mapping of telomere-associated sequences from *Hordeum vulgare* L. Mol Gen Genet 1992; 235:153-156.

130. Röder MS, Lapitan NLV, Sorrells ME et al. Genetic and physical mapping of barley telomeres. Mol Gen Genet 1993; 238:294-303.

131. Fan YS, Davis LM, Shows TB. Mapping small DNA sequences by fluorescence in situ hybridization directly on banded metaphase chromosomes. Proc Natl Acad Sci USA 1990; 87:6223-6227.

132. Dong H, Quick JS. Detection of a 2.6 kb single/low copy DNA sequence on chromosomes of wheat (*Triticum aestivum*) and rye (*Secale cereale*) by fluorescence in situ hybridization. Genome 1995; 38:246-249.

133. Leitch IJ, Leitch AR, Heslop-Harrison JS. Physical mapping of plant DNA sequences by simultaneous in situ hybridization of two differently labelled fluorescent probes. Genome 1991; 34:329-333.

134. Islam-Faridi MN, Mujeeb-Kazi A. Visualization of *Secale cereale* DNA in wheat germ plasm by fluorescent in situ hybridization. Theor Appl Genet 1995; 90:595-600.

135. Anamthawat-Jónsson K, Reader SM. Preannealing of total genomic DNA probes for simultaneous genomic in situ hybridization. Genome 1995; 38:814-816.

136. Arumuganathan K, Earle ED. Nuclear DNA content of some important plant species. Plant Mol Biol Rep 1991; 9:208-218.

137. Umehara, Y, Inagaki A, Tanoue H et al. Construction and characterization of a rice YAC library for physical mapping. Molecular Breeding 1995; in press.

138. Wang GL, Holsten T.E, Song WY et al. Construction of a rice bacterial artificial chromosome library and identification of clones linked to the Xa-21 disease resistance locus. Plant J 1995; 7:525-533.

139. Zhang HB, Choi S, Woo SS et al. Construction and characterization of two rice bacterial artificial chromosome libraries from the parents of a permanent recombinant inbred mapping population. Poster 276. Plant Genome III. San Diego, Calif. Jan. 15-19, 1995.

140. Song,W, Wang B, Wang GL et al. Construction of a rice cosmid library and identification of five cosmid clones candidate

bacterial blight resistance genes, *Xa*-21. Poster 218. Plant Genome III. San Diego, Calif. Jan. 15-19, 1995.

141. Devos KM, Gale MD. The use of random amplified polymorphic DNA markers in wheat. Theor Appl Genet 1992; 84:567-570.

142. Talbert LE, Blake NK, Chee PW et al. Evaluation of "sequence-tagged-site" PCR products as molecular markers in wheat. Theor Appl Genet 1994; 87:789-794.

143. Tragoonrung S, Kanazin V, Hayes PM et al. Sequence-tagged-site-facilitated PCR for barley genome mapping. Theor Appl Genet 1992; 84:1002-1008.

144. Nieto-Lopez RM, Blake TK. Russian wheat aphid resistance in barley: inheritance and linked molecular markers. Crop Sci 1994; 34:655-659.

145. Schondelmaier J, Martin R, Jahoor A et al. Microdissection and microcloning of the barley (*Hordeum vulgare* L.) chromosome 1HS. Theor Appl Genet 1993; 86:629-636.

146. Albani D, Côté MJ, Armstrong KC et al. PCR amplification of microdissected wheat chromosome arms in a simple 'single tube' reaction. Plant J 1994; 4:899-903.

147. Chen Q, Armstrong K. Characterization of a library from a single microdissected oat (*Avena sativa* L.) chromosome. Genome 1995; 38:706-714.

148. Wang ML, Leitch AR, Schwarzacher T et al. Construction of a chromosome-enriched *Hpa*II library from flow-sorted wheat chromosomes. Nucleic Acids Res 1992; 20:1897-1901.

149. Lisitsyn N, Lisitsyn N, Wigler M. Cloning the differences between two complex genomes. Science 1993; 259:946-951.

150. Delaney DE, Friebe BR, Hatchett JH et al. Targeted mapping of rye chromatin in wheat by representational difference analysis. Genome 1995; 38:458-466.

151. Dweikat I, Mackenzie S, Levy M. et al. Pedigree assessment using RAPD-DGGE in cereal crop species. Theor Appl Genet 1993; 85:497-505.

152. Penner GA, Bezte LG. Increased detection of polymorphism among randomly amplified wheat DNA fragments using a modified temperature sweep gel electrophoresis

(TSGE) technique. Nucleic Acids Res 1994; 22:1780-1781.

153. Rogowsky PM, Shepherd KW, Langridge P. Polymerase chain reaction based mapping of rye involving repeated DNA sequences. Genome 1992; 35:621-626.

154. Koebner RMD. Generation of PCR-based markers for the detection of rye chromatin in a wheat background. Theor Appl Genet 1995; 90:740-745.

155. Francis HA, Leitch AR, Koebner RMD. Conversion of a RAPD-generated PCR product, containing a novel dispersed repetitive element, into a fast and robust assay for the presence of rye chromatin in wheat. Theor Appl Genet 1995; 90:636-642.

156. Korzun VN, Börner A, Kartel NA. Construction and partial analysis of a *Pst*I DNA library for RFLP mapping of the rye (*Secale cereale* L.) genome. (In Russian) Genetika 1995; 31:767-772.

157. Röder MS, Plaschke J, König SU et al. Abundance, variability and chromosomal location of microsatellites in wheat. Mol Gen Genet 1995; 246:327-333.

158. Devos KM, Bryan GJ, Collins AJ et al. Application of two microsatellite sequences in wheat storage proteins as molecular markers. Theor Appl Genet 1995; 90:247-252.

159. Ma ZQ, Röder M, Sorrells ME. Frequencies, and sequence characteristics of di-, tri, and tetranucleotide microsatellites in wheat. Genome 1996 1996: 39;123-130.

160. Maroof MAS, Biyashev RM, Yang GP et al. Extraordinarily polymorphic microsatellite DNA in barley: species diversity, chromosomal locations, and population dynamics. Proc Natl Acad Sci USA 1994; 91:5466-70.

161. Becker J, Heun M. Barley microsatellites: allele variation and mapping. Plant Mol Biol 1995; 27:835-845.

162. Wu KS, Tanksley SD. Abundance, polymorphism and genetic mapping of microsatellites in rice. Mol Gen Genet 1993; 241:225-235.

163. Zhao XP, Kochert G. Phylogenetic distribution and genetic mapping of a (GGC)n microsatellite from rice (*Oryza sativa* L.). Plant Mol Biol 1993; 21:607-614.

164. Panaud O, Chen X, McCouch SR. Frequency of microsatellites in rice (*Oryza sativa* L.)

Genome 1995; 38: 1170-1176.

165. Bennetzen JL, Freeling M. Grasses as a single genetic system: genome composition, collinearity and compatibility. Trends Genet 1993; 9:259-261.

166. D'Ovidio R, Tanzarella OA, Porceddu E. Rapid and efficient detection of genetic polymorphism in wheat through amplification by polymerase chain reaction. Plant Mol Biol 1990; 15:169-171.

167. Weining S, Langridge P. Identification and mapping of polymorphisms in cereals based on the polymerase chain reaction. Theor Appl Genet 1991; 82:209-216.

168. Allard RW. Predictive methods for germplasm identification. *In* Stalker HT, Murphy JP, eds. Plant Breeding in the 1990s. Wallingford, UK; CAB International, 1992:119-146.

169. Lorenzen LL, Boutin S, Young N et al. Soybean pedigree analysis using map-based molecular markers: I. Tracking RFLP markers in cultivars. Crop Sci 1995; 35:1326-36.

170. O'Donoughue LS, Souza E, Tanksley SD et al. Relationships among North American oat cultivars based on restriction fragment length polymorphisms. Crop Sci 1994; 34:1251-1258.

171. Souza E, Sorrells ME. Prediction of progeny variation in oat from parental genetic relationships. Theor Appl Genet 1991; 82:233-241.

172. Röder MS, Sorrells ME, Tanksley SD. 5S ribosomal gene clusters in wheat: pulsed field gel electrophoresis reveals a high degree of polymorphism. Mol Gen Genet 1992; 232:215-220.

173. Kanazin V, Ananiev E, Blake T. The genetics of 5S rRNA encoding multigene families in barley. Genome 1993; 36:1023-1028.

CHAPTER 20

STATUS OF GENOME MAPPING IN THE TROPICAL GRAINS AND GRASSES (POACEAE)

Andrew H. Paterson, Sin-Chieh Liu, Yann-rong Lin and Charlene Chang

20.1. INTRODUCTION

Tropical grasses, characterized by C_4 photosynthesis which is a particularly efficient means of carbon fixation at high temperature, are among the most important agricultural crops. Many tropical grasses are cultivated to serve a wide range of needs, including food and feed production, animal forage, ornamental and sporting purposes, and erosion control. A few tropical grass species also enjoy a rich legacy of basic genetic research, such as maize and rice (which was addressed in the preceding chapter, to highlight comparative relationships established between temperate and tropical grasses).

We will focus this chapter on two tropical grass species cultivated for grain (maize and sorghum), and a third species cultivated for biomass (sugarcane). The comparative approach we will illustrate for simplifying genetic analysis of the complex genome of sugarcane, using a detailed understanding of genome organization and gene function from maize and sorghum, is likely to be extended in the near future to many tropical grasses which are cultivated as forages and turf, as well as additional grain crops.

20.2. ORIGIN AND DOMESTICATION OF MAIZE

Zea mays L., has historically served as an important genetic model for plants, often being thought of as the *Drosophila* or *E. coli* of the plant world. Beyond its major importance as a crop plant (with worldwide production valued near US $60 billion annually, about one third of this in the U.S. alone), maize played a key role in demonstrating the phenomenon of heterosis ("hybrid vigor"),[1-3] establishing the relationship between genetic variation, recombination, and chromosome transmission,[4] and served as a model for the first description of transposable genetic elements, yielding the Nobel Prize to Barbara McClintock (see chapter 13). A recent update of MaizeDB,[5] a public-sector database on the World Wide Web summarizing map-based information and tools for maize (http://www.agron.missouri.edu), includes descriptions of more than 3000 mapped genes and gene candidates. For a recent update regarding integration of the morphological and molecular maps, see reference 6.

Maize is of New World origin, with a center of diversity in Mexico and/or lowland Central America.[7] Although the ancestry of cultivated maize has long been debated, recent data from genome mapping (cited below) suggests that a relatively small number of mutations can account for the bulk of phenotypic variation between maize and teosinte, a weedy annual relative of maize. Both maize and teosinte share a common chromosome number (2n = 20), and exhibit strict bivalent pairing. Crosses between maize and teosinte are fertile.

Cultivation of maize may date to 5000 B.C. or earlier.[7] For the vast majority of its agricultural history, maize was grown as heterogeneous, open-pollinated populations. The discovery of hybrid vigor,[1-3] influenced cultivation of maize perhaps more than any other single crop, and has impelled development of hybrid production systems in many other crops. The transition from open-pollinated populations, to double-cross hybrids, and finally to single-cross hybrids, has been a major contributor to a greater than five-fold increase in maize yields during the period of 1930 to 1980,[8] and progress continues to be made. The joint effects of dramatic improvement in maize productivity, and the evolution of a hybrid seed industry, have greatly

stimulated activities in maize improvement, especially in the U.S.

20.3. GENETIC AND MOLECULAR MAPS FOR MAIZE

In view of both its economic importance, and its role as a genetic model, it is little surprise that maize was among the early crop plants to be subjected to molecular mapping. The first molecular map of maize was reported in 1986,[9] and a second in 1988.[10] Several additional maize maps are widely-used as primary maps by the research community,[11,12] and the various maps have been largely integrated by cross-mapping of DNA probes on different populations.[11,12] (We note that many additional low-density maps have been made for the purpose of identifying genes or QTLs in specific populations, many of which are cited below). MaizeDB[5] includes map information on about 2400 DNA probes, and 240 maps. The recombinational length of different maize maps varies widely, but appears to average approximately 2000 cM, or about 200 cM per chromosome.

20.3.1. PHENOTYPE MAPPING IN MAIZE

A large number of agriculturally-important traits have been subjected to genetic mapping in maize, and only a subset of key references will be cited here. The evolution of maize from teosinte, as briefly discussed above, has been the subject of a long and productive series of papers from Doebley and colleagues.[13-18] Inheritance of yield and its components has been the emphasis of research by Stuber and colleagues.[19-21] The molecular basis of classical long-term selection experiments for oil and protein composition of the maize kernel have recently been investigated,[22] as well as variation in seed characteristics of more elite populations.[23] Inheritance of plant height and flowering time have been investigated in detail,[24-29] and comparative relationships between these genes in maize and possible orthologs in sorghum and other taxa have been suggested.[30] A long list of additional characteristics have been, or are being mapped, including many disease and insect resistance genes, numerous morphological traits, and variation in response to a number of abiotic factors.[31,32]

20.3.2. Molecular Cloning in Maize

Because of the large physical size of the maize genome, gene isolation in maize has emphasized insertion mutagenesis by transposable elements (see chapter 13) rather than map-based cloning. The maize genome includes approximately 2500 Mb of DNA, or about 1 million nucleotides per centiMorgan (cM).[33] Although a largely-complete YAC library of the maize genome has been made,[34] few have undertaken a "chromosome walk" directly in maize. The large physical quantity of DNA in the genome is further complicated by the fact that a large portion of this DNA is comprised of repetitive elements, which are widely-distributed and would quickly interfere with a "chromosome walk."[35] Based on the finding that most of the maize genome is co-linear with the much smaller genomes of rice[36] and sorghum,[37-38] alternative strategies have been proposed[35-38] to map mutations in maize, walk across the gene in sorghum or rice, then isolate the transcript from maize cDNA libraries.

20.4. ORIGIN AND DOMESTICATION OF SORGHUM

Sorghum bicolor L., is widely-cultivated as an alternative to maize which is more tolerant of low inputs and semiarid conditions. Sorghum is particularly important in Africa and India, with the worldwide crop typically valued at about US$5 billion per year. Historically, sorghum has received less attention than maize as a genetic model, however, it has recently emerged as a facile tool for molecular analysis and gene isolation. Although maize and sorghum share a largely-common repertoire of genes,[37-39] the sorghum genome is comprised of only about 25% as much DNA

as the maize genome,[33] presumably due to a much smaller amount of repetitive DNA. Consequently, sorghum may prove to be a valuable system for map-based cloning of genes responsible for many aspects of grass growth and development.[38-40]

The Sorghum genus is of Old World origin, with at least two centers of diversity. The cultigen, *Sorghum bicolor*, is derived from robust members of a polymorphic African population which all share common ploidy (2n = 2x = 20). The wild population is referred to by the epithet *S. arundinaceum*,[41] however, crosses between cultivated and wild sorghums show a high degree of fertility, and normal cytology, contra-indicating the need for specific status. Although some investigators have subdivided this population into numerous species,[42] the view which has emerged from morphological and cytological analyses is that these various populations comprise a single species, which may be subdivided into several morphological races based largely on the architecture of the inflorescence.[43] A second center of diversity occurs in southeast Asia, including both a diploid (2n = 2x = 20) form, *Sorghum propinquum*, and a tetraploid or higher (2n = 4x = 40) form, *Sorghum halepense*.[44] The polyploid, which is perennial with aggressive rhizomes, has become widespread and is now one of the world's most noxious weeds, commonly referred to as "Johnson Grass." It is well-established that interspecific crosses between *S. bicolor* and *S. propinquum* are fertile, and this fact, together with morphological data, have long suggested that the polyploid is an interspecific hybrid. This view has recently received strong support from molecular data.[40]

Although the archaeological record provides limited documentation of the early history of sorghum cultivation, it was clearly cultivated in Oman by ca. 2500 B.C., and in (modern) Pakistan by about 1750 B.C.[45] Sorghum has become a staple food grain, and carbohydrate source for beer-making, throughout semi-arid regions of Africa and Asia. In the post-Columbian era, sorghum became established as a grain crop in drier parts of tropical America, and today it continues to be an important crop in many of these regions.[45] Various races of sorghum were introduced into the U.S. throughout the 19th century. Sorghum molasses was once a major product throughout the U.S., however, during the 20th century the primary use of sorghum in the U.S. has been as livestock feed.

Although sorghum domestication occurred in tropical Africa, scientific breeding of sorghum was pioneered largely in the U.S.. Sorghum domestication has not resulted in such striking morphological changes as the evolution of the maize ear, primarily involving elimination of shattering, reduction of tiller number, and increase in seed (grain) size and number. As was also true of maize domestication, cultivation in temperate climates required the identification of genotypes which flowered relatively independently of daylength, in contrast to the tropical forms which flower under relatively short days (usually 13 hours or less). The identification of dwarf varieties throughout the early part of the 20th century made it possible to harvest sorghum with a combine.[41] Scientific breeding is estimated to account for 28-34% of yield gains in sorghum, which averaged 95 kg per hectare per year, in the period of 1950-1980.[46] Although sorghum is largely self-pollinating in nature, the development of hybrid production systems in the late 1950s[41] are associated with a spurt of yield improvement, and modern sorghum cultivation in the U.S. largely uses hybrid cultivars.

20.5. GENETIC AND MOLECULAR MAPS FOR SORGHUM

The initiation of genetic mapping in sorghum was largely an offshoot of maize genetic mapping, using maize DNA probes to evaluate the comparative organization of maize and sorghum chromosomes.[39] Subsequent efforts focused increasingly on sorghum DNA clones, and to date at least

seven different genetic maps of sorghum have been developed.[37,39,47-51] Among these, only two[49,50] are "complete," with the number of linkage groups equal to the number of gametic chromosomes. At least 600 different DNA probes appear to have been mapped in sorghum, with many investigators using common probes derived from maize. While DNA polymorphism has been a factor constraining development of many sorghum molecular maps, an interspecific cross between *S. bicolor* and *S. propinquum* has proven especially facile, with a success rate of about 95% at detecting DNA polymorphisms. This is comparable to the rates of polymorphism which had previously been found only in outcrossing taxa such as maize and brassica.

20.5.1. PHENOTYPE MAPPING IN SORGHUM

As molecular maps became available, several groups quickly pursued mapping of agriculturally-important phenotypes. Plant height and flowering date, traits of primary importance to sorghum production, have been studied in considerable detail[26,30] and most of the genes identified by classical studies[52-54] have been placed on the molecular maps. Characteristics of fundamental importance to sorghum domestication, such as non-shattering habit, increased seed size, and daylength-insensitive flowering have been mapped, and the correspondence of the sorghum genes to those in maize and rice has been evaluated in detail.[30] A particularly interesting trait which has been mapped in sorghum is rhizomatousness, and associated tendency of sorghums to overwinter in sub-tropical and warm temperate climates[40]—this information may represent a starting point at understanding the molecular basis of rhizomatousness not only in Johnson Grass, bermuda grass, and other noxious weeds, but also in potentially valuable relatives of major crops such as *Saccharum spontaneum* and *Oryza longistaminata*. More recent efforts are focusing on genes associated with tolerance of pre- and post-flowering water deficit, resistance to major insect pests such as greenbug and midge, and resistance to numerous diseases.

20.5.2. MOLECULAR CLONING IN SORGHUM

In contrast to maize, the sorghum genome is of a physical size more amenable to map-based cloning. (In addition, no transposons have yet been discovered in sorghum—in principle, it would be possible to introduce exogenous transposons such as the Ac/Ds system of maize, but this has not yet been done). The sorghum genome is comprised of 760 Mb[33] or about 400 kb per centiMorgan, a distance amenable to map-based cloning. While no comprehensive study of the repetitive fraction of sorghum has been reported, anecdotal reports clearly indicate that sorghum enjoys a much smaller repetitive fraction than maize. Megabase DNA technology for sorghum has been developed, and a 3-genome-equivalent BAC library has been published.[55]

20.6. ORIGIN AND DOMESTICATION OF SUGARCANE

Saccharum spp., provides much of the world's sugar supply, and is important in many countries as a biomass/biofeedstock crop. The world sugarcane crop is valued near U.S.$20 billion/year, with less than 5% of this grown in the U.S. The genetic complexity of sugarcane, as reflected by high and unstable chromosome numbers (ranging from 2n = 36 to 2n = 170[56]), extensive chromosome duplication, a high rate of outcrossing,[57] and a large physical quantity of DNA per nucleus,[33] have made sugarcane a very difficult system in which to do even simple genetic studies. Sugarcane improvement remains one of the great challenges in plant breeding, and the productivity of modern sugarcane cultivars is a tribute to the skill, dedication, and diligence of a closely-knit nucleus of scientists, working mostly in the tropics. In this chapter, sugarcane will serve as an example of how a comparative approach can simplify genome analysis—using data and tools from well-characterized systems to overcome the challenges to genetic analysis of complex systems.

The early steps in the evolution of modern sugarcane appear to have occurred in Papua New Guinea, where native gardeners are believed to have selected clones of a native species, *Saccharum robustum*, with unusually sweet stalks for chewing raw. The resulting "noble canes" spread throughout the Pacific Islands and westward to continental Asia, overlapping the distribution of *S. spontaneum*, an aggressive pioneer of disturbed habitats.[45] Interspecific introgression of *S. spontaneum* germplasm into "noble canes" is postulated to be a key event in evolution of modern sugarcane cultivars.

20.7. GENETIC AND MOLECULAR MAPS FOR SUGARCANE

Not surprisingly, genetic mapping of sugarcane has proven complex. Sugarcane is generally intolerant of inbreeding, and most natural or synthetic populations are crosses between highly heterozygous genotypes. Almost invariably, three-generation pedigrees are not available, consequently genetic mapping has been based upon simplex DNA polymorphisms, or "single dose restriction fragments" (SDRFs).[58] Because sugarcane is autopolyploid, and pairing of a particular homolog is not limited to one partner but can be with any of 8-10 other partners, genetic maps of sugarcane describe individual homologs, rather than homologous pairs, and consequently a "complete" map is expected to have the zygotic chromosome number rather than the gametic chromosome number. To date, three molecular maps have been described,[59-61] including a total of about 600 genetic loci, however none of these maps are "complete," as the large number of chromosomes in sugarcane necessitates a correspondingly large number of DNA markers. Two of the maps are based upon a common population, and have been merged.[62]

20.7.1. PHENOTYPE MAPPING IN SUGARCANE

Mapping of phenotypes in sugarcane is just getting underway. A recent report shows that photoperiodic flowering of sugarcane appears to be controlled at least in part by a genetic locus which is homologous to the Ma-1 photoperiodic locus of sugarcane, as well as the Ppd homologous series of wheat and barley, and at least one QTL affecting photoperiodic flowering of maize.[38] A second report suggests possible associations between DNA markers and QTLs affecting stalk number and diameter, tasseled stalks, incidence of smut, pol (sugar content), and fiber percentage—however, significant marker x trait associations were found only at approximately the frequency which would be expected to occur at random (see pg. 358 of the cited publication), casting doubt on these findings.[63] Moreover, the findings are based on analysis of only 44 plants, an extremely small population for mapping QTLs. Several much larger-scale efforts to re-evaluate the molecular basis of these, and additional traits, are in progress.

20.4.3. MOLECULAR CLONING IN SUGARCANE

The physical size of the sugarcane genome is about 3000 Mb, modestly larger than that of maize. An estimate of the amount of DNA per centiMorgan is not yet available because a complete map has not yet been assembled—based on a conservative estimate of 5000 cM, sugarcane would have about 600 kb of DNA per centiMorgan, a value similar to sorghum.

The greatest obstacle to map-based cloning in sugarcane is likely to be the extent of genetic redundancy (polyploidy), rather than the physical distance from linked markers to genes. Traits which exhibit multiplex segregation may be difficult to place on the genetic map.[58] For traits which exhibit simplex segregation, chromosome walking will require that one discern megabase DNA clones from the target homolog, from the 6-12 or more homologs which might be present, for any particular chromosome. While megabase DNA libraries of diploid, inbreeding species can simply represent one homologous chromosome per pair, megabase DNA libraries in sugarcane must represent all homologous chromosomes in each set, in order to be confident of carrying the target allele. This is technically feasible—but involves considerable work.

An alternative approach to map-based cloning directly in sugarcane may be to use the smaller, diploid genome of sorghum as a guide. The comparative approach is discussed in more detail, below.

20.8. COMPARATIVE MAPPING OF TROPICAL GRASSES

Because such a large number of major crops are grasses, comparative mapping activity has been particularly active in the Poaceae. An abundance of data now demonstrates similarity in the order of genes along the chromosomes of several cultivated grasses[36-39] (also see numerous additional citations in the preceding chapter). Extensive conservation of gene order along the chromosomes of different grass taxa has led to the suggestion that grasses might be treated essentially as a "single genetic system,"[64] and indeed many DNA probes and megabase clones can be cross-utilized in diverse grass taxa.

Recently, the scope of comparative mapping has broadened, from analysis of chromosome organization, to comparison of the roles of specific genes in accounting for phenotypic variation in different grass species. Domestication of sorghum, maize and rice seems to have involved a largely-common set of genes, which were independently mutated to produce large-grained, non-shattering, day-neutral genotypes suitable to mechanized agriculture.[38] The parallels are not limited to these characteristics, but extend at least to plant height and flowering time,[26,30] and many additional traits are being evaluated in numerous labs. It remains unclear whether the mutations associated with the early steps in domestication correspond to those which account for phenotypic variation in the elite gene pool of crop cultivars—however, it is clear that the early steps in plant domestication involved a small number of genetic loci which are common to several grass species, and which may have been driven to fixation for new mutants rather quickly.

Comparative mapping offers the opportunity to shed new light on the evolution of grass genomes, specifically regarding the basal chromosome number for several grass species. It has been noted above, that maize and sorghum share a common chromosome number (2n = 2x = 10). However, early RFLP mapping studies of maize often showed DNA probes which mapped to two unlinked genetic loci, and the arrangements of duplicated DNA marker loci across the genome[65] supported prior isozyme data[66] in asserting that maize had recently undergone an extensive duplication of most, if not all chromosomes.

While the duplication of most maize chromosomes clearly pointed to n = 5 as the probable basal chromosome number for maize, molecular evidence regarding the basal chromosome number for sorghum has been unclear. Although sorghum occasionally shows multiple loci with individual DNA probes, a detailed comparison of maize and sorghum suggested that sorghum clearly exhibited less duplication than maize.[37] (It must be noted that this comparison was modestly biased, in that it relied upon use of maize DNA probes both in maize and in sorghum, however, subsequent studies based on sorghum DNA probes[48] tended to support this result). However, as complete genetic maps of sorghum were published, it became clear that those duplicate loci which were found showed patterns in the genome which were consistent with ancient duplication.[48,49] While past studies in sorghum were constrained by a relatively small number of DNA probes which detected multiple loci, a recent study (Lin, Chang, Liu, Schertz, Paterson, in preparation) has analyzed the distribution of duplicate loci from more than 100 DNA probes, distributed across the sorghum genome. Many regions of the sorghum genome appear to be duplicated, however, common gene order persists over relatively short chromosome segments, suggesting that the duplication event was ancient. Moreover, antiquity of the duplication event may account for the tendency of DNA probes to hybridize to one, rather than two loci in sorghum, more frequently than in maize.

Sugarcane, as a very close relative of sorghum,[67] should closely resemble the chromosome organization of sorghum. Although it remains to be confirmed that the patterns of chromosome duplication found in sorghum can be verified in sugarcane, it is clear that the evolutionary model for sugarcane may add at least one new twist. Based on evaluation of the periodicity in chromosome numbers of putatively euploid taxa within the various *Saccharum* species, the species *S. robustum* and *S. officinarum* appear to have basal chromosome numbers of 10, and the species *S. spontaneum* appears to have a basal chromosome number of eight.[56] While this dichotomy remains somewhat equivocal, it would suggest that since the divergence of sorghum and sugarcane, perhaps as little as five million years ago,[67] two chromosome fusions or chromosome losses may have occurred, accounting for evolution of the *S. spontaneum* karyotype.

20.8.1. SUGARCANE AS AN EXAMPLE OF THE POTENTIAL BENEFITS OF A COMPARATIVE APPROACH

The dilemma of "how to make a complete sugarcane map" may be simplified greatly by comparative analysis. It is well-established that among cultivated grasses, the closest relative of sugarcane is sorghum, which may be diverged from sugarcane for as little as five million years.[67] Partial comparative analysis of the chromosomes of sorghum and sugarcane, based on maize DNA probes, has showed a high degree of co-linearity between the two taxa.[61] A recent effort has been inaugurated to construct a "complete" comparative map for the chromosomes of sorghum and sugarcane, and to infer the genetic map of the

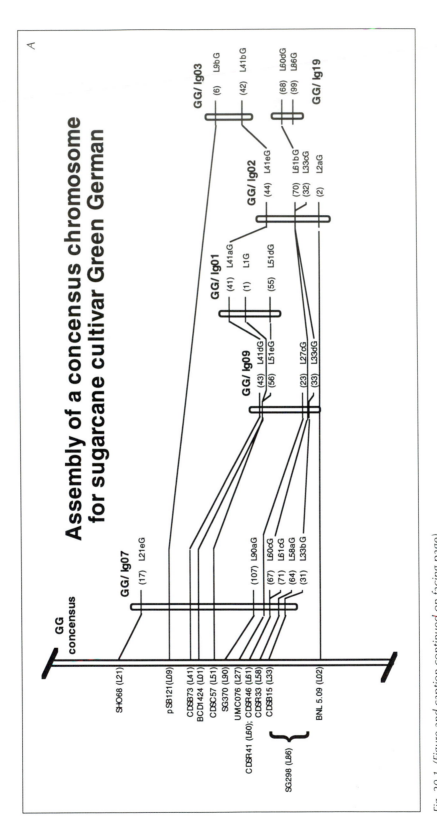

Fig. 20.1. (Figure and caption continued on facing page)

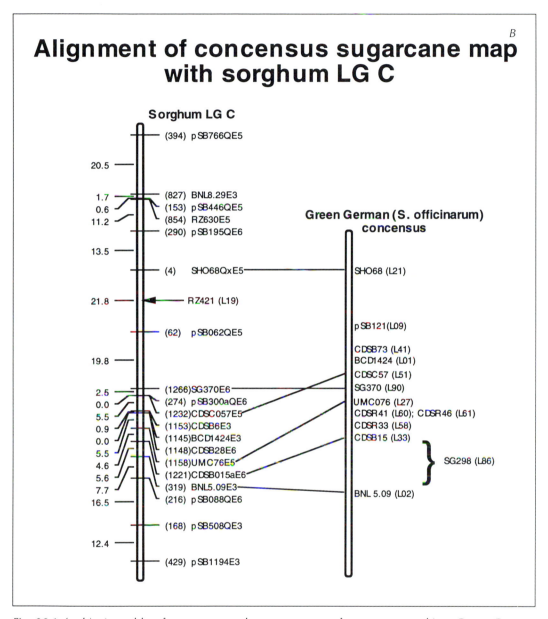

Fig. 20.1 (a, b). Assembly of a consensus chromosome map for sugarcane cultivar Green German (S. officinarum). The high level of polyploidy in the sugarcane genome, together with the fact that a sugarcane map based on single-dose restriction fragments describes each homolog, rather than each homologous set (see text), has the consequence that different homologs carry loci detected by different subsets of sugarcane DNA probes. By exploiting fortuitous probes which detect SDRF loci on several different homologs, the homologs can be aligned to one another, and the approximate order of all DNA markers can be inferred, as shown. This "consensus order," although it does not describe any single map of sugarcane, and does not accurately reflect the distances between DNA probes, serves as a useful abstraction for two purposes: (1) It may serve to guide future searches for DNA markers at specific locations on a particular homolog, to fill gaps in a linkage map or enrich for markers in the vicinity of a particular gene of interest. (2) It serves as a much simpler basis for comparative analysis of gene order in sugarcane and other taxa, as shown in part B of the figure.

basic homologous set of sugarcane chromosomes (Fig. 20.1). By mapping a common set of DNA probes at closely-spaced intervals across the chromosomes of sorghum and sugarcane, respectively, it will be possible to provide anchor loci in sorghum to simplify identification of homologous chromosomes in sugarcane. Further, by utilizing the subset of DNA probes which segregate at orthologous SDRF loci on different homologous chromosomes, it is possible to infer a consensus map order of DNA probes along the chromosomes of sugarcane, effectively reducing the sugarcane map to a set of 8-10 (depending on the species, see above) basic chromosomes. By this approach, it should be possible to align sugarcane linkage groups with the consensus map, and systematically identify DNA markers most likely to fill predicted "gaps" in the linkage groups, either based on the consensus sugarcane map, or based on the more detailed (see above) sorghum map. Moreover, since the sorghum map enjoys a growing level of integration with the maps of maize, wheat, rice, and other grasses.[36-39,61] Sugarcane will benefit from comparative information from these other sources as well.

20.8.2. A COMPARATIVE APPROACH MAKES MAP-BASED CLONING MORE TRACTABLE IN LARGE-GENOME GRASSES

Mounting evidence that the grass chromosomes contain a largely-common repertoire of genes, and differ by relatively few well-documented, chromosomal rearrangements, suggests that "chromosome walking" in large genome grasses such as maize and sugarcane might be guided by use of the smaller genomes of sorghum or rice (see preceding chapter).

The comparative approach to chromosome walking has several advantages. First, megabase DNA libraries of small genomes require fewer clones to be handled, to represent the population of genes in the grass genome. Second, individual clones contain a higher "concentration" of genes, as smaller genomes tend to have less repetitive DNA. Third, the repetitive DNA which is the greatest obstacle, the interspersed class, tends to be rapidly evolving, and by using megabase DNA tools from two or more species simultaneously, repeat arrays in one species might be overcome by successful "steps" in another species. Fourth, a comparative approach permits primary genetic mapping of a locus to be done in the taxon which shows the simplest inheritance of a particular phenotype—for example, a discrete locus imparting non-shattering habit to sorghum corresponds to QTLs affecting the trait in both rice and maize.[38] Clearly, the discrete locus can be mapped more precisely than the QTLs, accelerating progress toward cloning. (It is important to note that this idea is not new,[68] however, there are relatively few examples where it has previously been implemented.)

It is an inescapable conclusion that comparative analysis is revolutionizing the way in which we think about molecular analysis of grass genomes. This trend has spilled over from early work in the grasses and the nightshades (see chapter 21), to the crucifers (see chapter 16), legumes (see chapter 17), and many other taxa. Not only does this unified approach to crop genome analysis economize on tools and information, but it also affords the opportunity to pursue new learning opportunities such as comparative analysis of domestication in different grass taxa.[38] In the long term, comparative mapping will prove a particular boon to taxa which are less well-studied genetically, and which can quickly derive the benefit of enormous databases including decades of research results for maize, rice, and other models simply by using conserved DNA probes to align the genetic maps.

REFERENCES

1. Shull GH. The composition of a field of maize. Rep Am Breed Assoc 1908; 4:296-301.
2. Shull GH. The genotypes of maize. Am Nat 1911; 45:234-252.
3. East EM. Inbreeding in corn. Rep Conn

Agric Exp Stn 1907:419-428.

4. Creighton H. S., and B. McClintock. 1931. A correlation of cytological and genetical crossing-over in *Zea mays*. Proc Nat'l Acad Sci USA 17:492-497.

5. Polacco ML, D Hancock, P Byrne, G Yerk-Davis, EH Coe. 1996. MaizeDB, an Internet gateway to map-based information and tools. Proc Plant Genome IV, 1996:108.

6. Coe EH, D Hancock, S Lowalewski, M Polacco. Gene list and working maps. Maize Genetics Cooperative News 1995; 69: 191-267.

7. Goodman M. Maize. Chapter 37 in NW Simmonds, ed. Evolution of Crop Plants. Essex, UK: Longman Press, 1978:128-136.

8. Duvick D. Genetic contributions to yield gains of US hybrid maize, 1930 to 1980. In: Fehr W ed. Genetic Contributions to Yield Gains of Five Major Crop Plants. Madison WI: Crop Science Society of America, 1984:15-48.

9. Helentjaris T, M Slocum, S Wright, A Schaefer, J Nienhuis. Construction of genetic maps in maize and tomato using restriction fragment length polymorphisms. Theor Appl Genet 1986; 72:761-769.

10. Burr B, F Burr, KH Thompson, MC Albertsen, CW Stuber. Gene mapping with recombinant inbreds in maize. Genetics 1988; 118:519-526.

11. Beavis WD and D Grant. A linkage map based on information from four F2 populations of maize. Theor Appl Genet 1991; 82:636-644.

12. Gardiner J, EH Coe, S Melia-Hancock, DA Hoisington, S Chao. Development of a core RFLP map in maize using an immortalized F2 population. Genetics 1993; 134:917-30.

13. Doebley J, A Stec, J Wendel, and M Edwards. Genetic and morphological analysis of a maize-teosinte F2 population: implications for the origin of maize. Proc Nat Acad Sci USA 1990; 87:9888.

14. Doebley J and A Stec. Genetic analysis of the morphological differences between maize and teosinte. Genetics 1991; 129:285-295.

15. Doebley J and A Stec. Inheritance of the morphological differences between maize and teosinte: comparisons of results for two F2 populations. Genetics 1993; 134:559-70.

16. Dorweiler J, A Stec, J Kermicle, J Doebley. *Teosinte glume architecture-1*: a genetic locus controlling a key step in maize evolution. Science 1993; 262:233.

17. Doebley J, A Bacigalupo, A Stec. Inheritance of kernel weight in two maize-teosinte hybrid populations: Implications for crop evolution. J Hered 1994; 85:191-195.

18. Doebley J, A Stec, C Gustus. *teosinte branched 1* and the origin of maize: evidence for epistasis and the evolution of dominance. Genetics 1995; 141:333-346.

19. Edwards MD, CW Stuber and JF Wendel. Molecular-marker-facilitated investigations of quantitative-trait loci in maize. I. Numbers, genomic distribution and types of gene action. Genetics 1987; 116:113-125.

20. Stuber CW, MD Edwards and JF Wendel. Molecular-marker-facilitated investigations of quantitative-trait loci in maize. II. Factors influencing yield and its component traits. Crop Science 1987; 27:639-648.

21. Stuber CW, SE Lincoln, DW Wolff, T Helentjaris and ES Lander. Identification of genetic factors contributing to heterosis in a hybrid from two elite inbred lines using molecular markers. Genetics 1992; 132:823-839.

22. Rocheford TR. Chromosome regions associated with control of maize kernel composition. Proc Annu Corn Sorghum Ind Res Conf 1994; 49:239-249.

23. Schon CC, AE Melchinger, J Boppenmaier, E Brunklaus-Jung, RG Herrmann et al. RFLP mapping in maize: Quantitative trait loci affecting testcross performance of elite European flint lines. Crop Sci 1994; 34:378-389.

24. Beavis WD, D Grant, M Albertsen and R Fincher. Quantitative trait loci for plant height in four maize populations and their associations with qualitative loci. Theor Appl Genet 1991; 83:141-145.

25. Koester RP, PH Sisco and CW Stuber. Identification of quantitative trait loci controlling days to flowering and plant height in two near isogenic lines of maize. Crop Sci 1993; 33:1209-1216.

26. Pereira MG, M Lee and PJ Rayapati. Comparative RFLP and QTL mapping in sorghum and maize. Poster 169 in the Second

Internal Conference on the Plant Genome. New York: Scherago Internal, Inc., 1994.

27. Abler BSB, MD Edwards and CW Stuber. Isozymatic identification of quantitative trait loci in crosses of elite maize inbreds. Crop Sci 1991; 31:267-274.

28. Veldboom LR, M Lee and WL Woodman. Molecular marker-facilitated studies in an elite maize population: I. Linkage analysis and determination of QTL for morphological traits. Theor Appl Genet 1994; 88:7-16.

29. Zehr BE, JW Dudley, J Chojecki, MA Saghai-Maroof and RP Mowers. Use of RFLP markers to search for alleles in a maize population for improvement of an elite hybrid. Theor Appl Genet 1992; 83:903-11.

30. Lin YR, KF Schertz and AH Paterson. Comparative analysis of QTLs affecting plant height and maturity across the Poaceae, in reference to an interspecific sorghum population. Genetics 1995; 141:391-411.

31. Reiter RS, JG Coors, MR Sussman and WH Gabelman. Genetic analysis of tolerance to low-phosphorus stress in maize using restriction fragment length polymorphisms. Theor Appl Genet 1991; 82:561-568.

32. Causse M, J Rocher, S Pelleschi, Y Barriere, JL Prioul, D de Vienne. Sucrose phosphate synthase: An enzyme with heterotic activity correlated with maize growth. Crop Sci 1995; 35:995-1001.

33. Arumunganathan K, and ED Earle. Nuclear DNA content of some important plant species. Plant Mol Biol Rptr 1991; 9:208-18.

34. Edwards KJ, H Thompson, D Edwards, AD Saizieu, C Sparks, JA Thompson, AJ Greenland, M Eyers, W Schuch. Construction and characterization of a yeast artificial chromosome library containing three haploid maize genome equivalents. Plant Mol Biol 1992; 19:299-308.

35. Springer PS, Edwards KJ, Bennetzen JL. DNA class organization on maize *Adh1* yeast artificial chromosomes. Proc Natl Acad Sci USA 1994; 91:863-867.

36. Ahn S, and SD Tanksley. Comparative linkage maps of the rice and maize genomes. Proc Nat Acad Sci USA 1993; 90:7980-84.

37. Whitkus R, J Doebley, M Lee. Comparative genome mapping of sorghum and maize. Genetics 1992; 132:119-130.

38. Paterson AH, YR Lin, Z Li, KF Schertz, JF Doebley, SRM Pinson, SC Liu, JW Stansel, JE Irvine. Convergent domestication of cereal crops by independent mutations at corresponding genetic loci. Science 1995; 269:1714-1718.

39. Hulbert SH, TE Richter, JD Axtell, JL Bennetzen. Genetic mapping and characterization of sorghum and related crops by means of maize DNA probes. Proc Natl Acad Sci U S A 1990; 87:4251-4255.

40. Paterson AH, KF Schertz, YR Lin, SC Liu, YL Chang. The weediness of wild plants: Molecular analysis of genes influencing dispersal and persistence of johnsongrass, *Sorghum halepense* (L.) Pers. Proc Natl Acad Sci USA 1995; 92:6127-6131.

41. Doggett H. Sorghum. Chapter 34 in NW Simmonds (ed), Evolution of Crop Plants. Essex, UK: Longman Press, 1978:112-116.

42. Snowden JD. London: The Cultivated Races of Sorghum. 1936.

43. Harlan JR and JMJ deWet. A simplified classification of cultivated sorghum. Crop Sci 1972; 12:172-176.

44. Celarier RP. Cytotaxonomic notes on the subsection *Halepense* of the genus *Sorghum*. Bull Torrey Bot Club 1958; 85:49-62

45. Sauer J. Historical geography of crop plants: A select roster. Boca Raton, FL: Lewis Publishers, 1993.

46. Miller FR and Y Kebede. Genetic contributions to yield gains in sorghum, 1950 to 1980. Chapter 1 In: Fehr, WR (ed.), Genetic Contributions to Yield Gains of Five Major Crop Plants. Madison WI: Crop Science Society of America, 1984.

47. Binelli G, L Gianfranceschi, ME Pe, G Taramino, C Busso, J Stenhouse, E Ottaviano. Similarity of maize and sorghum genomes as revealed by maize RFLP probes. Theor Appl Genet 1992;84:10-16.

48. Chittenden LM, KF Schertz, YR Lin, RA Wing and AH Paterson. A detailed RFLP map of *Sorghum bicolor* x *S. propinquum*, suitable for high-density mapping, suggests ancestral duplication of Sorghum chromosomes or chromosomal segments. Theor Appl Genet 1994; 87:925-933.

49. Pereira MG, Lee, M, PJ Bramel Cox, W Woodman, JF Doebley, R Whitkus. Con-

struction of an RFLP map in sorghum and comparative mapping in maize. Genome 1994; 37:236-243.

50. Xu GW, CW Magill, KF Schertz, GE Hart. A RFLP linkage map of Sorghum bicolor. Theor Appl Genet 1995; (in press).

51. Ragab RA, S Dronavalli, MAS Maroof, YG Yu. Construction of a sorghum RFLP linkage map using sorghum and maize DNA probes. Genome 1995; (in press).

52. Quinby JR and RE Karper. The inheritance of three genes that influence time of floral initiation and maturity date in milo. Jour Amer Soc Agron 1945; 37:916-936.

53. Quinby JR and RE Karper. Inheritance of height in sorghum. Agron J 1954; 46:211-16.

54. Quinby JR. Fourth maturity gene locus in sorghum. Crop Sci 1966; 6:516-518.

55. Woo SS, J Jiang, BS Gill, AH Paterson, RA Wing. Construction and characterization of a bacterial artificial chromosome library of *Sorghum bicolor*. Nucl Acids Res 1994; 22:4922-4931.

56. Irvine JE. The circumscription of Saccharum officinarum. Baileya 1996; (in press).

57. Simmonds NW. Sugarcane. Chapter 32 In: NW Simmonds, ed. Evolution of Crop Plants. Essex, UK: Longman Press, 1978:104-107.

58. Wu KK, W Burnquist, ME Sorrells, TL Tew, PH Moore, SD Tanksley. The detection and estimation of linkage in polyploids using single-dose restriction fragments. Theor Appl Genet 1992; 83:294-300.

59. Al-Janabi SM, RJ Honeycutt, M McClelland, BWS Sobral. A genetic linkage map of Saccharum spontaneum L. 'SES208'. Genetics 1993; 134:1249-1260.

60. Da Silva J, W Burnquist, ME Sorrells, SD Tanksley. RFLP linkage map and genome analysis of Saccharum spontaneum. Genome 1993; 36:782-791.

61. Grivet L, A D'Hont, P Dufour, P Hamon, D Roques, JC Glaszmann. Comparative genome mapping of sugar cane with other species within the Andropogoneae tribe. Heredity 1994; 500-508.

62. Da Silva J, RJ Honeycutt, W Burnquist, SM Al-Janabi, ME Sorrells, SD Tanksley, BWS Sobral. *Saccharum spontaneum* genetic linkage map containing DNA markers. Molecular Breeding 1995; 1:165-179.

63. Sills GR, W Bridges, SM Al-Janabi, BWS Sobral. Genetic analysis of agronomic traits in a cross between sugarcane (*Saccharum officinarum* L.) and its presumed progenitor (*S. robustum* Brandes & Jesw. ex Grassl). Molecular Breeding 1995; 1:355-363.

64. Bennetzen JL and M Freeling. Grasses as a single genetic system: genome composition, collinearity and compatibility. Trends Genet 1993; 9:259-261.

65. Helentjaris T, D Weber and S Wright. Identification of the genomic locations of duplicate nucleotide sequences in maize by analysis of restriction fragment length polymorphism. Genetics 1988; 118:353-363.

66. Wendel JF, CW Stuber, MD Edwards, and MM Goodman. 1986. Duplicated chromosome segments in Zea mays L.: Further evidence from hexokinase isozymes. Theor Appl Genet 72:178-185.

67. Al-Janabi SM, M McClelland, C Peterson, BWS Sobral. Phylogenetic analysis of organellar DNA sequences in the Andropogoneae:Saccharinae. Theor Appl Genet 1994; 88:933-944.

68. Robertson DS. A possible technique for isolating genic DNA for quantitative traits in plants. J Theor Biol 1985; 117:1-10.

CHAPTER 21

STATUS OF GENOME MAPPING TOOLS IN THE TAXON SOLONACEAE

Klaus Pillen, Omaira Pineda, Candice B. Lewis and Steven D. Tanksley

The nightshade family (Solanaceae) consists of more than 90 genera and 2600 species. The family includes crop and garden plants, such as tomato (*Lycopersicon esculentum,* 2n=2x=24), potato (*Solanum tuberosum,* 2n=4x=48), tobacco (*Nicotiana tabacum,* 2n=4x=48), pepper (*Capsicum annuum,* 2n=2x=24), eggplant *(Solanum melongena,* 2n=2x=24) and petunia (*Petunia hybrida,* 2n=2x=14), as well as many poisonous plants. Most genera are of New World origin with centers of diversity located in Central and South America.

Due to the numerous data on genome organization in Solanaceous species, this chapter is further subdivided in three sections. In each section we summarize the current state of genome mapping of the genera *Lycopersicon, Solanum* and *Capsicum,* respectively.

21.1. GENOME MAPPING IN *LYCOPERSICON* (TOMATO)

21.1.1. DESCRIPTION OF THE TAXON

The genus *Lycopersicon* includes the cultivated tomato *(L. esculentum)* together with eight wild species. The wild species bear a wealth of genetic variability and since tomato is crossable with each wild species, the gene resources can be exploited for introgression of various traits of agronomic relevance into modern tomato cultivars.[1,2] The cultivated tomato can be crossed with all of the other species in the genus, however, embryo culture may be required with more distant species such as *L. peruvianum* or *L. chilense.*[3]

Genome Mapping in Plants, edited by Andrew H. Paterson. © 1996 R.G. Landes Company.

L. esculentum and its nearest relatives are autogamous while more distantly related species (e.g., *L. pennellii, L. peruvianum, L. hirsutum* and *L. chilense*) are self-incompatible. The self-incompatibility system in *Lycopersicon* is gametophytic, as is the case for most Solanaceous species. The system is controlled by one multi-allelic S locus which maps to chromosome 1.[4,5] Rick classified the *Lycopersicon* species into four groups according to their mating system (Table 21.1).[6]

Tomato centers of diversity are located in western South America and, in the case of *L. esculentum* var. *cerasiforme,* Mexico and Central America. *L. esculentum* var. *cerasiforme* is considered as the most likely ancestor of cultivated tomatoes.[7,8] Karyotypes of the *Lycopersicon* species are very similar, with little or no structural difference among species. However, tomato chromosomes can be distinguished cytologically in the pachytene stage of meiosis based on chromosome length, centromere position and heterochromatin distribution.[9,10]

21.1.2. GENETIC MAPS

Genetic linkage in tomato has been reported since the dawn of classical genetic studies at the beginning of the century.[11] Since then, natural and radiation-induced tomato mutants have been collected and the underlying genes have been systematically mapped.[12] By 1975, the morphological map of the tomato genome comprised 190 loci.[13] In addition, the exploitation of isozyme variation in tomato added 36 isozyme loci to the map.[14]

The number of mapped loci has increased considerably due to the introduction of restriction fragment length polymorphism (RFLP) markers.[15] Since the mid 1980s, genomic and cDNA sequences have been integrated into the existing tomato map.[16,17] Due to the low level of genetic variation among tomato cultivars, the linkage map was constructed using an F_2 population of the interspecific cross *L. esculentum* x *L. pennellii*. The current tomato RFLP map contains more than 1000 markers which are distributed over 1276 centi-Morgans (cM).[17,18] An updated tomato map including genes of known function or phenotype is presented in Figure 21.1. On average, markers are separated by less than 1 cM. However, the map still includes some gaps of up to 17 cM. Since most of the mapped markers in tomato are single or low copy sequences, the occurrence of gaps could be explained by the presence of chromosomal regions lacking single copy sequences. Alternatively, gaps could indicate the existence of hot spots for recombination in the tomato genome.

The haploid DNA content of tomato is approximately 950 million base pairs.[19] Thus, the average relationship between genetic and physical distance in tomato is equal to 750,000 base pairs (750 kb) per cM. The actual ratio of genetic and physical distance varies considerably depending on the chromosomal region under investigation. The tomato map reveals clusters of markers which are preferentially located in centromeric and telomeric regions. High resolution genetic and physical mapping

---

Table 21.1. Different mating systems in Lycopersicon species[6]

Mating system	Species
Autogamous	*L. esculentum, L. cheesmanii* and *L. parviflorum*
Facultative SC[1]	*L. pimpinellifolium, L. chmielewskii*
Facultative SC[1] and SI[2]	*L. hirsutum, L. pennellii, L. peruvianum*
Allogamous SI[2]	*L. chilense*

[1]SC: self-compatible
[2]SI: self-incompatible

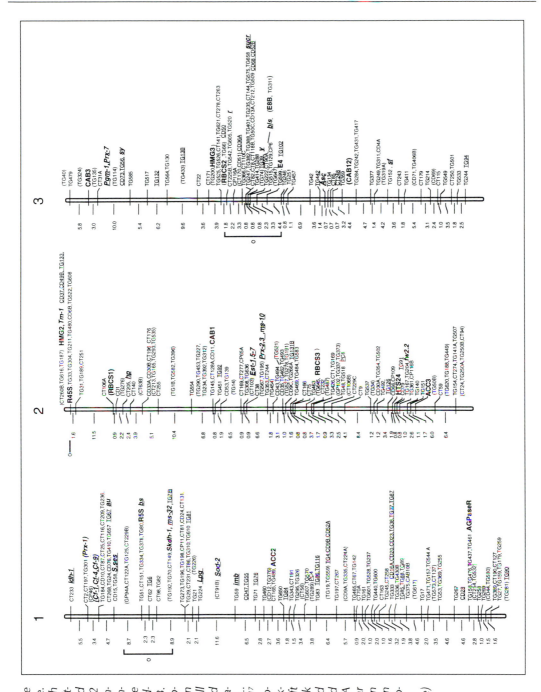

Fig. 21.1. Molecular linkage map of the tomato genome. Loci by tickmarks ordered with LOD > 3 with MapMaker software. Loci in bold correspond to known genes (see Table 21.2 for details). Loci following commas cosegregate. Markers enclosed in parentheses have been located to corresponding intervals with LOD < 3. Position of underlined loci approximated from placement on previously published maps. All other loci have been mapped directly within an F_2 population of 67 plants from L. esculentum x L. pennellii.[17] Approximate centromere positions are indicated by brackets and bars with 0s to the left of each chromosome.[123] Black boxes on chromosome 7 and 9 indicate precisely mapped centromere positions.[124] A major change in the molecular map is the flipped orientation of chromosome 7 based on assignment of markers to chromosome arms.[124]

(Figure continued on next page)

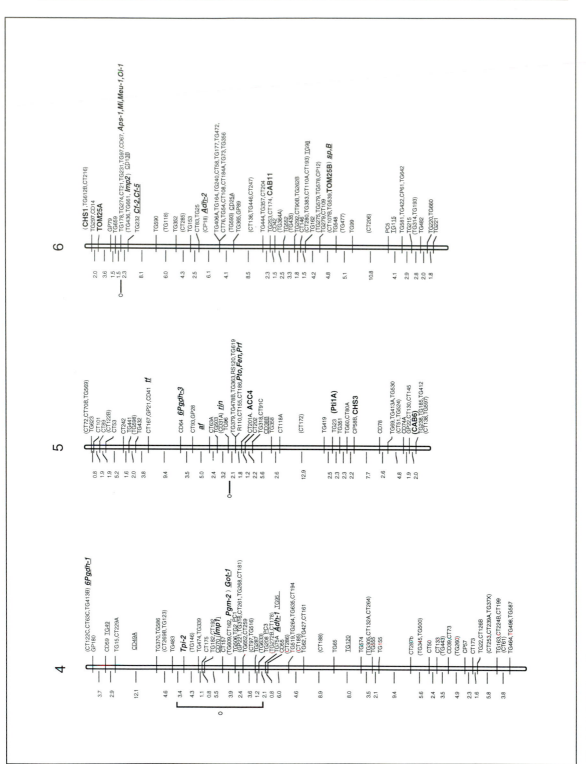

Fig. 21.1. (continued)

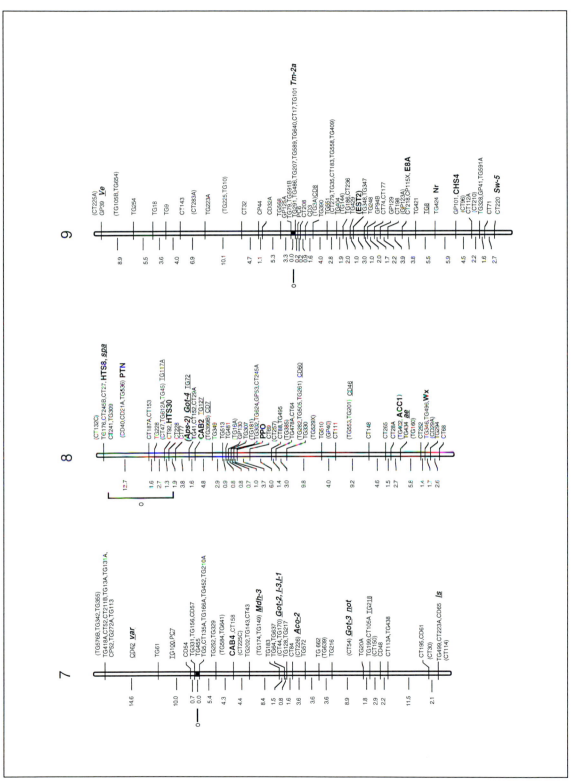

Fig. 21.1. (continued)

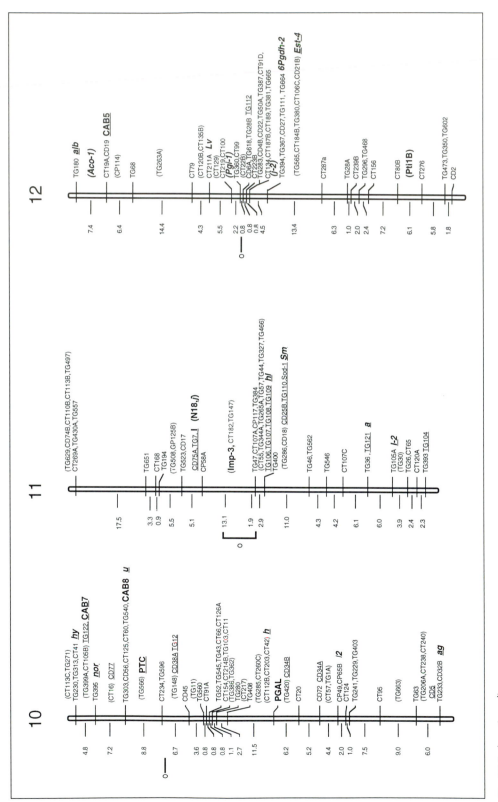

Fig. 21.1. (continued)

around the *Tm-2a* region, for example, indicate that in this area, which is located close to the centromere of chromosome 9, one cM corresponds to more than five million base pairs (Pillen, unpublished data).[20] In contrast, near the *I-2* locus, which is located in euchromatin distant from the centromere of chromosome 11, one cM is represented by only 43 kb.[21] Future research on physical mapping around target loci for map-based cloning will provide further information indicating whether the cited values of reduced and enhanced frequencies of recombination are the extremes in the tomato genome.

Genetic recombination between any two markers may vary considerably depending on several factors. (1) In general, the recombination frequency is higher in female than in male tomato gametes.[22,23] (2) The genetic diversity between the parents of a cross can also affect recombination. Recombination frequencies observed in interspecific crosses are often reduced when compared to the same frequencies in an intraspecific cross. For instance, the comparison of a *L. peruvianum* intraspecific map with the original *L. esculentum* x *L. penellii* map revealed on average a 10% increase in chromosome length for the intraspecific map.[23] (3) The recombination frequency in interspecific crosses also depends on the backcross or selfing generation used for mapping. Advanced generations often reveal less recombination between markers than early generations. This was demonstrated by comparison of interval lengths between BC_1, BC_3 and BC_5 in a *Solanum pennellii* x *Lycopersicon esculentum* cross.[24,25] The author showed that the recombination frequency of five intervals in BC_3 and BC_5 generations diminished between 70 and 48% compared to the BC_1 values. Similar results are reported for mapping markers associated with a QTL for dry weight accumulation on chromosome 1 (S. Bentolila, personal communication). In this instance, a five-fold reduction of recombination frequency in a BC_2F_3 generation was detected when compared

with the original F_2 generation of the *L. esculentum* x *L. pennellii* cross.

The current tomato map is thought to be fairly complete for two reasons: (1) All molecular and classical markers could be mapped to linkage groups indicating that no loci failed to link up with the map; (2) The sub-telomeric repeat TGRI, approximately 150 kb distant to most tomato chromosomal ends, has been genetically mapped to four tomato chromosomal ends.[26] In all cases, TGRI mapped less than 10 cM from the most terminal molecular marker of the respective linkage group. This supports the hypothesis that each linkage group most likely represents the entire corresponding tomato chromosome.

For tagging new monogenic characters and mapping quantitative trait loci, a set of 200 anchor loci, evenly distributed throughout the entire tomato genome, has been identified and can be requested from our lab. However, the construction of high resolution maps around target genes for map-based cloning approaches will require the selective enrichment of sub-cM regions with additional molecular markers. The isolation of additional markers, tightly linked to a target gene, can be achieved by bulked segregant analysis (BSA) or screening with near-isogenic lines (NILs).[27-29]

21.2. GENOME ORGANIZATION

The tomato genome is comprised of approximately 78% single copy sequences, as evaluated under high stringency hybridization conditions.[30] The remaining part consists of repetitive DNA of which four major classes have been characterized. Ribosomal DNA represents the most abundant repetitive DNA family and comprises approximately 3% of the tomato genome. Both 5S and 45S rDNA are tandemly repeated, containing 1000 and 2300 copies, and map to single loci on chromosome 1 and 2, respectively.[16,31,32]

The highest number of repeats is reported for TGRI, a 162 bp tandem repeat with 77,000 copies in the tomato genome.

Together with a 7 bp telomeric repeat, TGRI is present at 20 of 24 tomato telomeres.[26,33,34] In situ hybridization also confirmed that TGRI is present at 20 tomato chromosome ends and, in addition, a few centromeric and interstitial sites.[35] Non-telomeric sites have also been discovered for the 7 bp telomere repeat (Presting et al, in preparation). All mapped interstitial repeats were placed near the centromeres on 8 of the 12 tomato chromosomes. In one instance, the interstitial repeat was part of the marker cluster which, based on 8000 gametes examined, mapped closest to the centromere of chromosome 9. Two further repeats, TGRII and TGRIII, are less abundant (4200 and 2100 copies, respectively) and, like TGRI, are present in *Lycopersicon* species and in *Solanum lycopersicoides*.[33] Whereas TGRII is apparently randomly distributed among the tomato chromosomes with an average spacing of 133 kb, TGRIII is predominantly clustered in the vicinity of centromeres.

Broun and Tanksley characterized minisatellites as a third class of repeats in tomato.[36] The minisatellite repeat units vary from 100 to 3000 bp and occur on average approximately every 4 Mb.

Recently, much attention in genome mapping has been directed to microsatellites with repeat units of 2-4 bp. In tomato, a probe which detects GACA microsatellites could be used to differentiate a set of six *Lycopersicon* species and 15 tomato cultivars.[37] The copy number and, thus, the size of microsatellite-containing restriction fragments is highly variable between closely related tomato varieties. Since microsatellites can be screened by PCR they are potentially useful in automated genotyping assays of tomato populations. Four independent studies on the localization of GATA or GACA microsatellites showed that these repeats are predominantly associated with tomato centromeres (Broun and Tanksley, in press; Eshed and Zamir, personal communication).[38,123] Thus, the mapping of GATA and GACA repeats may also help to understand the structure of tomato centromeric regions.

21.3. LOCALIZATION OF TOMATO CENTROMERES

The approximate map position of the centromere is now known for each tomato chromosome (see Fig. 21.1). For chromosomes 1 and 2 the centric positions have been identified by RFLP mapping and by in situ hybridization with 5S rDNA and 45S rDNA, respectively.[32,39] The centromeres of chromosome 3 and 6 have been located on the integrated molecular-classical map and by deletion mapping.[40-42] Since there is evidence that the potato/tomato inversions on chromosomes 5, 9, 10, 11 and 12 involve entire chromosome arms, the respective centromeres are most likely located at the inversion breakpoints (Fig. 21.2).[17] Likely map positions for the centromeres of chromosomes 4 and 8 were predicted based on the relationship among the cytological, genetic and molecular tomato maps.[10,17] A more precise localization of the centromeres of chromosomes 7 and 9 have been achieved by RFLP hybridization and dosage analysis of telo-, secondary and tertiary trisomic stocks.[124] In both instances, the centromeres are located within a cluster of tightly linked markers. By constructing high resolution maps for both centromeric regions, markers in these clusters could be genetically separated and the centromeres placed between them. The precise localization of the centromeres of chromosome 7 and 9 represent the first step towards the characterization and ultimate isolation of tomato centromeres.

21.4. GENETIC MAPPING OF TOMATO PHENOTYPES

In tomato a wealth of mapping information for defined characters has been accumulated since the advent of molecular markers. Mapped phenotypes can be classified into two categories: (1) monogenic characters, including morphological markers and resistance genes; and (2) loci underlying quantitative traits (QTL).

21.4.1. MONOGENIC CHARACTERS

The introgression of monogenic traits from wild species has been greatly facilitated by means of marker-assisted selection using isozyme and, later, DNA markers to select positive genotypes. In an early example of gene introgression by means of marker-assisted selection, the isozyme *Aps-1* was used to select for the tightly-linked nematode resistance gene *Mi*.[43,44] A further example for marker-assisted selection in tomato is the use of the isozyme *Prx-2* to select for male-sterile progeny.[45]

Table 21.2 summarizes morphological markers and genes of known function mapped relative to molecular markers. The exact genome position of the genes can be found in Figure 21.1. Several of these linkages are being exploited for marker assisted selection in tomato breeding. For instance, the powdery mildew resistance gene *Lv* has been localized on tomato chromosome 12 by means of mapping with molecular markers.[46] The authors further demonstrated that in mildew resistant tomato lines approximately half of chromosome 12 still contained DNA from the wild species *L. chilense* which was used to introduce *Lv*. Currently, the linked molecular markers are used to select recombinant tomato offspring from these lines in order to significantly reduce the *Lv* bearing *L. chilense* segment on chromosome 12.

21.4.2. QUANTITATIVE TRAITS

A wealth of data has been accumulated for the mapping of quantitative trait loci (QTL) from various interspecific tomato crosses.[47-58] In most of the QTL studies yield components were examined, however, in three cases QTLs for quantitative resistance or salt tolerance have been reported.[49,56,58]

One of the more extensive QTL studies published for tomato was conducted on progeny from a cross between *L. esculentum* and *L. cheesmanii*. Three-hundred and eighty F_2 progeny were investigated at three different locations in California and Israel.[54] A total of 29 putative QTLs affecting fruit size, pH and soluble solids concentration were detected on 11 of 12 tomato chromosomes. Four QTLs were significant in all three environments whereas

15 QTLs (~50%) could be detected at only one location. These findings suggest that a majority of QTLs are environment-dependent. However, major QTLs showing high LOD scores were likely to be present in at least two environments simultaneously. Consequently, major QTLs are most likely the best candidates for introgression into modern tomato cultivars.

The authors also studied the value of QTL mapping to predict the progeny phenotype which is an important measure in breeding programs. They found that the predictive value of the F_2 QTL data was inversely correlated with the heritability of the trait. The F_3 progeny phenotype for traits of low heritability like soluble solids concentration (h^2=0.15) could be predicted more than twice as accurately using F_2 DNA marker data than using F_2 phenotype data. The ratio of prediction is reduced to less than one if the trait under investigation shows high heritability like fruit size (h^2=0.45). Accordingly, marker-assisted selection using QTL-linked DNA markers is especially valuable in breeding for quantitative traits with low heritability.

QTL studies in tomato have also shed light on the basis of transgressive variation.[55] QTL mapping in the interspecific cross *L. esculentum* x *L. pennellii* revealed that for all quantitative traits examined QTLs with allelic effects opposite of that predicted by the parental phenotype could be observed. This result illustrates that molecular markers can be exploited to identify QTL alleles from wild species with a potential to improve agronomic traits in cultivars, even though the wild species do not display the phenotype in question. This was also demonstrated by Eshed and Zamir.[57] The authors selected tomato introgression lines (IL) from a *L. esculentum* x *L. pennellii* cross which represent putative QTL regions for fruit yield and soluble solids content, respectively. The largest effect of wild tomato germplasm on these traits was shown for chromosome 5. A heterozygous IL containing a *L. pennellii* segment of chromosome 5 displayed a 50% increase in soluble solids content when grown in wide spacing.

Fine mapping of QTLs has also advanced in tomato. Applying a strategy called substitution mapping, Paterson and co-authors could narrow down the size of introgressed QTL segments from *L. chmielewskii* into tomato nearly isogenic lines (NILs) to as little as 3 cM.[53] Introgressions can also be employed to measure the effect of small introgressed regions on quantitative traits.[57] Continuous selection of recombinants in successive generations will result in even smaller introgressed segments. These new NILs will then be useful in closing the gap between linkage mapping and physical mapping of QTLs and might be employed for a map-based cloning approach of single QTLs.[29]

21.5. PHYSICAL MAPPING AND MAP-BASED CLONING

Among crop plants, tomato is ideally suited for map-based cloning of target genes without knowledge of the gene product. (1) The tomato genome map is well-populated with more than 1000 markers supporting the localization of any target gene within a short period of time. (2) Tomato is easy to self-pollinate and cross and a single tomato plant can yield a large segregating population needed for high resolution mapping. (3) The tomato genome is relatively small (950 Mb) compared with other crop plants. Tomato high molecular weight DNA libraries, contained in yeast artificial chromosomes (YAC) can be employed to walk along the recombination points of the high resolution map.[59,60] (4) Finally, an *Agrobacterium* transformation system is well established for tomato which can be used to examine the candidate gene action through complementation in transgenic tomato plants.[61] In addition, Van Eck and coauthors recently reported the stable integration of a 80 kb YAC into the tomato genome by means of particle bombardment of cell cultures.[62] This strategy could be useful to further facilitate complementation experiments with DNA segments known to include a candidate gene.

A further advantage of tomato in regard to its potential for gene cloning is the availability of a transposon tagging system.[63-68]

Both gene cloning strategies have been successfully applied to clone resistance genes in tomato. The map-based cloning approach yielded the identification of the *Pto* gene, which confers resistance to *Pseudomonas syringae*.[69] The *Pto* gene product shows homology to a serine/threonine protein kinase and, thus, is most likely involved in a signal transduction pathway which ultimately triggers the resistance response. However, in contrast to resistance genes isolated from *Arabidopsis* and tobacco, the *Pto* peptide lacks leucine rich repeats (LRR) or other subsequences which could act as a receptor domain in recognition of the pathogen.[70,71]

The map-based cloning of the *Pto* gene revealed a second protein kinase which is encoded by *Fen* and causes sensitivity to the insecticide Fenthion. The two genes share 80% protein identity and are physically linked.[72] Interestingly, the activity of *Pto* and *Fen* are simultaneously controlled by a gene designated *Prf* (for *Pseudomonas* resistance and fenthion sensitivity) and all three genes are located at the same chromosomal position.[73] This observation might indicate the existence of multigene clusters in the tomato genome encoding multiple components of a plant defense response. Similar findings in tomato are reported for clustering of *Cf-1, Cf-4, Cf-9* on chromosome 1 and a major resistance cluster containing *Cf-2, Cf-5, Mi, Meu-1* and *Ol-1* on chromosome 6 (Fig. 21.1).

The transposon tagging approach resulted in the cloning of the *Cf-9* gene, which is responsible for race-specific tomato resistance to *Cladosporium fulvum*.[74] The *Cf-9* gene encodes a putative membrane-anchored extracytoplasmic glycoprotein. Its extracellular leucine rich repeats most likely act as a receptor which is involved in pathogen recognition.

Presumably, the cloning of *Pto* and *Cf-9* signals the onset for further isolation of

Table 21.2. Genes of known function or phenotype that have been mapped onto the molecular map of tomato/potato (Chr = Chromosome; morph = morphological marker)

Gene	Type	Product/phenotype	Chr	Reference
6Pgdh-1	isozyme	6phosphogluconate dehydrogenase	4	16, 125
6Pgdh-2	isozyme	6phosphogluconate dehydrogenase	12	125
6Pgdh-3	isozyme	6phosphogluconate dehydrogenase	5	86, 125
a	morph	anthocyaninless	11	17, 126
ae	morph	entirely anthocyaninless	8	17, 126
af	morph	anthocyanin free	5	17, 126
ag	morph	anthocyanin gainer	10	17, 126
alb	morph	albescent	12	17, 126
ACC1	RFLP	ACC synthase	8	127
ACC2	RFLP	ACC synthase	1	127
ACC3	RFLP	ACC synthase	2	127
ACC4	RFLP	ACC synthase	5	127
Aco-1	isozyme	aconitase	12	16, 128
Aco-2	isozyme	aconitase	7	16, 128
Adh-1	isozyme	alcohol dehydrogenase	4	16, 128
Adh-2	isozyme	alcohol dehydrogenase	6	17, 129
AGPaseR	RFLP	regulatory subunit of ADP-glucose pyrophosphorylase	1	130, Lewis and Tanksley, unpubl.
Aps-1	isozyme	acid phosphatase	6	17, 128
Aps-2	isozyme	acid phosphatase	8	17, 128
Asc	morph	resistance to *Alternaria alternata*	3	131, 132
au	morph	aurea	1	133
B	morph	beta carotene	6	17, 126
bls	morph	baby leaf syndrome	3	134
bs	morph	brown seed	1	133
CAB1	RFLP	chlorophyll a/b binding polypeptide	2	16, 31
CAB2	RFLP	chlorophyll a/b binding polypeptide	8	16, 31
CAB3	RFLP	chlorophyll a/b binding polypeptide	3	16, 31
CAB4	RFLP	chlorophyll a/b binding polypeptide	7	17, 135
CAB5	RFLP	chlorophyll a/b binding polypeptide	12	17, 135
CAB6	RFLP	chlorophyll a/b binding polypeptide	5	17, 136
CAB7	RFLP	chlorophyll a/b binding polypeptide	10	17, 137
CAB8	RFLP	chlorophyll a/b binding polypeptide	10	17, 138
CAB11	RFLP	chlorophyll a/b binding polypeptide	6	139
CAB12	RFLP	chlorophyll a/b binding polypeptide	3	139
Cf-1	morph	resistance to *Cladosporium fulvum*	1	133
Cf-2	morph	resistance to *Cladosporium fulvum*	6	77, 140, 141
Cf-4	morph	resistance to *Cladosporium fulvum*	1	141,142
Cf-5	morph	resistance to *Cladosporium fulvum*	6	77, 140, 141
Cf-9	morph	resistance to *Cladosporium fulvum*	1	141, 142
CHS1	RFLP	chalcone synthase	6	Drew and Goldberg, pers. comm.
CHS3	RFLP	chalcone synthase	5	Drew and Goldberg, pers. comm.
CHS4	RFLP	chalcone synthase	9	Drew and Goldberg, pers. comm.
D	morph	potato flower color (red anthocyanins)	2	82
E4	RFLP	ethylene inducible polypeptide	3	17, 143
E8A	RFLP	ethylene inducible polypeptide	9	17, 143

(Continued on next page)

Table 21.2. (Continued)

Gene	Type	Product/phenotype	Chr	Reference
E8B	RFLP	ethylene inducible polypeptide	3	17, 143
Est-1	isozyme	esterase	2	16, 128
Est-2	isozyme	esterase	9	17, 128
Est-4	isozyme	esterase	12	16, 128
Est-5	isozyme	esterase	2	16, 128
Est-6	isozyme	esterase	2	16, 128
Est-7	isozyme	esterase	2	16, 128
F	morph	potato flower color	10	82
Fen	morph	fenthion sensitivity	5	144
fw2.2	morph	fruit weight	2	145
Got-1	isozyme	glutamate oxaloacetate transaminase	4	128
Got-2	isozyme	glutamate oxaloacetate transaminase	7	16, 128
Got-3	isozyme	glutamate oxaloacetate transaminase	7	16, 128
Got-4	isozyme	glutamate oxaloacetate transaminase	8	128
h	morph	hairs absent	10	17, 126
hl	morph	hairless	11	17, 126
HMG2	RFLP	HMG CoA reductase	2	Narita and Gruissem, in prep.
HMG3	RFLP	HMG CoA reductase	3	Narita and Gruissem, in prep.
hp	morph	high pigment	2	Yen and Giovannoni, unpubl.
HTS8	RFLP	heat shock transcription factor	8	146
HTS24	RFLP	heat shock transcription factor	2	146
HTS30	RFLP	heat shock transcription factor	8	146
hy	morph	homogeneous yellow	10	17, 126
I	morph	resistance to *Fusarium oxysporum* race 1	11	Eshed, pers. comm.
I-1	morph	resistance to *Fusarium oxysporum* race 1	7	147
I-2	morph	resistance to *Fusarium oxysporum* race 2	11	21, 78, 148
I-3	morph	resistance to *Fusarium oxysporum* race 3	7	149, 150
Idh-1	isozyme	isocitrate dehydrogenase	1	16
imb	morph	imbecilla	1	133
Imp1	RFLP	Inositol monophosphatase	4	Keddie and Gruissem, in prep., Lewis and Tanksley, unpubl.
Imp2	RFLP	Inositol monophosphatase	6	Keddie and Gruissem, in prep., Lewis and Tanksley, unpubl.
Imp3	RFLP	Inositol monophosphatase	11	Keddie and Gruissem, in prep., Lewis and Tanksley, unpubl.
j	morph	jointless	11	151
j-2	morph	jointless	12	Wing, unpubl.
l2	morph	lutescent-2	10	17, 126
Lpg	morph	Lapageria	1	133
ls	morph	lateral supressor	7	152
Lv	morph	resistance to *Leveillula taurica*	12	46
Mdh-3	isozyme	malate dehydrogenase	7	17
Meu-1	morph	resistance to aphid *Macrosiphum euphorbiae*	6	153

(continued on facing page)

Table 21.2. (Continued)

Gene	Type	Product/phenotype	Chr	Reference
Mi	morph	resistance to root knot nematodes	6	140, 154, 155
ms-10	morph	male sterility	2	45
ms-32	morph	male sterility	1	133
N18	morph	resistance to tobacco mosaic virus	11	71, Lewis and Tanksley, unpubl.
nor	morph	non-ripening	10	17, 76, 126
not	morph	notabilis	7	124, 156
Nr	morph	never ripe	9	17, 126
Ol-1	morph	resistance to *Oidium lycopersicum*	6	157
P	morph	potato flower color (blue anthocyanins)	11	82
PGAL	RFLP	polygalacturonidase	10	17, 158
Pgi-1	isozyme	phosphoglucoisomerase (cytosolic)	12	16, 128
Pgm-1	isozyme	phosphoglucomutase (plastid)	3	16
Pgm-2	isozyme	phosphoglucomutase (cytosolic)	4	16, 128
PPO	RFLP	polyphenol oxidase	8	17, 159
Prf	morph	locus controlling *Pseudomonas* resistance and fenthion sensitivity	5	73
Prx-1	isozyme	peroxidase	1	16, 128
Prx-2	isozyme	peroxidase	2	16, 128
Prx-3	isozyme	peroxidase	2	16, 128
Prx-7	isozyme	peroxidase	3	16, 128
PSC	morph	potato purple skin color	10	88
PTC	RFLP	phytochrome	10	160
Pti1A	RFLP	serine-threonine protein kinase	5	Martin et al, submitted, Lewis and Tanksley, unpubl.
Pti1B	RFLP	serine-threonine protein kinase	12	Martin et al, submitted, Lewis and Tanksley, unpubl.
PTN	RFLP	patatin (tuber storage protein), potato	8	36, 97
Pto	morph	resistance to *Pseudomonas syringae*	5	144, 161
R	morph	potato cortex pigmentation	10	82
r	morph	tomato yellow fruit flesh	3	134
Ro	morph	potato tuber shape	10	82
rin	morph	ripening inhibitor	5	17, 76
R45s	RFLP	45s ribosomal RNA	2	16, 31
R5s	RFLP	5s ribosomal RNA	1	32, Lewis and Tanksley, unpubl.
RBCS1	RFLP	ss ribulose bisphosphate carboxylase	2	16, 31
RBCS2	RFLP	ss ribulose bisphosphate carboxylase	3	16, 31
RBCS3	RFLP	ss ribulose bisphosphate carboxylase	2	16, 31
S	morph	self-incompatibility	1	4, 162
ses	morph	semisterilis	1	133
sf	morph	solanifolia	3	134
Skdh-1	isozyme	shikimic acid dehydrogenase	1	16
Sm	morph	resistance to *Stemphylium*	11	163
Sod-2	isozyme	superoxide dismutase	1	17
sp	morph	self-pruning	6	52
spa	morph	sparsa	8	17, 126

(continued on next page)

Table 21.2. (Continued)

Gene	Type	Product/phenotype	Chr	Reference
sucr	morph	sucrose accumulator	3	134
Sw-5	morph	resistance to tomato spotted wilt virus	9	Brommonschenkel et al, in press
sy	morph	sunny	3	134
tf	morph	trifoliate	5	17, 126
Tm-1	morph	resistance to tobacco mosaic virus	2	164
Tm-2a	morph	resistance to tobacco mosaic virus	9	165, Pillen et al, in prep.
TOM25A	RFLP	ripening related	6	158
TOM25B	RFLP	ripening related	6	158
Tpi-2	isozyme	triose phosphate isomerase	4	16, 128
u	morph	uniform ripening	10	52, 158
var	morph	variabilis	7	124, 166
Ve	morph	resistance to *Verticillium*	9	167, 168
Wx	morph	waxy (potato)	8	88
y	morph	yellow flesh (potato)	3	86

tomato genes without prior knowledge of their gene products. The presence of detailed genetic maps and the construction of YAC contigs, which include DNA markers tightly flanking a target gene, are primary criteria for the possible success of a map-based cloning approach. Recently, the strategy of map-based cloning has been slightly modified, placing more emphasis on the selective enrichment of the target gene region with molecular markers, and on high resolution mapping.[29] A list of tomato genes which meet the outlined criteria is compiled in Table 21.2. Likely gene candidates to be cloned in the near future are those controlling morphological traits like jointless *(j)*, nonripening *(nor)* and ripening inhibitor *(rin)* and resistance genes like *Cf-2/Cf-5*, *I-2*, *Sw-5* (Brommonschenkel et al, in press), *Mi* (Ganal et al, unpublished data) and *Tm-2a* (Pillen et al, unpublished data).[75-78]

21.6. GENOME MAPPING IN *SOLANUM* (POTATO)

21.6.1. DESCRIPTION OF THE TAXON

Potato (*Solanum tuberosum* L.) is one of the major crops in human nutrition, ranking fourth in world production after wheat, corn and rice. It is superior to all other major crops in protein production per unit area and time.[79] The cultivated potato is the most important food crop of the *Solanaceae* family, section Tuberarium, consisting of a number of species and species hybrids of which approximately 225 are tuber-bearing. The basic chromosome number is x = 12 and the species ranges from diploids (2n = 2x = 24) to hexaploids (2n = 6x = 72). The cultivated tetraploid is the most widely grown and has the highest clonal diversity.

The center of origin of the potato is believed to be the Andean Region of South America between Southern Peru and Northern Bolivia. It is assumed that potato was taken to Spain in the 1570s and then to England in the 1590s.[80] The native potato cultivars that were introduced were day-length sensitive with tuber formation inhibited by a day-length longer than 12 hours. After many generations of natural selection in Europe, they adapted to the long summer days of this region.[81] By the late 18th and early 19th century, this long-day adaptation was complete, allowing potato cultivation in Central and Eastern Europe. Potatoes were introduced to the North American colonies in 1691. They were taken to India, China, Japan and some parts of Africa in the 1700s. They

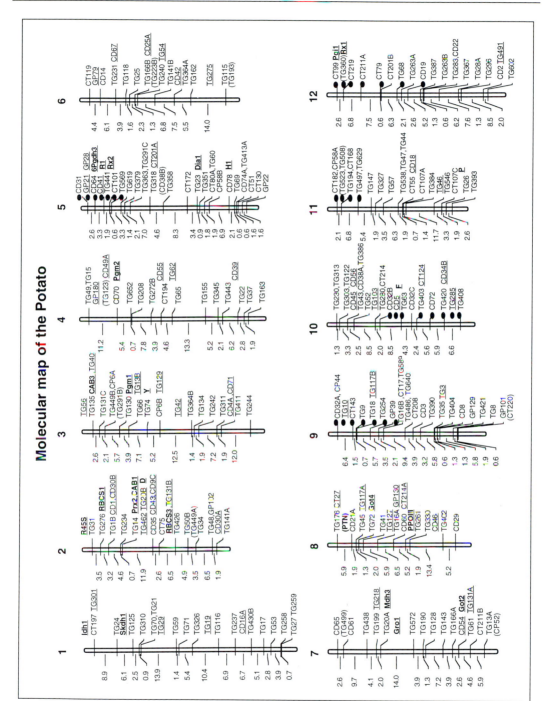

Molecular map of the Potato

Fig. 21.2. Molecular linkage map of the potato genome. Loci by tick marks ordered with LOD >2. Loci following commas cosegregate. Markers enclosed in parentheses have been located to corresponding intervals with LOD < 2. Position of underlined loci approximated from placement on previously published maps. All other loci directly mapped on backcross family N263 of 155 plants (S. tuberosum x S. berthaultii x S. berthaultii). Dots indicate markers involved in inversions that differentiate potato from tomato.[17]

also appeared in New Zealand by the 1760s.[81] Today potatoes are grown in temperate, subtropical and tropical regions as one of the most important world crops.

The cultivated potato (*S. tuberosum L.*) is an annual plant about 30-100 cm tall and is vegetatively propagated through tubers. It is a highly heterozygous tetraploid with $2n=4x=48$ chromosomes. At the diploid level, potato species are obligatory outbreeders because of gametophytic self-incompatibility. At the tetraploid level, potato can be self-compatible and some degree of spontaneous self-fertilization can be found. However at both, the diploid and tetraploid level, it is rarely possible to obtain homozygous lines because severe inbreeding depression is observed.[82]

Potato breeding is a difficult job, because of the tetrasomic inheritance of characters, the high heterozygosity, self incompatibility and male-sterility in many cultivars.[83] Conventional methods of potato breeding involve selection, crossing for recombination and the use of mutations in order to achieve all the traits that should be combined in a variety.[84] These traits can be grouped in three different categories:[84]

(1) Yield and all the factors controlling it, as well as characteristics for good growth, harvest and storage.

(2) Resistance to environmental factors, pests and diseases.

(3) Quality based upon the needs of the specific industry that a particular breeding program serves.

Most of the selection in breeding programs is limited to natural variation. This is time consuming and has a very limited efficiency. For example, a typical program, can take up to 14 years to develop and release a variety (starting with > 500,000 seedlings in the first year).[85]

21.6.2. GENETIC MAPS

The first genetic map of potato was generated by Bonierbale and colleagues using genomic and cDNA clones from tomato in an interspecific cross between *S. phureja* x (*S. tuberosum* x *S. chacoense*) ($2n=2x=24$).[86] A total of 134 DNA and isozyme markers were mapped in 12 linkage groups corresponding to the 12 potato chromosomes and covering 606 cM. Three paracentric inversions in chromosomes 5, 9 and 10 were detected in relation to the tomato chromosomes. The other nine chromosomes were found to be homosequential.[86] Double crossovers were very rare in the potato cross and a large number of the individuals failed to exhibit a single crossover in the majority of their chromosomes.[87]

Gebhardt and colleagues published a second genetic map of potato using potato cDNA clones in a cross between *S. tuberosum* clones.[88] One-hundred and forty one loci were mapped covering 690 cM of the *S. tuberosum* genome on 12 linkage groups corresponding to the 12 potato chromosomes. These two potato maps were aligned with the tomato map using 230 probes from tomato and potato.[89] Three-hundred and four loci were mapped and the similarity between the tomato and potato genomes was confirmed. The comparison between RFLP maps obtained from different segregating backgrounds showed conservation of marker order but differences in recombination rates.

The most recent genetic map of potato from our laboratory was published by Tanksley and coauthors.[17] A population of 155 plants was obtained from a cross between *S. tuberosum* and *S. berthaultii*, backcrossed to *S. berthaultii,* and cDNA and genomic probes were used. The three previously reported paracentric inversions were confirmed (in chromosomes 5, 9, 10), as well as two new paracentric inversions in chromosomes 11 and 12. This potato map consists of 684 cM, approximately one-half that of tomato. These results confirmed earlier reports of reduced recombination in potato compared with tomato.[86] A current map of potato is presented in Figure 21.2. Although not all of the molecular markers mapped in tomato have been mapped in potato, a large number of them have been mapped to allow the deduction of the position of the remaining markers based on the homosequentiality of the two genomes.[17] More than 1400 markers are available for potato, which makes this species to one of the most fully covered plant genomes. Jacobs and coauthors published a new genetic map of potato consisting of molecular markers as well as isozymes and morphological markers.[90] They constructed separate maps for the female and male parents, and the combined map is composed of 175 molecular markers, 10 morphological markers and 8 isozyme markers with a total length of 1120 cM.

21.7. GENETIC MAPPING OF POTATO PHENOTYPES

The utility of RFLP maps of potato has already been demonstrated by the localization of economically important genes. These mapped traits have been classified into two classes: (1) Monogenic characters and (2) Quantitative traits.

21.7.1. MONOGENIC CHARACTERS

In potato breeding there are several important monogenic traits that are frequently transferred from one genetic background to another. By identifying tightly linked markers to a gene of interest, marker assisted selection in potato breeding programs will be enhanced. Several resistance genes as well as other traits of agronomic interest have been mapped in potato.

A major dominant gene *Gro1* conferring resistance to pathotype Ro1 of *Globodera rostochiensis* was mapped using RFLP markers and positioned to the short arm of chromosome 7.[91] Two segregating populations were used to map two major genes controlling extreme resistance to potato virus X (PVX).[92] The resistance gene *Rx1* was mapped to the distal end of chromosome 12 and *Rx2* was mapped to an intermediate position on chromosome 5. A F$_1$ segregating population between a dihaploid line carrying the *R1* locus (that confers vertical resistance to all races of *Phytophthora infestans*) and a susceptible line was used to map the position of the *R1* gene. Using RFLPs the *R1* locus was mapped to chromosome 5 in an interval

between GP 21 and GP 179 in very close proximity to the *Rx2* locus.[93] The *H1* gene from *S. tuberosum* ssp. *andigena* confers resistance to the root cyst nematode *Globodera rostochiensis* pathotypes Ro1 and Ro4. Two independent groups have mapped the *H1* gene to the long arm of chromosome 5 approximately 60 cM from the *Rx2* locus.[94,95] Using a dihaploid segregating population, a second locus *R3* conferring resistance to *P. infestans* was mapped to a distal position of chromosome 11.[96]

Other important agronomic traits have also been mapped. Ten to fifteen copies of the gene responsible for the major storage protein of potato were mapped to a single locus at the end of the long arm of chromosome 8.[97] Three different loci for flower color have been mapped.[82] The locus *P*, predictor of blue anthocyanins, was mapped on chromosome 11 near TG 30. Locus *F*, involved in flower specific expression of genes, was mapped to chromosome 10, near TG 63. Locus *D*, involved in biosynthesis of red anthocyanins, was mapped on chromosome 2, near TG 20. Pigmentation of the cortex, probably locus *R*, appeared to be located on chromosome 10 at approximately the same position as the skin color locus involved in pigmentation of the epidermis.[82] Other morphological traits that have been mapped are: tuber skin color on chromosome 10, the *Ro* locus involved in tuber shape on chromosome 10 and purple skin color, the *PSC* locus, on chromosome 10.[82,88]

21.7.2. QUANTITATIVE TRAITS

Mapping of several quantitative traits has also been pursued in potato. Loci for resistance to two races of *P. infestans* have been mapped in a F$_1$ progeny from a cross between non-inbred diploid parents.[98] Eleven chromosome segments on nine potato chromosomes were identified with at least one race-specific QTL locus. Two mapped regions correspond to segments on chromosomes 5 and 12 to which *R1*, conferring resistance to *P. infestans*, and *Rx1* and *Rx2*, conferring resistance to PVX, have been mapped. Resistance to another species of the potato cyst nematode, *Globodera pallida* pathotypes *Pa2* and *Pa3*, from *S. spegazinii* has been mapped. A major locus, *Gpa*, was mapped to chromosome 5 and two minor loci were mapped to chromosomes 4 and 7.[99] Two QTLs involved in disease resistance to *Globodera rostochiensis* pathotype *Ro1*, were mapped to chromosomes 10 and 11 of the potato genome.[100]

One of the most comprehensive studies dealing with QTLs for insect resistance was that of Bonierbale and colleagues.[101] Biochemical assays that measure the enzymatic browning activity (MEBA) and the concentration of polyphenol oxidase (PPO) in type A trichomes, and the production of sugar esters by the type B trichomes were used to screen a segregating population between *S. tuberosum* x *S. berthaultii*. These traits are believed to be associated with insect resistance. QTLs were identified for type A and B trichomes, as well as for PPO, MEBA and concentration of sugar esters. The QTLs for concentration of sugar esters were located on chromosome 1, 2, 4 and 5. QTLs for type B trichome density were mapped to chromosome 5 and 11. Type A density QTLs were mapped to chromosomes 4, 6 and 10. Two QTLs for browning activity were located on chromosomes 6 and 10 and three QTLs for PPO were mapped to chromosomes 2, 5 and 8.[101]

The segregating population (*S. tuberosum* x *S. berthaultii)* used by Bonierbale and colleagues was also used to study QTLs associated with earliness of tuberization. Ten QTLs on eight different chromosomes were mapped. Earliness of tuberization was favored by *S. tuberosum* alleles at seven of the QTLs and by *S. berthaultii* alleles at three of the QTLs.[85] Using a diploid population from *S. tuberosum* haploid x (*S. chacoense* x *S. phureja*), 13 genetic markers representing six QTLs were found to be associated with chip color. The most important QTL explained 43.5% of the phenotypic variation for this trait.[102]

21.7.3. MARKER-ASSISTED SELECTION

Marker-assisted selection by means of oligonucleotide primers and PCR technology can be a rapid and inexpensive early generation screening tool for potato breeders. Niewöhner and colleagues have developed allele-specific oligonucleotide primers to screen for the presence of two independent resistance genes to the potato cyst nematode *Globodera rostochiensis*.[103] The inheritance of *Gro1* and *H1* could be followed in crosses of diploid potato. When tested in unrelated tetraploid varieties, the specific allele for *Gro1* was not correlated with the resistance phenotype and the allele indicative of the *H1* presence was correlated in only 2% of the varieties screened. Work is being done to try to develop a reliable allele-specific marker to screen for the presence of the resistance gene in tetraploid breeding materials carrying the *H1* gene (Pineda et al, unpublished results).

21.8. GENOME MAPPING IN *CAPSICUM* (PEPPER)

The genus *Capsicum* includes the garden pepper *(C. annuum)* which is grown worldwide and is considered the second most cultivated vegetable species after tomato in the third world. The main center of origin of *Capsicum* is located in South America, where 22 out of approximately 27 *Capsicum* species are endemic.[104] Pepper shares the basic chromosome number (x=12) with most nightshade species. However, its haploid DNA content is approximately four times larger than tomato.[105]

Many genes have been characterized in pepper, particularly those for fruit characters and disease resistance.[106] A set of primary trisomics was used to localize 10 genes to pepper chromosomes, however, no linkage could be found among these genes.[107] The first isozyme linkage map was developed by Tanksley.[108] The map included 14 isozyme loci which were placed on four linkage groups.

The first RFLP map of pepper was published by Tanksley and coauthors and further extended by Prince and coauthors.[109,110] Both maps were based on interspecific crosses between *C. annuum* and *C. chinense* and include mainly tomato cDNA and genomic probes as markers. The authors found that most of the tomato clones hybridized with pepper DNA. However, the linear order of these markers on pepper chromosomes has been widely modified relative to tomato.

The current pepper genome map represents a total size of 720 cM including 19 linkage groups and 192 molecular markers.[110,111] Three linkage groups could be assigned to corresponding pepper chromosomes due to morphological and isozyme markers that had been previously assigned to trisomic chromosomes.[108,112] Based on mapped tomato clones in pepper, an estimated minimum of 31% of the pepper genome is conserved relative to tomato. The order of markers in the pepper genome indicates that at least 15 chromosome breakage events (inversions and translocations) have occurred since the divergence of pepper and tomato from a common ancestor. The excessive rearrangement of pepper linkage groups relative to tomato is in contrast with the high degree of genome conservation between potato and tomato.

Since almost all *Capsicum* species harbor a haploid complement of 12 chromosomes and the current linkage map of pepper contains 19 linkage groups, it is evident that several gaps still exist in the current pepper map. In order to fill these chromosome gaps in pepper and to join unlinked pepper linkage groups, it might be necessary to use probes originated from pepper DNA libraries rather than rely exclusively on tomato probes in future mapping efforts.

Although only a small number of morphological markers have been placed on the current pepper linkage map, two QTLs for multiple flowers per node (*mf1* and *mf2*) could be placed on linkage group 10.[110,111] *Mf1* and *mf2* explain 28 and 39% of the quantitative variation, respectively. The gene action of *mf1* was characterized as

recessive for the *C. chinense* allele. A previous QTL study with isozyme markers revealed five QTLs for the same trait.[113] At least two QTLs mapped to different linkage groups than *mf1* and *mf2*.

Recently, a third molecular map of pepper has been constructed and is based on three doubled-haploid populations derived from an intraspecific *C. annuum* cross.[114] The new map comprises 85 markers distributed over 14 linkage groups and covers approximately 820 cM. The authors included RAPD markers in the new pepper map in addition to 49 RFLP markers from the two previous pepper maps.[109,110] Based on the common RFLP markers, 4 of the 14 linkage groups could be assigned to the previous pepper linkage groups. In accordance with the previous pepper maps, the new map is regarded as incomplete since: (1) 14 markers failed to show linkage to any other marker; and (2) the number of linkage groups still exceeds the haploid set of pepper chromosomes. In addition to the fruit pungency gene *c* which was already assigned to the old linkage group 12, two other genes have been located in the new pepper map. The *L* gene, conferring resistance to tobacco mosaic virus (TMV) in pepper, mapped to linkage group 4 and the *up* gene, controlling the upright fruit habit, mapped to linkage group 6.[114]

Currently, three additional pepper genome maps derived from the interspecific crosses *C. annuum* x *C. chinense, C. annuum* x *C. frutescens* and *C. chinense* x *C. frutescens* are in progress (Kyle et al, in preparation). So far, each map contains more than 150 tomato and pepper markers which are distributed over 17-18 linkage groups. Moreover, the authors carried out a QTL study for resistance to cucumber mosaic virus (CMV) using a *C. annuum* x *C. frutescens* and an intraspecific *C. annuum* cross. Altogether, they could localize 11 independent QTLs which are associated with quantitative resistance to CMV (Kyle et al, in preparation).

Besides the construction of pepper linkage maps, molecular markers have been applied to study genetic variation in the genus *Capsicum*. Isozyme and RFLP markers proved to be useful to discriminate between *Capsicum* accessions from Mexico and to order these accessions in dendograms relative to their genetic distance.[115,116] A further RFLP study of genetic variation within *C. annuum* var. *annuum* demonstrated that related pepper cultivars could be distinguished based on RFLP markers.[117] This finding has been confirmed by Prince and coauthors.[118] Using RFLP and RAPD markers, the authors could distinguish between 21 *Capsicum* accessions and four *C. annuum* cultivars, respectively.

DNA markers are also practical for marker-assisted selection in pepper breeding. An informative RFLP marker has been utilized to distinguish between two parental lines of a commercial hybrid cultivar of pepper.[119] This RFLP marker can be applied to assess the genetic purity of the pepper hybrid progeny. More recently, the RFLP assay has been converted into a PCR assay which is more suitable for large-scale quality-control of hybrid pepper varieties.[120]

21.9. PROSPECTS

During the past 10 years a wealth of genome information and resources have accumulated for Solanaceous species. Shared probe homologies and conserved linkage maps (especially between tomato and potato) have accelerated genome mapping research in the nightshade family.

In the future, these maps are likely to assist in the introgression of linked genes from wild species into cultivars. A procedure designated "advanced backcross breeding" is currently applied in our lab to facilitate the introgression of desirable QTL alleles from wild tomato species into breeding lines.[121,122] The selection of genotypes based on linked DNA markers not only expedites the breeding process but it also allows a much higher precision of gene introduction, excluding undesirable linkage drag, than was feasible in conventional backcross breeding programs.

In the case of quantitative traits, it is now possible to dissect the underlying genes and study their individual phenotypic effects in an isogenic background. In the future, the isogenic lines containing a single QTL interval might be a starting point to isolate the corresponding QTL genes by means of map-based cloning.

While the cloning of QTLs still awaits realization, the cloning of single resistance genes in tomato has already been accomplished and the pool of cloned genes is expected to increase further in the near future. The nature of the identified resistance genes suggests that they are part of a signal transduction pathway which starts with recognition of the pathogen and ultimately leads to a resistance response. The cloning of additional resistance genes will reveal whether the involvement in signal transduction is a general feature of resistance genes or whether different resistance strategies are established in plant-pathogen interactions.

REFERENCES

1. Rick CM. Potential genetic resources in tomato species: Clues from observations in native habitats. In: Srb AM, ed. Genes, Enzymes and Populations. New York: Plenum Press, 1973; 255-269.
2. Rick CM. Natural variability in wild species of *Lycopersicon* and its bearing on tomato breeding. Genet Agr 1976; 30:249-59.
3. Soost RW. Progenies from sesquidiploid F₁ hybrids of *Lycopersicon esculentum* and *L. peruvianum*. J Hered 1958; 49:208-213.
4. Tanksley SD and Loaiza-Figueroa F. Gametophytic self-incompatibility is controlled by a single major locus on chromosome 1 in *Lycopersicon peruvianum*. Proc Natl Acad Sci USA 1985; 82:5093-5096.
5. Rivers BA, Bernatzky R, Robinson SJ et al. Molecular diversity at the self-incompatibility locus is a salient feature in natural populations of wild tomato *Lycopersicon peruvianum*. Mol Gen Genet 1993; 238:419-27.
6. Rick CM. Genetic resources in *Lycopersicon*. In: Nevins DJ and Jones RA, eds. Plant Biology, Vol. 4, Tomato Biotechnology, New York: Alan R. Liss Inc., 1987; 17-26.

7. Jenkins JA. The origin of the cultivated tomato. Econ Bot 1948; 2:379-392.

8. Rick CM. Tomato. In: Simmonds NW, ed. Evolution of Crop Plants. London and New York: Longman, 1976; 268-273.

9. Barton DW. Pachytene morphology of the tomato chromosome complement. Am J Bot 1950; 37:639-643.

10. Khush GS, Rick CM. Cytogenetic analysis of the tomato genome by means of induced deficiencies. Chromosoma 1968; 23:452-84.

11. Jones DF. Linkage in *Lycopersicon*. Am Nat 1917; 51:608-621.

12. Butler L. The linkage map of tomato. J Hered 1952; 43:25-25.

13. Rick CM. The tomato. In: King RC, ed. The Handbook of Genetics. Vol. 2, New York: Plenum Press, 1975; 247-280.

14. Tanksley SD and Bernatzky R. Molecular markers for the nuclear genome of tomato. In: Nevins DJ and Jones RA, eds. Plant Biology, Vol. 4, Tomato Biotechnology, New York: Alan R. Liss Inc., 1987; 37-44.

15. Botstein D, White RL, Skolnick M et al. Construction of a genetic linkage map in man using restriction fragment length polymorphisms. Am J Hum Genet 1980; 32:314-331.

16. Bernatzky R, Tanksley SD. Toward a saturated linkage map in tomato based on isozymes and random complementary DNA sequences. Genetics 1986; 112:887-898.

17. Tanksley SD, Ganal MW, Prince JP et al. High density molecular linkage maps of the tomato and potato genomes. Genetics 1992; 132:1141-1160.

18. Tanksley SD. Linkage map of the tomato *(Lycopersicon esculentum)* (2n=24). In: O'Brien SJ, ed. Genetic Maps, Sixth Edition. Cold Spring Harbor, NY: Cold Spring Harbor Laboratory Press, 1993; 6.39-6.60.

19. Arumuganathan K, Earle ED. Nuclear DNA content of some important plant species. Plant Mol Biol Rep 1991; 9:208-218.

20. Ganal MW, Young ND, Tanksley SD. Pulsed field gel electrophoresis and physical mapping of large DNA fragments in the *Tm-2a* region of chromosome 9 in tomato. Mol Gen Genet 1989; 215:395-400.

21. Segal G, Sarfatti M, Schaffer MA et al. Correlation of genetic and physical struc-ture in the region surrounding the *I-2 Fusarium oxysporum* resistance locus in tomato. Mol Gen Genet 1992; 231:179-185.

22. DeVicente MC, Tanksley SD. Genome-wide reduction in recombination of backcross progeny derived from male versus female gametes in an interspecific cross of tomato. Theor Appl Genet 1991; 83:173-178.

23. Van Ooijen W, Sandbrink JM, Vrielink M et al. An RFLP linkage map of *Lycopersicon peruvianum*. Theor Appl Genet 1994; 89:1007-1013.

24. Rick CM. Controlled introgression of chromosomes of *Solanum pennellii* into *Lycopersicon esculentum:* Segregation and recombination. Genetics 1969; 62:753-768.

25. Rick CM. Further studies on segregation and recombination in backckross derivatives of a tomato species hybrid. Biol Zentralblatt 1972; 91:209-220.

26. Ganal MW, Broun P, Tanksley SD. Genetic mapping of tandemly repeated telomeric DNA sequences in tomato *Lycopersicon esculentum*. Genomics 1992; 14:444-448.

27. Michelmore RW, Paran I, Kesseli RV. Identification of markers linked to disease-resistance genes by bulked segregant analysis. A rapid method to detect markers in specific genomic regions by using segregating populations. Proc Natl Acad Sci USA 1991; 88:9828-9832.

28. Giovannoni JJ, Wing RA, Ganal MW et al. Isolation of molecular markers from specific chromosomal intervals using DNA pools from existing mapping populations. Nucleic Acids Res 1991; 19:6553-6558.

29. Tanksley SD, Ganal MW, Martin GB. Chromosome landing: A paradigm for map-based gene cloning in plant species with large genomes. Trends in Genet 1995; 11:63-68.

30. Zamir D, Tanksley SD. Tomato genome is comprised largely of fast-evolving low copy number sequences. Mol Gen Genet 1988; 213:254-261.

31. Vallejos CE, Tanksley SD, Bernatzky R. Localization in the tomato genome of DNA restriction fragments containing sequences homologous to the ribosomal RNA 45S, the major chlorophyll a-b binding polypeptide

and the ribulose bisphosphate carboxylase genes. Genetics 1986; 112:93-106.

32. Lapitan NLV, Ganal MW, Tanksley SD. Organization of the 5S ribosomal RNA genes in the genome of tomato. Genome 1991; 34:509-514.

33. Ganal MW, Lapitan NLV, Tanksley SD. A molecular and cytogenetic survey of major repeated DNA sequences in tomato Lycopersicon esculentum. Mol Gen Genet 1988; 213:262-268.

34. Broun P, Ganal MW, Tanksley SD. Telomeric arrays display high levels of heritable polymorphism among closely related plant varieties. Proc Natl Acad Sci USA 1992; 89:1354-1357.

35. Lapitan NLV, Ganal MW, Tanksley SD. Somatic chromosome karyotype of tomato based on in situ hybridization of the TGRI satellite repeat. Genome 1989; 32:992-998.

36. Broun P, Tanksley SD. Characterization of tomato DNA clones with sequence similarity to human minisatellites 33.6 and 33.15. Plant Mol Biol 1993; 23:231-242.

37. Rus-Kortekaas W, Smulders MJM, Arens P et al. Direct comparison of levels of genetic variation in tomato detected by a GACA-containing microsatellite probe and by random amplified polymorphic DNA. Genome 1994; 37:375-381.

38. Arens P, Odinot P, Van Heusden AW et al. GATA- and GACA-repeats are not evenly distributed throughout the tomato genome. Genome 1995; 38:84-90.

39. Tanksley SD, Miller JC, Paterson AH et al. Molecular mapping of plant chromosomes. In: Gustafson JP, Appels R, eds. Chromosome Structure and Function, New York: Plenum Press, 1988; 157-172.

40. Koorneef M, Bade J, Hanhart C et al. Characterization and mapping of a gene controlling shoot regeneration in tomato. Plant J 1993; 31:131-141.

41. Van der Biezen EA, Overduin B, Nijkamp HJJ et al. Integrated genetic map of tomato chromosome 3. Rep Tomato Genet Coop 1994; 44:8-10.

42. Van Wordragen MF, Weide R, Liharska T et al. Genetic and molecular organization of the short arm and pericentromeric region of tomato chromosome 6. Euphytica 1994;

79:169-174.

43. Rick CM, Fobes JF. Association of an allozyme with nematode resistance. Rep Tomato Genet Coop 1974; 24:25.

44. Medina-Filho HP. Linkage of Aps-1, Mi and other markers on chromosome 6. Rep Tomato Genet 1980; 30:26-28.

45. Tanksley SD, Zamir D. Double tagging of a male-sterile gene in tomato using a morphological and enzymatic marker gene. Hort Science 1988; 23:387-388.

46. Chunwongse J, Bunn TB, Crossman C et al. Chromosomal localization and molecular marker tagging of the powdery mildew resistance gene (Lv) in tomato. Theor Appl Genet 1994; 89:76-79.

47. Tanksley SD, Medina-Filho H, Rick CM. Use of naturally occurring enzyme variation to detect and map genes controlling quantitative traits in an interspecific backcross of tomato. Heredity 1982; 49:11-25.

48. Osborn TC, Alexander DC, Fobes JF. Identification of restriction fragment length polymorphisms linked to genes controlling soluble solids content in tomato fruit. Theor Appl Genet 1987; 73:350-356.

49. Nienhuis J, Helentjaris T, Slocum M et al. Restriction fragment length polymorphism analysis of loci associated with insect resistance in tomato. Crop Sci 1987; 27:797-803.

50. Weller JI, Soller M, Brody T. Linkage analysis of quantitative traits in an interspecific cross of tomato Lycopersicon esculentum x Lycopersicon pimpinellifolium by means of genetic markers. Genetics 1988; 118:329-340.

51. Tanksley SD, Hewitt J. Use of molecular markers in breeding for soluble solids content in tomato. A re-examination. Theor Appl Genet 1988; 75:811-823.

52. Paterson AH, Lander ES, Hewitt JD et al. Resolution of quantitative traits into Mendelian factors, using a complete linkage map of restriction fragment length polymorphism. Nature 1988; 335:721-726.

53. Paterson AH, DeVerna JW, Lanini B et al. Fine mapping of quantitative trait loci using selected overlapping recombinant chromosomes in an interspecies cross of tomato. Genetics 1990; 124:735-742.

54. Paterson AH, Damon S, Hewitt JD et al. Mendelian factors underlying quantitative

traits in tomato. Comparison across species, generations and environments. Genetics 1991; 127:181-198.

55. DeVicente MC, Tanksley SD. QTL analysis of transgressive segregation in an interspecific tomato cross. Genetics 1993; 134:585-596.

56. Foolad MR, Jones RA. Mapping salt-tolerance genes in tomato (*Lycopersicon esculentum*) using trait-based marker analysis. Theor Appl Genet 1993; 87:184-192.

57. Eshed Y, Zamir D. Introgressions from *Lycopersicon pennellii* can improve the soluble solids yield of tomato hybrids. Theor Appl Genet 1994; 88:891-897.

58. Danesh D, Aarons S, McGill GE et al. Genetic dissection of oligogenic resistance to bacterial wilt in tomato. Mol Plant-Microbe Interact 1994; 7:464-471.

59. Martin GB, Ganal MW, Tanksley SD. Construction of a yeast artificial chromosome library of tomato and identification of cloned segments linked to two disease resistance loci. Mol Gen Genet 1992; 233:25-32.

60. Nakata K, Tanaka H, Ito T et al. Construction and some characterization of a yeast artificial chromosome library from DNA of a tomato line having four disease resistance traits. Bioscience Biotechnology and Biochemistry 1993; 57:1790-1792.

61. McCormick S, Niedermeyer J, Fry J et al. Leaf disc transformation of cultivated tomato (*Lycopersicon esculentum*) using *Agrobacterium tumefaciens*. Plant Cell Rep 1986; 5:81-84.

62. Van Eck JM, Blowers AD, Earle ED. Stable transformation of tomato cell cultures after bombardment with plasmid and YAC DNA. Plant Cell Rep 1995; 14:299-304.

63. Yoder JI, Palys J, Alpert K et al. *Ac* transposition in transgenic tomato plants. Mol Gen Genet 1988; 213:291-296.

64. Osborne BI, Corr CA, Prince JP et al. *Ac* transposition from a T-DNA can generate linked and unlinked clusters of insertions in the tomato genome. Genetics 1991; 129:833-844.

65. Healy J, Corr C, Deyoung J et al. Linked and unlinked transposition of a genetically marked dissociation element in transgenic tomato. Genetics 1993; 134:571-584.

66. Schmitz G, Theres K. A self-stabilizing *Ac* derivative and its potential for transposon tagging. Plant Journal 1994; 6:781-786.

67. Charng YC, Pfitzner AJP. The firefly luciferase gene as a reporter for in vivo detection of *Ac* transposition in tomato plants. Plant Science 1994; 98:175-183.

68. Carroll BJ, Klimyuk VI, Thomas CM et al. Germinal transpositions of the maize element dissociation from T-DNA loci in tomato. Genetics 1995; 139:407-420.

69. Martin GB, Brommonschenkel SH, Chunwongse J et al. Map-based cloning of a protein kinase gene conferring disease resistance in tomato. Science 1993; 262:1432-1436.

70. Bent AF, Kunkel BN, Dahlbeck D et al. *RPS2* of *Arabidopsis thaliana:* A leucine-rich repeat class of plant disease resistance genes. Science 1994; 265:1856-1860.

71. Whitham S, Dinesh-Kumar SP, Choi D et al. The product of the tobacco mosaic virus resistance gene *N:* Similarity to Toll and the interleukin-1 receptor. Cell 1994; 78:1101-1115.

72. Martin GB, Frary A, Wu T et al. A member of the tomato *Pto* gene family confers sensitivity to Fenthion resulting in rapid cell death. Plant Cell 1994; 6:1543-1552.

73. Salmeron JM, Barker SJ, Carland FM et al. Tomato mutants altered in bacterial disease resistance provide evidence for a new locus controlling pathogen recognition. Plant Cell 1994; 6:511-520.

74. Jones DA, Thomas CM, Hammond-Kosack KE et al. Isolation of the tomato *Cf-9* gene for resistance to *Cladosporium fulvum* by transposon tagging. Science 1994; 266:789-793.

75. Zhang HB, Martin GB, Tanksley SD et al. Map-based cloning in crop plants: Tomato as a model system II. Isolation and characterization of a set of overlapping yeast artificial chromosomes encompassing the jointless locus. Mol Gen Genet 1994; 244:613-621.

76. Giovannoni JJ, Noensie EN, Ruezinsky DM et al. Molecular genetic analysis of the *ripening-inhibitor* and *non-ripening* loci of tomato: A first step in genetic map-based cloning of fruit ripening genes. Mol Gen Genet 1995;

248:195-206.

77. Dixon MS, Jones DA, Hatzixanthis K et al. High-resolution mapping of the physical location of the tomato *Cf-2* gene. Mol Plant-Microbe Interact 1995; 8:200-206.

78. Ori N, Paran I, Aviv D et al. A genomic search for the gene conferring resistance to *Fusarium* wilt in tomato. Euphytica 1994; 79:201-204.

79. Rowe RC. Potato health management: A holistic approach. In: Rowe RC ed. Potato Health Management. APS Press, 1993:3-10.

80. Brown CR. Modern evolution of the cultivated potato gene pool. In: Vayda ME, Park WD, eds. The Molecular and Cellular Biology of the Potato. Redwood Press Ltd, 1990:1-12.

81. Hawkes JG. Origins of cultivated potatoes and species relationships. In: Bradshaw JE, Mackay GR, eds. Potato Genetics, Cambridge: University Press, 1994:3-42.

82. Van Eck HJ. Localisation of morphological traits on the genetic map of potato using RFLP and isozyme markers. PhD dissertation. Dept Plant Breeding, Wageningen: Agricultural University, 1995:146.

83. Bajaj YPS. Biotechnology and 21st century potato. In: Bajaj YPS, ed. Biotechnology in Agriculture and Forestry 3. Berlin, Heidelberg: Springer-Verlag, 1987:3-22.

84. Ross H. Potato breeding-problems and perspectives. In: Advances in Plant Breeding. Supplement to Journal of Plant Breeding 13, 1986:132.

85. Plaisted RL, Bonierbale MW, Yencho GC et al. Potato improvement by traditional breeding and opportunities for new technologies. In: Belknap WR, Vayda ME, Park WD, eds. The Molecular and Cellular Biology of the Potato, Biddles Ltd, 1994:1-20.

86. Bonierbale MW, Plaisted RL, Tanksley SD. RFLP maps based on a common set of clones reveal modes of chromosomal evolution in potato and tomato. Genetics 1988; 120:1095-1103.

87. Bonierbale MW, Ganal MW, Tanksley SD. Applications of restriction fragment length polymorphisms and genetic mapping in potato breeding and molecular genetics. In: Vayda ME, Park WD, eds. The Molecular and Cellular Biology of the Potato. Red-wood Press Ltd, 1990; 13-24.

88. Gebhardt C, Ritter E, Debener T et al. RFLP analysis and linkage mapping in *Solanum tuberosum.*. Theor Appl Genet 1989; 78:65-75.

89. Gebhardt C, Ritter E, Barone A et al. RFLP maps of potato and their alignment with the homoeologous tomato genome. Theor Appl Genet 1991; 83:49-57.

90. Jacobs JM, Van Eck HJ, Arens P et al. A genetic map of potato (*Solanum tuberosum*) integrating molecular markers, including transposons, and classical markers. Theor Appl Genet 1995; 91:289-300.

91. Barone A, Ritter E, Schachtschabel U et al. Localization by restriction fragment length polymorphism mapping in potato of a major dominant gene conferring resistance to the potato cyst nematode *Globodera rostochiensis*. Mol Gen Genet 1990; 224:177-182.

92. Ritter E, Debener T, Barone A et al. RFLP mapping on potato chromosomes of two genes controlling extreme resistance to potato virus X (PVX). Mol Gen Genet 1991; 227:81-85.

93. Leonards-Schippers C, Gieffers W, Salamini F et al. The *R1* gene conferring race-specific resistance to *Phytophthora infestans* in potato is located on potato chromosome 5. Mol Gen Genet 1992; 233:278-283.

94. Gebhardt C, Mugniery D, Ritter E et al. Identification of RFLP markers closely linked to the *H1* gene conferring resistance to *Globodera rostochiensis* on potato. Theor Appl Genet 1993; 85:541-544.

95. Pineda O, Bonierbale MW, Plaisted RL et al. Identification of RFLP markers linked to the *H1* gene conferring resistance to the potato cyst nematode *Globodera rostochiensis*. Genome 1993; 36:152-156.

96. El-Kharbotly A, Leonards-Schippers C, Huigen DJ et al. Segregation analysis and RFLP mapping of the *R1* and *R3* alleles conferring race-specific resistance to *Phytophthora infestans* in progeny of dihaploid potato parents. Mol Gen Genet 1994; 242:749-754.

97. Ganal MW, Bonierbale MW, Roeder MS et al. Genetic and physical mapping of the patatin genes in potato and tomato. Mol Gen Genet 1991; 225:501-509.

98. Leonards-Schippers C, Gieffers W, Schafer-Pregl R et al. Quantitative resistance to *Phytophthora infestans* in potato: A case study for QTL mapping in an allogamous plant species. Genetics 1994; 137:68-77.

99. Kreike CM, De Koning JRA, Vinke JH et al. Quantitatively inherited resistance to *Globodera pallida* is dominated by one major locus in *Solanum spegazzinii*. Theor Appl Genet 1994; 88:764-769.

100. Kreike CM, De Koning JRA, Vinke JH et al. Mapping of loci involved in quantitatively inherited resistance to the potato cyst nematode *Globodera rostochiensis* pathotype Ro1. Theor Appl Genet 1993; 87:464-470.

101. Bonierbale MW, Plaisted RL, Pineda O et al. QTL analysis of trichome-mediated insect resistance in potato. Theor Appl Genet 1994; 87:973-987.

102. Douches D, Freyre R. Identification of genetic factors influencing chip color in diploid potato (*Solanum* spp). American Potato J 1994; 71:581-590.

103. Niewöhner J, Salamini F and Gebhardt C. Development of PCR assays diagnostic for RFLP marker alleles closely linked to alleles *Gro1* and *H1*, conferring resistance to the root cyst nematode *Globodera rostochiensis* in potato. Mol Breeding 1995; 1:65-78.

104. Hunziker AT. South American *Solanaceae:* A synoptic survey. In: Hawkes JG, Lester RN, Skelding AD, eds. Biology and Taxonomy of the *Solanaceae.* New York: Academic Press, 1979; 49-85.

105. Galbraith DW, Harkins KR, Maddox JM et al. Rapid flow cytogenetic analysis of the cell cycle in intact plant tissues. Science 1983; 220:1049-1051.

106. Lippert LF, Bergh BO, Smith PG. Gene list for the pepper. J Hered 1965; 56:3-34.

107. Pochard E, Dumas de Vaulx R. Localization of *vy2* and *fa* genes by trisomic analysis. Capsicum Newsl 1982; 11:18-19.

108. Tanksley SD. Linkage relationships and chromosomal locations of enzyme-coding genes in pepper, *Capsicum annuum.* Chromosoma 1984; 89:352-360.

109. Tanksley SD, Bernatzky R, Lapitan NL et al. Conservation of gene repertoire but not gene order in pepper and tomato. Proc Natl Acad Sci USA 1988; 85:6419-6423.

110. Prince JP, Pochard E, Tanksley SD. Construction of a molecular linkage map of pepper and a comparison of synteny with tomato. Genome 1993; 36:404-417.

111. Tanksley SD Prince JP, Kyle MM. Linkage map of pepper *(Capsicum annuum)* (2n=24). In: O'Brien SJ, ed. Genetic Maps, Sixth Edition. Cold Spring Harbor, NY: Cold Spring Harbor Laboratory Press, 1993; 6.220-6.227.

112. Pochard E. Localization of genes in *Capsicum annuum* L. by trisomic analysis. Ann Amelior Plantes 1977; 27:255-266.

113. Tanksley SD, Iglesias-Olivas J. Inheritance and transfer of mutiple-flower character from *Capsicum chinense* into *Capsicum annuum.* Euphytica 1984; 33:769-777.

114. Lefebvre V, Palloix A, Caranta C et al. Construction of an intraspecific integrated linkage map of pepper using molecular markers and doubled-haploid progenies. Genome 1995; 38:112-121.

115. Loaiza-Figueroa F, Ritland K, Laborde-Cancino JA et al. Patterns of genetic variation of the genus *Capsicum (Solanaceae)* in Mexico. Plant Syst Evol 1989; 165:159-188.

116. Prince JP, Loaiza-Figueroa F, Tanksley SD. Restriction fragment length polymorphism and genetic distance among Mexican accessions of *Capsicum*. Genome 1992; 35:726-732.

117. Lefebvre V, Palloix A, Rives M. Nuclear RFLP between pepper cultivars (*Capsicum annuum* L.). Euphytica 1993; 71:189-199.

118. Prince JP, Lackney VK, Angeles C et al. A survey of DNA polymorphism within the genus *Capsicum* and the fingerprinting of pepper cultivars. Genome 1995; 38:224-31.

119. Livneh O, Nagler Y, Tal Y et al. RFLP analysis of a hybrid cultivar of pepper *Capsicum annuum* and its use in distinguishing between parental lines and in hybrid identification. Seed Sci Technol 1990; 18:209-214.

120. Livneh O, Vardi E, Stram Y et al. The conversion of a RFLP assay into PCR for the determination of purity in a hybrid pepper cultivar. Euphytica 1992; 62:97-102.

121. Tanksley SD and Nelson JC. Advanced backcross QTL analysis: A method for the simultaneous discovery and transfer of valu-

able QTLs from unadapted germplasm into elite breeding lines. Theor Appl Genet 1995; in press.

122. Tanksley SD, Grandillo S, Fulton TM et al. Advanced backcross QTL analysis in a cross between an elite processing line of tomato and its wild relative *L. pimpinellifolium.* Theor Appl Genet 1995; in press.

123. Grandillo S, Tanksley SD. Genetic analysis of RFLPs, GATA microsatellites and RAPDs in a cross between *L. esculentum* and *L. pimpinellifolium.* Theor Appl Genet 1995; in press.

124. Frary A, Presting GG, Tanksley SD. Molecular mapping of the centromeres of tomato chromosomes 7 and 9. Mol Gen Genet 1995; in press

125. Tanksley SD, Kuehn G. Genetics, subcellular localization and molecular characterization of 6-phosphogluconate dehydrogenase isozymes in tomato. Biochem Genet 1985; 23:442-454.

126. Rick CM. Tomato linkage survey. Rep Tomato Genet Coop 1980; 30:2-17.

127. Rottmann WH, Peter GF, Oeller PW et al. 1-Aminocylopropane-1-carboxylate synthase in tomato is encoded by a multigene family whose transcription is included during fruit and floral senescence. J Mol Biol 1991; 222:937-961.

128. Tanksley SD, Rick CM. Isozymic gene linkage map of the tomato: Applications in genetics and breeding. Theor Appl Genet 1980; 57:161-170.

129. Tanksley SD, Jones RA. Effect of stress on tomato ADH's: Description of a second ADH coding gene. Biochem Genet 1981; 19:397-409.

130. Iglesias AA, Barry GF, Meyer C et al. Expression of the potato tuber ADP-glucose pyrophosphorylase in *Escherichia coli.* J Biol Chem 1993; 268:1081-1086.

131. Witsenboer HMA, Van de Griend EG, Tiersma JB et al. Tomato resistance to *Alternaria* stem canker localization in host genotypes and functional expression compared to non-host resistance. Theor Appl Genet 1989; 78:457-462.

132. Van der Biezen EA, Glagotskaya T, Overduin B et al. Inheritance and genetic mapping of resistance to *Alternaria alternata*

f.sp. *lycopersici* in *Lycopersicon pennellii.* Mol Gen Genet 1995; 247:453-461.

133. Balint-Kurti PJ, Jones DA, Jones JDG. Integration of the classical and RFLP linkage maps of the short arm of tomato chromosome 1. Theor Appl Genet 1995; 90:17-26.

134. Chetelat RT, DeVerna JW, Bennett AB. Introgression into tomato (*Lycopersicon esculentum*) of the *L. chmielewskii* sucrose accumulator gene (*sucr*) controlling fruit sugar composition. Theor Appl Genet 1995; 91:327-333.

135. Pichersky E, Hoffman NE, Malik VS et al. The tomato *Cab-4 and Cab-5* genes encode a second type of CAB polypeptides localized in photosystem II. Plant Mol Biol 1987; 9:109-120.

136. Pichersky E, Hoffman NE, Bernatzky R et al. Molecular characterization and genetic mapping of DNA sequences encoding the type I chlorophyll *a/b*-binding polypeptide of photosystem I in *Lycopersicon esculentum* (tomato). Plant Mol Biol 1987; 9:205-216.

137. Pichersky E, Tanksley SD, Piechulla B et al. Nucleotide sequence and chromosomal location of *Cab-7:* the tomato gene encoding the type II chlorophyll *a/b*-binding polypeptide of photosystem I. Plant Mol Biol 1988; 11:69-71.

138. Pichersky E, Brock TG, Nguyen D et al. A new member of the CAB gene family: structure, expression and chromosomal location of *Cab-8:* the tomato gene encoding the type III chlorophyll *a/b*- binding polypeptide of photosystem I. Plant Mol Biol 1989; 12:257-270.

139. Schwartz E, Shen D, Aebersold R et al. Nucleotide sequence and chromosomal location of *Cab11* and *Cab12:* the genes for the fourth polypeptide of photosystem I light harvesting antenna (LHCI). FEBS Lett 1991; 280:229-234.

140. Dickinson MJ, Jones DA, Jones JDG. Close linkage between the *Cf-2, Cf-5* and *Mi* resistance loci in tomato. Mol Plant-Microbe Interact 1993; 6:341-347.

141. Jones DA, Dickinson MJ, Balint-Kurti PJ et al. Two complex resistance loci revealed in tomato by classical and RFLP mapping of the *Cf-2, Cf-4, Cf-5* and *Cf-9* genes for

resistance to *Cladosporium fulvum*. Mol Plant-Microbe Interact 1993; 6:348-57.

142. Balint-Kurti PJ, Dixon MS, Jones DA et al. RFLP linkage analysis of the *Cf-4* and *Cf-9* genes for resistance to *Cladosporium fulvum* in tomato. Theor Appl Genet 1994; 88:691-700.

143. Lincoln JE, Cordes S, Read E et al. Regulation of gene expression by ethylene during *Lycopersicon esculentum* (tomato) fruit development. Proc Natl Acad Sci USA 1987; 84:2793-2797.

144. Carland FM, Staskawicz BJ. Genetic characterization of the *Pto* locus of tomato. Semidominance and cosegregation of resistance to *Pseudomonas syringae* pathovar tomato and sensitivity to the insecticide Fenthion. Mol Gen Genet 1993; 239:17-27.

145. Alpert KB, Grandillo S, Tanksley SD. *fw2.2:* A major QTL controlling fruit weight is common to both red and green-fruited tomato species. Theor Appl Genet; in press.

146. Scharf K, Rose S, Zott W et al. Three tomato genes code for heat stress transcription factors with a region of remarkable homology to the DNA-binding domain of the yeast HSF. EMBO J 1990; 9:4495-4501.

147. Sarfatti M, Abu-Abied M, Katan J et al. RFLP mapping of *I-1* a new locus in tomato conferring resistance against *Fusarium oxysporum* f. sp. *lycopersici* race 1. Theor Appl Genet 1991; 82:22-26.

148. Sarfatti M, Katan J, Fluhr R et al. An RFLP marker in tomato linked to the *Fusarium oxysporum* resistance gene *I-2*. Theor Appl Genet 1989; 78:755-9.

149. Bournival BL, Scott JW, Vallejos CE. An isozyme marker for resistance to race 3 of *Fusarium oxysporum* f. sp. *lycopersici* in tomato. Theor Appl Genet 1989; 78:489-494.

150. Tanksley SD, Costello W. The size of the *L. pennellii* chromosome 7 segment containing the *I-3* gene in tomato breeding lines measured by RFLP probing. Rep Tomato Genet Coop 1991; 41:60.

151. Wing RA, Zhang HB, Tanksley SD. Map-based cloning in crop plants: Tomato as a model system: I. Genetic and physical mapping of jointless. Mol Gen Genet 1994; 242:681-688.

152. Schumacher K, Ganal MG, Theres K. Genetic and physical mapping of the *lateral suppressor (ls)* locus in tomato. Mol Gen Genet 1995; 246:761-766.

153. Kaloshian I, Lange WH, Williamson VM. An aphid resistance locus is tightly linked to the nematode resistance gene *Mi* in tomato. Proc Natl Acad Sci USA 1995; 92:622-625.

154. Messeguer R, Ganal M, DeVicente MC et al. High resolution RFLP map around the root knot nematode resistance gene *Mi* in tomato. Theor Appl Genet 1991; 82:529-36.

155. Klein-Lankhorst R, Rietveld P, Machiels B et al. RFLP markers linked to the root knot nematode resistance gene *Mi* in tomato. Theor Appl Genet 1991; 81:661-667.

156. Boynton JE and Rick CM. Linkage tests with mutants of Stubbe's groups I, II, III and IV. Rep Tomato Genet Coop 1965:24-27.

157. Van der Beek G, Pet G, Lindhout P. Resistance of powdery mildew (*Oidium lycopersicum*) in *Lycopersicon hirsutum* is controlled by an incompletely dominant gene *Ol-1* on chromosome 6. Theor Appl Genet 1994; 89:467-473.

158. Kinzer SM, Schwager SJ, Mutschler MA. Mapping of ripening-related or -specific cDNA clones of tomato (*Lycopersicon esculentum*). Theor Appl Genet 1990; 79:489-496.

159. Newman SM, Eannetta NT, Yu H et al. Organization of the tomato polyphenol oxidase gene family. Plant Mol Biol 1993; 21:1035-1051.

160. Lissemore JL, Colbert JT, Quail PH. Cloning of cDNA for phytochrome from etiolated *Cucurbita* and coordinate photoregulation of the abundance of two distinct phytochrome transcripts. Plant Mol Biol 1987; 8:485-496.

161. Martin GB, Williams JGK, Tanksley SD. Rapid identification of markers linked to a *Pseudomonas* resistance gene in tomato by using random primers and near-isogenic lines. Proc Natl Acad Sci USA 1991; 88:2336-2340.

162. Bernatzky R. Genetic mapping and protein product diversity of the self-incompatability locus in wild tomato *Lycopersicon peruvianum*.

Biochem Genet 1993; 31:173-184.

163. Behare J, Laterrot H, Sarfatti M et al. Restriction fragment length polymorphism mapping of the *Stemphylium* resistance gene in tomato. Mol Plant-Microbe Interact 1991; 4:489-492.

164. Levesque H, Vedel F, Mathieu C et al. Identification of a short rDNA spacer sequence highly specific to a tomato line containing the *Tm-1* gene introgressed from *Lycopesicon hirsutum*. Theor Appl Genet 1990; 80:602-8.

165. Young ND, Zamir D, Ganal MW et al. Use of isogenic lines and simultaneous probing to identify DNA markers tightly linked to the *Tm-2a* gene in tomato. Genetics 1988; 120:579-586.

166. Reeves AF, Zobel RW, Rick CM. Further tests with mutants of the Stubbe Series I, II, III and IV. Rep Tomato Genet Coop 1968; 32-34.

167. Zamir D, Bolkan H, Juvik JA et al. New evidence for placement of Ve- the gene for resistance to *Verticilium* race 1. Rep Tom Genet Coop 1993; 43:51-52.

168. Kawchuk LM, Lynch DR, Hachey J et al. Identification of a codominant amplified polymorphic DNA marker linked to the *Verticillium* wilt resistance gene to tomato. Theor Appl Genet 1994; 89:661-664.

CHAPTER 22

GENOME MAPPING IN GYMNOSPERMS: A CASE STUDY IN LOBLOLLY PINE (*PINUS TAEDA* L.)

David B. Neale and Ronald R. Sederoff

22.1. INTRODUCTION

Genome mapping is a powerful experimental technique for studying the genetics and evolution of gymnosperms, an old and successful group of higher plants. Gymnosperms, defined by the absence of an ovary wall which encloses the seeds of angiosperms, are now represented by less than a thousand total species.[10] The gymnosperms are predominantly woody perennials, are highly diverse in size and structure and were the dominant vascular plants during the Mesozoic. The earliest gymnospermous seed plants were found in the late Devonian and the earliest appearance of conifers was at the end of the Carboniferous. The modern families of conifers became apparent in fossils of the Mesozoic.

Although gymnosperms have largely been replaced by the more recently evolved and diverse angiosperms, gymnosperms are distributed widely throughout the world, including vast regions of northern temperate forests where they dominate the landscape. Gymnosperms represent the most important source of wood and fiber for the developed countries of the world. Among the gymnosperms are the oldest, largest and tallest of earth's organisms. Bristlecone pine (*Pinus aristata* Engelm.) approaches 5000 years of age, the giant sequoia (*Sequoiadendron giganteum* (Lindl. Buchholz)) is the largest living organism and the coastal redwoods (*Sequoia sempervirens* (D. Don) Endl.) can exceed 100 meters in height.

Genome Mapping in Plants, edited by Andrew H. Paterson. © 1996 R.G. Landes Company.

Pines (*Pinus*) as a genus were already well differentiated by the Cretaceous.[43] The secondary centers of speciation and diversity developed during the Eocene and continued through the Pliestocene in regions like the North American southwest and Japan. In North America, tree populations migrated with the ebb and flow of glaciers, while in Europe, east-west mountain ranges reduced migration and consequently, also reduced diversity. At the present time, pines are widely distributed in the Northern Hemisphere. Pines are found from the polar region to the tropics and dominate the natural vegetation in many regions.[13] There are 100 or more species which include many commercially important forest tree species.

Because of its wide range, rapid growth and versatility for wood products, loblolly pine (*Pinus taeda* L.) is the principal commercial pine species in the United States and one of the world's most important forest trees. It grows along the Coastal Plain and the Piedmont from Delaware to eastern Texas. Its range is limited, at least in part, by its requirement for moderate temperatures and plentiful rainfall. Well over a billion loblolly pine seedlings are planted in the southeastern U.S. each year.

The long generation times, high levels of genetic diversity and the absence of defined inbred lines or cultivars appeared to be formidable barriers to genomic mapping in forest trees.[48] In addition, most conifers have large genome sizes, typically ten-fold larger than angiosperms and 200 times larger than the model plant *Arabidopsis thaliana*. However, the high levels of genetic diversity and heterozygosity of conifers have provided large numbers of excellent DNA markers and genomic mapping is now being carried out in dozens of academic and industrial laboratories around the world. The application of molecular markers analyses to existing pedigrees and to individual trees has made it possible to obtain a high level of genetic information related to genome evolution, wood properties, disease resistance and growth traits in a remarkably short period of time. Molecular marker systems have exploited

variation in restriction sites (RFLP markers), oligonucleotide target sites for PCR using arbitrary primers (RAPD markers) and defined sequences for PCR amplification (microsatellites).

Why construct genetic maps for forest trees? As the world continues to grow in population and as the level of technology increases throughout the world, there will be an ever increasing demand for wood and paper products. At the same time, it becomes more important to protect and conserve the dwindling resources of the natural forests and thus produce larger amounts of wood on a decreasing area of land. It will be necessary to use improvements in silviculture and genetic technology for the supply of wood products to keep pace with the demand. Gymnosperms, particularly conifers, provide all of the world's softwood. Softwoods are distinguished by their simpler wood cell structure and by their long wood fibers and are used for many applications. These include a wide diversity of paper products, pulp, composite wood products and the many uses of the wood itself.

22.2. PINE GENOME ORGANIZATION

Genome organization in conifers appears to be highly conserved in ploidy level and chromosome number.[48] Most conifers are diploid. The haploid chromosome numbers for the major families of conifers are 12 for the Pinaceae and 11 for the Taxodiaceae and the Cupressaceae. There are a few exceptions such as Douglas-fir (*Pseudotsuga menziesii* (Mirb.) Franco) (n=13) and the hexaploid coast redwood (n=33). Sax and Sax[58] identified 12 metacentric chromosomes in pine that were very similar in size and morphology. Recent studies of pine chromosomes using confocal fluorescence microscopy and in situ hybridization have begun to differentiate the chromosomes.[17] The average number of chiasmata observed in pine pollen mother cells was 2.5 chiasmata per bivalent.[59] Total map distance should correspond to half the number of chiasmata, which should be approximately 1500 cM.

DNA contents in conifers are considerably higher than is typical for other higher plants.[5,68] The most recent and more reliable estimates of genome size in conifers have been determined by laser flow cytometry and show a range of DNA contents within the pines of 19.9 pg to 31.7 pg.[68] The genome content of loblolly pine is 22 pg per haploid genome. DNA reassociation analysis of eastern white pine (*Pinus strobus* L.) indicated that about 76% of the sequences were in a moderately or highly repeated fraction.[35] Early attempts at genetic mapping in pine based on RFLP analysis with cDNA probes showed that gene family sizes in conifers may be significantly larger than in other groups of plants.[4,16] The lipid transfer protein, for example, has at least 14 members in loblolly pine,[33] whereas just a few copies are reported for angiosperm species. One mechanism that could have led to this significant amount of gene family amplification is retrotransposition. Sederoff et al[60] discovered a highly repeated element in pine called *IFG*. Kossack[34] cloned and sequenced this element and determined that it was a retrotransposable element that exists in at least 100,000 copies in the pine genome. It is possible that this same element has also been responsible for amplification of structural genes by reverse transcription of mRNA templates and integration into new regions of the genome. Kvarnheden et al[37] recently showed that pseudogenes of the *cdc2* gene family in spruce exist and that the structure of these pseudogenes suggests an origin by retrotransposition.

22.3. GENETIC MAPS

Genomic mapping is being conducted in only one group of Gymnosperms, the conifers; we are not aware of mapping research in the Cycads, Gnetum, or Ginkgo. There are ten families in the order Coniferae but most genetic mapping is being done in the Pinaceae. There is one report from the Taxodiaceae (*Cryptomeria japonica* D. Don)[45] and one from the Taxaceae (*Taxus brevifolia* Nutt.).[20] Within the Pinaceae nearly all published reports have come from the genus *Pinus*, although there are reports from the *Picea*,[7,66] and projects are underway in the *Pseudotsuga* and *Larix*. Genetic maps have been published for a number of commercially important pine species including slash pine (*Pinus elliottii* Engelm.),[50] longleaf pine (*P. palustris* Mill.),[49] maritime pine (*P. pinaster* Ait.),[55,56] sugar pine (*P. lambertiana* Dougl.),[14] and Turkish red pine (*P. brutia* Ten.).[28] In this paper, however, we will only review in detail genetic mapping in loblolly pine. Because of easy electronic access to genome databases, no genetic maps will be presented in this paper. All genetic maps referred to in this paper can be viewed via the Dendrome World Wide Web (http://s27w007.pswfs.gov) and Gopher (s27w007.pswfs.gov) servers or by obtaining a copy of the TreeGenes Database (dendrome@s27w007.pswfs.gov).

Genetic mapping in loblolly pine began in the late 1980s and initially two basic approaches were used to construct maps. Neale and coworkers at the Institute of Forest Genetics (IFG) in Placerville, California used RFLP markers and three-generation outbred pedigrees for constructing maps whereas the Forest Biotechnology Group at North Carolina State University (NCSU) used RAPD markers and segregating megagametophytes from individual trees. In recent years, however, both groups have used a combination of marker types and pedigree structures as the specific objectives of map construction grew and diversified.

22.3.1. RFLP MAPS

Devey et al[15] constructed the first RFLP map for loblolly pine. This map was based on segregation data from 95 progeny of a single, three-generation outbred pedigree. Two types of RFLP mapping probes were used: (1) complementary DNA (cDNA) probes derived from loblolly pine seedlings and (2) genomic DNA probes from loblolly pine.[16] Seventy-three RFLP loci and two isozyme loci were positioned onto 20 linkage groups of two or more loci:

17 RFLP and 4 isozyme loci were unlinked. This map was constructed with the linkage program GMendel 2.0[40] which can estimate linkages from phase-unknown data. A single sex-averaged map was constructed, no attempt was made to construct maternal-and paternal-specific maps due to the limited amount of data.

The second loblolly pine RFLP map was also constructed using a three-generation outbred pedigree.[21] This pedigree was selected for mapping quantitative trait loci (QTL) influencing wood specific gravity (see Phenotype Mapping below). Statistically independent maps were constructed for the maternal and paternal parents from the segregation data of 177 progeny using the linkage program JoinMap 1.4.[63] The maternal map included 87 loci positioned onto 17 linkage groups and the paternal map included 75 loci on 23 linkage groups. There were 26 loci that mapped to both the maternal and paternal maps. Because maternal and paternal maps were constructed and linked pairs of loci were common to both maps, Groover et al[22] were able to test for sex-related differences in meiotic recombination between the maternal and paternal parent. They found a slightly (26%) higher rate of recombination in the male parent.

22.3.2. RAPD Maps

The conifer megagametophyte is haploid and genetically identical to the egg. This system provides an exceptional opportunity for linkage analysis. Heterozygous loci in mother trees produce two types of megagametophytes each carrying one of the two alleles and segregate in a 1:1 ratio. Two-locus linkage is inferred when gametic ratios deviate from the Mendelian ratio of 1:1:1:1. Linkage relationships among allozyme loci have been determined using this system in many conifers.[11,12,18,23,24,31,46] Allozyme linkage analyses provided an initial view of conifer genome organization even though only a small number of loci could be examined.

The large genome sizes of conifers and the small DNA content of the megagameto-phyte precluded the use of RFLP markers on individual megagametophytes. RFLP analysis in conifers requires approximately 10 μg of DNA to make a single Southern blot, whereas megagametophytes from most conifers contain only a few nanograms of DNA. The advent of anonymous PCR-based DNA markers[69,75] made it possible to conduct linkage analyses in conifers using the small amounts of DNA that can be isolated from individual megagametophytes.[66] Anonymous PCR-based markers require no prior sequence information and can be analyzed with only a few nanograms of DNA. Genetic maps based on RAPDs were quickly constructed for a number of conifers.[7,20,28,49,50,66]

The first loblolly pine RAPD map was constructed by O'Malley et al[52] for genotype 7-56, a tree widely used as a parent in many breeding programs and one of the grandparents in the three-generation pedigree used for one of the RFLP maps of loblolly pine.[15] The current RAPD map has 232 RAPD markers that are grouped on the map at a criteria of LOD greater than four with a recombination fraction less than 0.25. The map has 15 linkage groups with a total map distance between ordered markers of 1067 cM. A framework map has been generated with 91 markers at an interval support of a LOD greater than three. At this interval support, the order of framework markers is roughly 1000 times more probable than the next most likely order. One-hundred-twenty-one additional markers can be assigned to the map but the order cannot be assigned within the interval at the interval support criterion. Fourteen markers are unordered in five small linkage groups containing two to four markers and six markers remain unlinked.

In addition to genotype 7-56, two other genotypes have been mapped in some detail as part of a study to investigate the genetic basis for resistance to fusiform rust disease.[71] These two genotypes are related; one is the open-pollinated seed parent of the other. One is known as genotype 10-5, which has been widely used as a standard check in assays for resistance and susceptibility to fusiform rust disease. The daughter of 10-5 that has been mapped is the genotype designated 152-231. Moderate density maps have been constructed for both of these genotypes. For genotype 10-5, 91 RAPD primers were used to identify and map 295 segregating polymorphisms. For the daughter, genotype 152-231, 58 primers were used to map 227 segregating polymorphic DNA fragments. Markers from both genotypes were assigned to linkage groups and ordered using a LOD criterion of greater than five and a recombination fraction of less than 0.25. The genomic map of 10-5 is 1727 cM in length, with an average density of one marker per 5.9 cM.[71] The map for tree genotype 152-231 has a total length of 1587 cM with an average density of one marker per 6.8 cM. The two maps were constructed independently using different primers, however, many of the primers were common to both maps. As a result there is a high degree of relationship between both maps and at least half of the linkage groups of one map are syntenic with the other map by inspection.

22.3.3. RELIABILITY AND REPEATABILITY OF LOCUS ORDERING IN RAPD MAPS IN PINE

Plomion et al[56] have constructed two genomic maps from an individual tree of maritime pine (*Pinus pinaster* Ait.) using a common set of 263 RAPD markers chosen for their validity by inheritance and repeatability. Each map was constructed from marker analysis of an independent set of 62 individuals, one set from a population of selfed seed and one from open pollinated seed of the same individual. Comparision of the two framework maps showed the same order for 98% of the mapped markers. A bootstrap method indicated that the rate of errors in marker order was representative for the data set. The differences in locus order were consistent with expected variation due to chance.

22.3.4. CONSENSUS MAPS

Genetic maps are constructed from individual mapping populations and are specific to that population. In effect, seven different loblolly pine maps have been constructed if the male and female RFLP maps for two crosses and the three RAPD maps are considered. To a limited extent, these maps share some common pedigree and genetic markers such that it will be possible to merge all seven maps into an integrated map for loblolly pine. Such a consensus map is an artificial construction and the position of genes reflects an average across maps, but will nevertheless be useful for summarizing map information in the species loblolly pine. To begin, Sewell and Neale[61] merged the four independent RFLP maps. This map contains a total of 349 markers in 19 linkage groups and a total map distance of 1105 cM. This map was constructed from: (1) 240 loci from loblolly pine cDNA probes, (2) 20 loci from loblolly pine genomic DNA probes, (3) 20 loci from cDNA or genomic DNA probes from other species, (4) 11 isozyme loci and (5) 58 RAPD loci. Approximately 60 of these loci were common to one or more of the four RFLP maps and were thus used for merging the maps with JoinMap 1.4.[63] The IFG and NCSU labs are in the process of identifying RAPD markers common to these maps such that they can be used to integrate all maps. More common markers may yet needed to successfully merge these maps.

22.4. PHENOTYPE MAPPING IN LOBLOLLY PINE

The considerable interest in genetic mapping in the forestry community is in part based on its potential for use in marker-aided breeding.[48,65,74] Trees have long generation times and traits of economic interest are often difficult and expensive to measure and have low heritabilities. Before marker-aided breeding can be applied, however, it must be shown that quantitative trait loci (QTL) for such traits can be mapped. In pine, three traits of economic interest have been the subject of genetic mapping experiments: (1) resistance to fusiform rust (*Cronartium quercuum* (Berk.) Miyabe ex Shirai f. sp. *fusiforme*), (2) wood specific gravity (WSG) and (3) early height growth.

22.4.1. FUSIFORM RUST

Fusiform rust is the major disease of loblolly pine and the economic cost of susceptibility to rust has been estimated to be between U.S. $36 and $192 million per year.[57] The genetic basis of resistance was not understood despite 37 years of work directed to this question. Genomic mapping has shown that resistance in at least some families is conferred by a single region of the genome. The region is two map units from the most closely linked RAPD marker which confers complete resistance to a common race of the pathogen.[36,71,72] The resistance locus in the pedigree of loblolly pine family 10-5 is dominant and was found to confer resistance over three generations, both in the greenhouse and in the field. The resistant genotype was found to confer resistance in the field to natural inoculum at two test sites, suggesting that the common fungal genotype interacts with the resistance gene identified in the 10-5 pedigree. It was further determined that standard assays for rust resistance and susceptibility can have a significant number of escapes. It is likely that the variable frequency of escapes, due to a variety of environmental factors, was the basis for the apparent quantitative variation in expression of resistance that confounded genetic analysis for a great many years. Genomic analysis, as applied to disease resistance in forest trees, may become a powerful tool to understand the genetic basis of complex traits in natural systems, just as genetic analysis has been used for the analysis of complex traits in humans and other animals.[38]

22.4.2 WOOD SPECIFIC GRAVITY

Wood specific gravity is a trait of considerable economic importance in loblolly pine.[42,77] WSG is the primary determinant of lumber strength and pulp yield. WSG has moderate to high heritability, suggesting that genes of major effect could be detected in mapping experiments.[74] Groover et al[21] mapped QTLs for WSG in a three-generation pedigree. Using the simple single-factor ANOVA approach to QTL mapping, they found five QTLs on five different linkage groups that contained QTLs for WSG. Because some of the RFLP loci linked to QTLs were segregating for three or four alleles (i.e., a fully informative marker), it could be shown that some of the QTLs were also segregating for more than two alleles among the parents. Two of the QTLs showed genotype x environment interaction but there was no evidence found for digenic epistasis among QTLs. The efficiency of marker breeding in highly heterozygous and outbred conifers will be enhanced if tightly-linked markers flanking QTLs can be identified. Kiehne et al[30] developed a strategy to identify such markers for loblolly pine WSG QTLs that is based on the approach of Giovannoni et al.[19] Four DNA pools, based on the nonrecombinant progeny genotypes of fully-informative RFLP markers linked to a WSG QTL, were screened with a large number of RAPD primers. Nine new RAPD markers were identified which mapped to an interval containing the WSG QTL.

22.4.3. EARLY HEIGHT GROWTH

Genomic mapping has been used to investigate the genetic basis of early height growth in maritime pine under controlled growth conditions.[54] F2 plants were grown in a greenhouse under intense fertilization and periods of continuous light during growth seasons. QTLs were identified for: (1) total height at two years, (2) length of shoot cycles, (3) number of stem units and (4) mean stem unit length. Shoot elongation has been identified as a criterion for early selection because it is strongly correlated with field performance at eight years in loblolly pine.[8,9,73] QTL detected in any one year of growth were not necessarily detected in any other year, suggesting that specific genes may be differentially expressed in different years of growth. These results could provide important insights into the genetic basis of juvenile mature correlations in forest trees.

22.5. MAP-BASED CLONING IN PINE

Map-based or positional cloning of genes from plants is a powerful technique that is being employed extensively in *Arabidopsis thaliana* because of its small genome size and abundance of phenotypic mutants. To a lesser extent this approach is being used in agronomic crops such as tomato and rice. The initial targets for map-based cloning have been disease resistance genes and a number of successes have been reported.[6,27,39,41,44,70] Map-based cloning in loblolly pine and other conifers will be difficult using current technologies due to their extremely large genomes. One centiMorgan in pine may be as much as four megabases,[48] although this has never been experimentally determined. Nevertheless, new technologies are certain to develop and in at least one case there is a serious effort to clone a disease resistance gene from pine that is characterized only by its phenotype and genetic map position. Devey et al[14] recently mapped a dominant gene for resistance to white pine blister rust (*Cronartium ribicola* Fisch.) in sugar pine. The most tightly linked RAPD marker was 0.9 cM from the resistance gene R. A single marker linked at this distance in pine would not be sufficient to begin a physical mapping experiment to clone R. Subsequent efforts using a large number of RAPD primers (1200+) and large sample sizes to detect rare recombinants have yielded additional and more tightly linked markers (Neale, unpublished data). However, flanking markers linked to within 0.1 cM of R, that would justify the construction and screening of YAC or BAC libraries, have not yet been identified. Nevertheless, the precise genetic definition of the map location of R will have significant value for whatever approach to cloning R is ultimately employed.

22.6. DIRECTIONS FOR FUTURE RESEARCH

Genetic marker development will continue to be an important area of research for conifer genome mapping. The ideal genetic marker for use in highly heterozygous and outbred conifers is yet to be developed.[47] RFLP markers are highly informative (codominant and multiallelic) and repeatable but they are slow and difficult to develop and apply. RAPD markers are easy and fast to apply and develop, but are much less informative (dominant and diallelic). A new type of marker, amplified fragment length polymorphisms (AFLPs),[76] offers the potential to assay large numbers of markers very quickly. These markers are codominant but are only diallelic. For this reason they will be useful for fine-structure mapping in defined crosses but will not meet the needs of applications requiring a high level of marker informativeness. Microsatellite, or simple sequence repeat (SSR), markers are highly informative and easy to apply and would be extremely useful markers for most applications in conifers. These markers are, however, difficult and very expensive to develop. Nevertheless, SSRs are being developed for a few conifer species.[62] Harry and Neale[25] are taking a different approach to developing highly informative markers for loblolly pine and other conifers. The ends of cDNAs are sequenced and primers are designed for PCR amplification of genomic DNA. The amplicons are digested with restriction enzymes to reveal polymorphisms within the amplicon or are separated by density gradient gel electrophoresis (DDGE) to reveal site changes as length variants. This approach will lead to highly informative, PCR-based markers and also reveal variations in structural gene loci. These markers are, however, very difficult and slow to develop.

A second important area of conifer genome research will be the identification of large numbers of structural gene loci through high-throughput cDNA sequencing and mapping. Already this approach has identified scores of genes from human brain,[1,2] *Arabidopsis*,[26,51] rice,[67] corn,[29] and *Brassica*.[53] In loblolly pine, Kinlaw et al[32] have sequenced 200+ cDNAs that have previously been used as mapping probes.[15,21] Approximately 30% of the cDNAs were assigned tentative identities based on database comparisons. Both the IFG and NCSU labs are sequencing 1000+ cDNAs from loblolly pine secondary xylem to identify some of the genes involved in the production and differentiation of this important tissue. Attempts will be made to map these cDNAs either by RFLP or PCR-based approaches.

Will sequenced cDNAs then be the ultimate genetic markers? Although anonymous markers can be easily obtained and are useful markers, it is desirable in the long term to obtain the location of functional genes themselves. This approach is needed to complement and develop the power of quantitative genetics to define and locate functional genes for complex traits. Fine-structure mapping of complex traits could ultimately lead to the identification of the genes determining phenotypes. The combined mapping of quantitative trait loci and structural gene loci will lead to a profound understanding of the growth, physiology and evolution of forest trees that is needed to meet the increasing demands for both conservation of forests and the production of wood, pulp and paper products in the decades to come.

REFERENCES

1. Adams MD, Dubnick M, Kerlavage AR, Moreno R, Kelley JM, Utterback TR, Nagle JW, Fields C and Venter JC. Sequence identification of 2,375 human brain genes. Nature 1992; 355:632-634.
2. Adams MD, Kelley JM, Gocayne JD et al. Complementary DNA sequencing: Expressed sequence tags and human genome project. Science 1991; 242:1651-1656.
3. Adams WT and Joly RJ. Linkage relationships among twelve allozyme loci in loblolly pine. J Hered 1980; 71:199-202.
4. Ahuja MR, Devey ME, Groover AT, Jermstad KD and Neale DB. Mapped DNA probes from loblolly pine can be used for restriction fragment length polymorphism mapping in other conifers. Theor Appl Genet 1994; 88:279-282.
5. Arumuganananthan K and Earle ED. Nuclear DNA content of some important plant species. Plant Mol Biol Rep 1991; 9:208-218.
6. Bent AF, Kunkel BN, Dalbeck D et al. RPS2 of Arabidopsis thaliana: A leucine-rich repeat class of plant disease resistance genes. Science 1994; 265:1860-1866.
7. Binelli G and Bucci B. A genetic linkage map of *Picea abies* Karst., based on RAPD markers, as a tool in population genetics. Theor Appl Genet 1994; 88:283-288.
8. Bridgwater FE. Shoot elongation patterns of loblolly pine families selected for contrasting growth potential. For Sci 1990; 36:641-656.
9. Bridgwater FE, Williams CG, Campbell RG. Pattern of leader elongation in loblolly pine families. For Sci 1985; 31:933-944.
10. Chamberlin CJ. Gymnosperms. Structure and evolution. Chicago: Univ Chicago Press, 1933.
11. Cheliak WM and Pitel JA. Inheritance and linkage of allozymes in Larix laricina Silvae Genet 1985; 34:142-148.
12. Conkle MT. Isozyme variation and linkage in six conifer species. In: Proceedings of the Symposium on Isozymes of North American Forest Trees and Forest Insects. Technical Coordinator. M.T. Conkle. USDA For Ser Gen Tech Rep PSW-48. 1981:11-17.
13. Critchfield WB and Little EL. Geographic Distribution of the Pines of the World. USDA/Forest Service Micellaneous Publication 991, 1966:97.
14. Devey ME, Delfino-Mix A, Kinloch BB and Neale DB. Efficient mapping of a gene for resistance to white pine blister rust in sugar pine. Proc Natl Acad Sci USA 1995; 92:2066-2070.
15. Devey ME, Fiddler TA, Liu B-H, Knapp SJ and Neale DB. An RFLP linkage map for loblolly pine based on a three-generation outbred pedigree. Theor Appl Genet 1994; 88:273-278.
16. Devey ME, Jermstad KD, Tauer CG and Neale DB. Inheritance of RFLP loci in a loblolly pine three-generation pedigree. Theor Appl Genet 1991; 83:238-242.

17. Doudrick RL, Heslop-Harrison JS, Nelson CD, Schmidt T, Nance WL and Schwarzacher T. The karyotype of Pinus elliottii var. elliottii using patterns of fluorescent in situ hybridization and fluorochrome banding. J Hered 1995; 86:289-96.

18. El-Kassaby YA, Sziklai O and Yeh FC. Linkage relationships among 19 polymorphic allozyme loci in coastal Douglas-fir (*Pseudotsuga menziesii* var. menziesii). Can J Genet Cytol 1982; 24:101-108.

19. Giovannoni JJ, Wing RA, Ganal MW and Tanksley SD. Isolation of molecular markers from specific chromosomal intervals using DNA pools from existing mapping populations. Nucleic Acids Res 1991; 19:6553-6558.

20. Gocmen B, Jermstad KD, Neale DB and Kaya Z. A partial genetic linkage map of *Taxus brevifolia* Nutt. using random amplified polymorphic DNA markers. Can J For Res 1995; in press.

21. Groover A, Devey M, Fiddler T, Lee J, Megraw R, Mitchell-Olds T, Sherman B, Vujcic S, Williams C and Neale D. Identification of quantitative trait loci influencing wood specific gravity in an outbred pedigree of loblolly pine. Genetics 1994; 138:1293-1300.

22. Groover AT, Williams CG, Devey ME, Lee JM and Neale DB. Sex-related differences in meiotic recombination frequency in Pinus taeda. J Hered 1995; 86:157-158.

23. Guries RP, Friedman ST and Ledig FT. A megagametophyte analysis of linkage in pitch pine (*Pinus rigida* Mill.). Heredity 1978; 40:309-314.

24. Harry, D.E. Inheritance and linkage of isozyme variants in incense-cedar. J Hered 1986; 77:261-266.

25. Harry DE and DB Neale. Developing codominant PCR markers in pines. Proceedings of the 22nd Southern Forest Tree Improvement Conference, Atlanta, GA, June 14-17, 1993:249-256.

26. Hofte H, Desprez T, Amselem J, Chiapello H, Caboche M, Moisan A, Jourjon M, Charpenteau J, Berthomieu P, Guerrier D, Giraudat J, Quigley F, Thomas F, Yu D, Mache R, Raynal M, Cooke R, Grellet F, Delseny M, Parmentier Y, Marcillac G, Gigot C, Fleck J, Phillips G, Axelos M, Bardet C, Tremousaygue D and Lescure B. An inventory of 1152 expressed sequence tags obtained by partial sequencing of cDNAs from Arabidopsis thaliana. Plant J 1993; 4:1051-1061.

27. Jones DA, Thomas CM, Hammond-Kosack KE, Balint-Kurti PJ and Jones JDJ. Isolation of the tomato *Cf-9* gene for resistance to *Cladosporium fulvum* by transposon tagging. Science 1994; 266:789-793.

28. Kaya Z and Neale DB. Linkage mapping in Turkish red pine (*Pinus brutia* Ten.) using random amplified polymorphic DNA (RAPD) genetic markers. Silvae Genet 1995; 44:110-116.

29. Keith CS, Hoang DO, Barrett BM, Feigelman B, Nelson MC, Thai H and Baysdorfer C. Partial sequence analysis of 130 randomly selected maize cDNA clones. Plant Phys 1993; 101:329-332.

30. Kiehne KL, Sewell MM and Neale DB. Saturation mapping of quantitative trait loci associated with wood specific gravity in loblolly pine using bulked DNAs based on genotypic classes of a full-informative RFLP marker. San Diego CA: Poster Abstract, Plant Genome III, 1995.

31. King JN and Dancik BP. Inheritance and linkage of isozymes in white spruce (*Picea glauca*) Can J Genet Cytol. 1983; 25:430-36.

32. Kinlaw CS, Baysdorfer C, Gladstone E, Harry DE, Sewell MM and Neale DB. Loblolly pine cDNA sequence analysis. San Diego CA: Poster Abstract, Plant Genome III, 1995.

33. Kinlaw CS, Gerttula SM and Carter MC. Lipid transfer protein genes of loblolly pine are members of a complex gene family. Plant Mol Biol 1994; 26:1213-1216.

34. Kossack D. The IFG copia-like element: Characterization of a transposable element present in high copy number in *Pinus* and a history of the pines using IFG as a marker. Davis, CA: Ph.D. thesis, University of California, 1989.

35. Kriebel HB. DNA seqence components of the *Pinus* strobus nuclear genome. Can J For Res 1985; 15:1-4.

36. Kuhlman EG, Amerson HV and Wilcox PL. Recent research on fusiform rust dis-

ease. In Proc. 4th IUFRO Rusts of Pines Working Party Conf., Tsukuba, Japan, 1995:17-21.

37. Kvarnheden A, Tandre K and Engström P. A *cdc2* homologue and closely related processed retropseudogenes from Norway spruce. Plant Mol. Biol. 1995; 27:391-403.

38. Lander ES and Schork NJ. Genetic dissection of complex traits. Science 1994; 265:2037-2048.

39. Lawrence GJ, Finnegan EJ, Ayliffe MA and Ellis JG. The *L6* gene for flax rust resistance is related to the Arabidopsis bacterial resistance gene *RPS2* and the tobacco viral resistance gene *N*. The Plant Cell 1995; 7:1195-1206.

40. Liu BH and Knapp SJ. GMENDEL: A program for Mendelian segregation and linkage analysis of individual or multiple progeny populations using log-likelihood ratios. J Hered 1990; 8:407.

41. Martin GB, Brommonschenkel SW, Chunwongse J, Frary A, Ganal MW, Spivy R, Wu T, Earle ED and Tanksley SD. Map-based cloning of a protein kinase gene conferring disease resistance in tomato. Science 1993; 262:1432-1436.

42. Megraw RA. Wood quality factors in loblolly pine: Influence of tree age, position in tree and cultural practice on wood specific gravity, fiber length and fibril angle. Atlanta GA: TAPPI Press, Technology Park, 1985:88.

43. Millar CI. Impact of the Eocene on the evolution of *Pinus* L. Ann. Missouri Bot Gard 1993; 80:471-498.

44. Mindrinos M, Katagiri F, Yu G-L and Ausubel FM. The *A. thaliana* disease resistance gene *RPS2* encodes a protein containing a nucleotide-binding site and leucine-rich repeats. Cell 1994; 79:1089-109.

45. Mukai Y, Suyama Y, Tsumura Y, Kawahara T, Yoshimaru H, Kondo T, Tomaru N, Kuramoto N and Murai M. A linkage map for sugi (*Cryptomeria japonica*) based on RFLP, RAPD and isozyme loci. Theor Appl Genet 1995; 90:835-840.

46. Neale DB and Adams WT. Inheritance of isozyme variants in seed tissues of balsam fir (*Abies balsamea*). Can J Bot 1981; 59:1285-1291.

47. Neale DB and Harry DE. Genetic mapping in forest trees: RFLPs, RAPDs and beyond. AgBiotech News and Infor 1994; 6:107N-114N.

48. Neale DB and Williams CG. Restriction fragment length polymorphism mapping in conifers and applications to forest tree genetics and tree improvement. Can J For Res 1991; 21:545-554.

49. Nelson CD, Kubisiak TL, Stine M and Nance WL. A genetic linkage map of longleaf pine (*Pinus palustris* Mill.) based on random amplifed polymorphic DNAs. J Hered 1994; 85:433-439.

50. Nelson CD, Nance WL and Doudrick RL. A partial genetic linkage map of slash pine (*Pinus elliottii* Engelm. var. *elliottii*) based on random amplified polymorphic DNAs. Theor Appl Genet 1993; 87:145-151.

51. Newman T, de Bruijn FJ, Green P, Keegstra K, Kende H, McIntosh L, Ohlrogge J, Raikhel N, Somerville S, Thomashow M, Retzel E and Somerville C. Genes galore: A summary of methods for accessing results from large-scale partial sequencing of anonymous Arabidopsis cDNA clones. Plant Physiol 1995; 106:1241-1255.

52. O'Malley DM, Grattapaglia D, Chaparro JX, Wilcox PL, Amerson HV, Liu B-H, Whetten R, McKeand S, Kuhlman EG, McCord S, Crane B and Sederoff R. Molecular markers, forest genetics and tree breeding. 22nd Stadler Genetics Symposium, 1995; (in press).

53. Park YS, Kwak JM, Kwon O, Kim YS, Lee DS, Cho MJ, Lee HH and Nam HG. Generation of expressed sequence tags of random root cDNA clones of *Brassica napus* by single-run partial sequencing. Plant Physiol 1993; 103:359-370.

54. Plomion C. These. Presente devant L'Ecole Nationale Surperierure Agronomique de Rennes pour obtenir le Titre de Docteur de L'ENSAR. Cartographie et Determinism Genetique de la Hauteur Juvenile Chez le Pin Maritime (Pinus pinaster Ait.), en Condition de Croissance Acceleree, 1995:172.

55. Plomion C, Bahrman N, Durel CE and O'Malley DM. Genomic mapping in maritime pine (*Pinus pinaster*) using RAPD and protein markers. Heredity, 1995;

74:661-668.

56. Plomion C, O'Malley DM and Durel CE. Genomic analysis in maritime pine (*Pinus pinaster*). Comparison of two RAPD maps using selfed and open-pollinated seeds of the same individual. Theor Appl Genet 1995; 90:1028-1034.

57. Pye JM, Wagner JE, Holmes TP and Cubbage FW. Economic evaluation of fusiform rust research and protection. 23rd Southern Forest Tree Improvement Conference, 1995:32.

58. Sax R and Sax HJ. Chromosome number and morphology in conifers. J Arnold Arbor Harv Univ, 1933; 14:356-375.

59. Saylor LC. Karyotype analysis of the genus *Pinus* - subgenus *Pinus*. Silvae Genet 1972; 19:155-163.

60. Sederoff R, Stomp A-M, Gwynn G, Ford E, Loopstra C, Hodgskiss P and Chilton WS. Application of recombinant DNA techniques to pines: A molecular approach to genetic engineering in forestry. In: Bonga JM and DurzanCell DJ, ed. Tissue Culture in Forestry, 1987:314-329.

61. Sewell MM and Neale DB. Genetic mapping in loblolly pine. In: Proceedings of the 23rd Southern Forest Tree Improvement Conference, Asheville, N.C. June 19-22, 1995.

62. Smith DN and Devey ME. Occurance and inheritance of microsatellites in *Pinus radiata*. Genome 1994; 37:977-983.

63. Stam P. Construction of integrated genetic linkage maps by means of a new computer package: Join Map Plant J 1993; 3:739-744.

64. Staskawicz BJ, Ausubel FM, Baker BJ, Ellis JG and Jones JDG. Molecular genetics of plant disease resistance. Science 1995; 268:661-667.

65. Strauss SH, Lande R and Namkoong G. Limitations of molecular-marker-aided selection in forest tree breeding. Can J For Res 1992; 22:1050-1061.

66. Tulsieram LK, Glaubitz JC, Kiss G and Carlson JE. Single tree genetic linkage mapping in conifers using haploid DNA from magagametophytes. Bio/Tech 1992; 10:686-690.

67. Uchimiya H, Kidou S, Shimazaki T, Aotsuka S, Takamatsu S, Nishi R, Hashimoto H, Matsubayashi Y, Kidou N, Umeda M and Kato A. Random sequencing of cDNA libraries reveals a variety of expressed genes in cultured cells of rice (*Oryza sativa* L.). Plant J 1992; 2:1005-1009.

68. Wakamiya I, Newton RJ, Johnston JS and Price HJ. Genome size and environmental factors in the genus *Pinus*. Am J Bot 1993; 80:1235-1241.

69. Welsh J and McClelland M. Finger-printing genomes using PCR with arbitrary primers. Nulceic Acids Res 1990; 18:7213-7218.

70. Whitham S, Dinesh-Kumar SP, Choi D, Hehl R, Corr C and Baker B. The product of the tobacco mosaic virus resistance gene N. Similarity to Toll and the interleukin-1 receptor. Cell 1994; 78:1-20.

71. Wilcox P. Genetic dissection of fusiform rust resistance in loblolly pine. PhD Dissertation. Raleigh, NC, USA: North Carolina State University, 1995.

72. Wilcox P, Amerson HV, Kuhlman G, Liu B-H, O'Malley DM and Sederoff RR. Genomic mapping of resistance to fusiform rust disease in loblolly pine. submitted to Proc Natl Acad Sci USA.

73. Williams CG. Influence of shoot ontogeny on juvenile-mature correlations in loblolly pine. For Sci 1987; 33:411-422.

74. Williams CG and Neale DB. Conifer wood quality and marker-aided selection: a case study. Can J For Res 1992; 22:1009-1017.

75. Williams JGK, Kubelik AR, Livak KJ, Rafalski JA and Tingey SV. DNA polymorphisms amplified by arbitrary primers are useful as genetic markers. Nucleic Acids Res 1990; 18:6531-6535.

76. Zabeau M. Selective restriction fragment amplification: a general method for DNA fingerprinting. European Patent Application. Publication No. 0 534 858 A1, 1993.

77. Zobel BJ and van Buijtenen JP. Wood Variation: Its Causes and Control. Berlin Heidelberg: Springer-Verlag, 1989:363.

INDEX